CRC MECHANICAL ENGINEERING SERIES

Series Editor Frank A. Kulacki, University of Minnesota

Published

Entropy Generation Minimization
  Adrian Bejan, Duke University

The Finite Element Method Using MATLAB
  Young W. Kwon, Naval Postgraduate School
  Hyochoong Bang, Korea Aerospace Research Institute

Mechanics of Composite Materials
  Autar K. Kaw, University of South Florida

Viscoelastic Solids
  Roderic Lakes, University of Iowa

Nonlinear Analysis of Structures
  M. Sathyamoorthy, Clarkson University

Practical Inverse Analysis in Engineering
  David M. Trujillo, Trucomp
  Henry R. Busby, Ohio State University

To be Published

Fundamentals of Environmental Discharge Modeling
  Lorin R. Davis, Oregon State University

Mechanics of Solids and Shells
  Gerald Wempner, Georgia Institute of Technology
  Demosthenes Talaslidis, Aristotle University of Thessalonika

Mathematical and Physical Modeling of Materials Processing Operation
  Olusegun Johnson Ileghusi, Northeastern University
  Manabu Iguchi, Osaka University
  Walter E. Wahnsiedler, Alcoa Technical Center

MECHANICS
of
FATIGUE

Vladimir V. Bolotin
Russian Academy of Sciences

CRC Press
Boca Raton London New York Washington, D.C.

Library of Congress Cataloging-in-Publication Data

Bolotin, V. V. (Vladimir Vasil ´evich)
   Mechanics of fatigue / Vladimir V. Bolotin.
       p.   cm. -- (Mechanical engineering)
   Includes bibliographical references and index.
   ISBN 0-8493-9663-8 (alk. paper)
   1. Materials--Fatigue--Testing. 2. Contact mechanics.
I. Series: Mechanical engineering (CRC Press).
TA418.38.B65 1998
620.1′126—dc21   98-29151
                                                                        CIP

No claim to original U.S. Government works
International Standard Book Number 0-8493-9663-8
Library of Congress Card Number 98-29151
Printed in the United States of America 1 2 3 4 5 6 7 8 9 0
Printed on acid-free paper

CONTENTS

PREFACE

This book is dedicated to the fatigue of materials and structural components of engineering systems. The term *fatigue* is understood in a broad sense, including crack nucleation and growth up to the final failure under cyclic and/or sustained loads and actions. In addition to the classic high-cycle and low-cycle fatigue, such phenomena as fatigue associated with creep, corrosion fatigue, stress corrosion cracking, etc., are also discussed.

Fatigue and related phenomena are the most frequent causes of failures in engineering systems, beginning from minor failures that result in the interruption of operation and the accompanying economic loss up to major accidents with disastrous consequences.

Fatigue is a subject of extensive study by engineers in various branches of industry as well as by researchers in the mechanics of solids and material science. As a complement to already published works, this book discusses the subject in a more theoretical way. Fatigue and all kinds of delayed fracture are considered here from the unified viewpoint of the mechanics of solids and structures. As with other centered positions, this approach cannot be comprehensive. However, it allows deeper insight into fatigue phenomena using the terms and concepts more appropriate for mechanical, civil, naval, and aeronautical engineers. Such an approach develops more detailed planning of fatigue tests, their statistical treatment and interpretation, as well as more efficient use of the known experimental data. In addition, this approach presents new analytical and numerical tools for the analysis of structural safety and reliability, as well as for life-time prediction and other important aspects of an engineer's activities.

The core of this book is built upon the synthesis of micro- and macromechanics of fracture. By micromechanics, the author means both the modeling of mechanical phenomena on the level of material structure (i.e., on the level of grains, fibers, microinclusions, microcracks and micropores) and the continuous approach based on the use of certain internal field parameters characterizing the dispersed microdamage. This is so-called continuum damage mechanics. As for macromechanics of fracture, the version used here is that developed by the author and called *analytical fracture mechanics*. This term means that the system cracked body-loading or loading device is considered as a mechanical system, and the tools of analytical (rational) mechanics are applied thoroughly to describe the crack propagation until the final failure.

The book consists of ten chapters. Chapter 1 is of an introductory character. It contains preliminary information on fatigue and engineering methods for the design of machines and structures against failures caused by fatigue.

Fatigue crack nucleation is the subject of Chapter 2. As this early stage of fatigue is controlled by factors of distinctly probabilistic nature, probabilistic models are widely used to describe the crack nucleation and early growth. Both microstructural and continuous models are discussed here. The emphasis is upon the prediction of statistical scattering and size effect during the early stages of fatigue damage before a macroscopic crack begins to grow through the bulk of the material. The initial stage of crack growth into the depth of material is also considered here in the framework of a probabilistic model.

The general approach to the theory of fatigue crack propagation is presented in Chapter 3. Here the basic concepts of analytical fracture mechanics are discussed which, in fact, present a generalization of the standard fracture mechanics to multiparameter problems in the framework of analytical mechanics. Fatigue crack growth is treated as a result of interaction between the generalized forces in the system cracked body-loading and the accumulation of dispersed damage in the body. Both the damage in the vicinity of crack tips and far field damage are considered. Due to the relationship between driving and resistance generalized forces, a crack propagates either continuously or in a step-wise pattern. To reduce the system of governing equations to ordinary differential equations related to those widely used in engineering design, a special approach is developed called "quasistationary approximation." Chapter 3 is necessary for understanding the later contents of the book, which are dedicated to different types of fatigue and related phenomena in a variety of materials, for various crack and body geometries, loading and environmental conditions.

Chapter 4 is dedicated to fatigue crack growth in linear elastic materials that are subject to dispersed damage. In general, this damage affects all material properties. The main factor is the influence of dispersed damage on the resistance against fracture. The inclusion of this effect is sufficient to describe fatigue crack growth. The subject of this chapter is the simplest crack of mode I in an unbounded body under cyclic loading called Griffith's fatigue crack in this book. This model is used for a systematic study of the influence of material and loading parameters, as well as of initial crack conditions on the fatigue crack growth rate. Presented in the standard form of crack growth rate diagrams, the results allow discussion in terms of common engineering approaches. Even within the scope of this simple model, adaptability and flexibility of the suggested theory are demonstrated.

The same model of material behavior is considered in Chapter 5 where more complex problems are studied. For example, two- and multiple-parameter fatigue cracks, such as elliptical mode I and initially oblique cracks, are analyzed. A special study is performed to predict the trajectories of mixed mode cracks propagating in materials with randomly distributed local heterogeneities.

Fatigue cracks in elastoplastic material are considered in Chapter 6. The thin plastic zone model is used here to predict fatigue crack growth accounting for plastic strain. Both the influence of dispersed damage on resistance against fracture and stress distribution within the plastic zone are considered. Effects inherent to elastoplastic fracture, such as the crack growth retardation due to overloading, are discussed in the framework of the proposed theory. The quasistationary approximation is also discussed to comment upon patterns of low-cycle fatigue, in particular, to make a sensible modeling of damage accumulation within the process zone.

Fatigue and related phenomena in hereditary solids are discussed in Chapter 7. Two types of materials are considered: 1) linear visco-elastic solids modeling the behavior of most polymer materials at small deformations; 2) nonlinear hereditary solids used as models for metals at elevated temperatures. Both cyclic and slowly varying sustained loads are significant for hereditary solids. A number of temporial characteristic time parameters are involved in this class of problems. Depending on the ratio between the parameters, a certain combination of mechanisms is dominant. In particular, creep affected crack growth is considered in this chapter, along with cyclic fatigue in materials with hereditary compliance properties.

The objective of Chapter 8 is to demonstrate that the general theory of fatigue crack growth is also applicable when environmental factors are involved. Corrosion fatigue and stress corrosion cracking are considered from a purely mechanical view-point not addressing internal chemical mechanisms. The equations from the previous chapters are complemented with equations describing the damage produced by environmental agents. Special attention is given to the equations describing an agent's penetration in the crack. It is demonstrated by numerical modeling that most effects observed in environmentally affected fatigue can be described in the framework of the proposed theory.

Damage and fatigue in composite materials are the subjects of the concluding two chapters of the book. Generally, composites as specific structural materials deserve a special monograph concerning their fatigue. Chapter 9 is dedicated to a special type of composites, unidirectional fiber composites with ductile matrix and brittle, initially continuous fibers. Tension along fibers is considered, and two modes of structural failure are studied. The first is fracture due to the loss of integrity as a result of dispersed damage accumulation. The second is the by multiple cracking accompanied by fibers debonding resulting in the formation of "brush-like" cracks. In contrast to Chapters 4 through 8, where the dispersed damage is described in the framework of continuum damage mechanics, the micromechanical approach is systematically used to predict the damage of fiber composites.

Laminate composites are the subject of Chapter 10. The typical pattern of their failure is delamination because of the damage of comparatively weak interlayers. This pattern becomes more complicated when buckling of the delaminated portion is involved. Here it is necessary to take into account the interaction between buckling, damage of the interlayer, and the general energy balance in the system. Various types of delamination in composite plates and shells are considered in this chapter. Analysis of this interrelationship carries a number of complexities. For example, damage in the interlayer affects its compliance that, in turn, affects the crack growth rate.

As it is seen from the above summary, the book covers a broad topic concerning damage, fatigue and fracture of engineering materials and structural components. The book is written by a researcher in mechanics of solids and structures. The reader will not find any mention of, for example, the diffusion of crystal lattice defects or equations for the reaction of metal with humid air. This is beyond the scope of the present study. This book is mainly oriented to readers who deal (at least on the level of application) with mechanical aspects of fracture and fatigue, or conduct research in fracture mechanics, structural safety, mechanics of composites and other modern branches of mechanics of solids and structures.

The bibliography to this book consists of three groups of references: 1) the references to publications used directly are included; 2) most of the monographs and manuals are cited that cover fracture and fatigue in a broader sense than could be done in a single monograph; 3) references are included of representative publications which, being somewhat removed from the author's approach, must be considered as milestones in this topic. The complete bibliography on fatigue contains thousands of items and cannot be totally covered. The author apologizes beforehand for exclusions which may be a result of limited exposure to Western literature. It must also be stressed that references to publications which are now considered as classic in the field or frequently referenced, are usually omitted. They are replaced with names and dates in the text, e.g., Griffith (1920).

The writing of this book was a laborious task that took almost five years. The author is indebted to many people who, directly or indirectly, assisted in its writing and publication. Most of the results included in this book have been presented by the author both in Russia and abroad. Discussions with many colleagues at seminars and symposia as well as the author's experience have proven extremely valuable and stimulating. Colleagues at the Moscow Power Engineering Institute (Technical University) and the Mechanical Engineering Research Institute of the Russian Academy of Sciences have aided in all the stages of writing this book, beginning from the pioneering numerical experimentation through to setting the final text and preparing final figures. These colleagues are too numerous to mention here, but most of them are listed in the references of this book. Last, but by no means least, the author is indebted to his wife, Kira S. Bolotina. She not only provided a perfect working environment to complete this book (as most scientists' wives do) but read, amended and typed the primary version of this book.

The study underlying some parts of this book was supported by the Russian Foundation of Basic Research (grant 96–01–01488) and the International Science Foundation (grant RLS300).

Chapter 1

INTRODUCTION

1.1 Fatigue of Materials and Related Phenomena

In a narrow sense, the term fatigue of materials and structural components means damage and fracture due to cyclic, repeatedly applied stresses. In a wide sense, it includes a large number of phenomena of delayed damage and fracture under loads and environmental conditions. The systematic study of fatigue was initiated by Wöhler, who in the period 1858-1860 performed the first systematic experimentation on damage of materials under cyclic loading. In particular, Wöhler introduced the concept of the fatigue curve, i.e., the diagram where a characteristic magnitude of cyclic stress is plotted against the cycle number until fatigue failure. Up to now, the Wöhler curve has been used widely in the applied structural analysis. At the same time, fatigue and related phenomena are considered as a subject of mechanics of solids, material science, as well as that of basic engineering.

It is expedient to distinguish between high-cycle (classic) and low-cycle fatigue. If plastic deformations are small and localized in the vicinity of the crack tip while the main part of the body is deformed elastically, then one has high-cycle fatigue. If the cyclic loading is accompanied by elasto-plastic deformations in the bulk of the body, then one has low-cycle fatigue. Usually we say low-cycle fatigue if the cycle number up to the initiation of a visible crack or until final fracture is below 10^4 or $5 \cdot 10^4$ cycles.

The mode of damage and final fracture depends on environmental conditions. At elevated temperatures plasticity of most materials increases, metals display creep, and polymers thermo-plastic behavior. At lower temperatures plasticity of metals decreases, and brittle fracture becomes more probable. If a structural component is subjected both to cyclic loading and variable thermal actions, mixed phenomena take place, such as creep fatigue, creep accelerated by vibration, and thermo-fatigue. The combination of fatigue and corrosion is called corrosion fatigue. It is a type of damage typical for metals interacting with active media, humid air, etc. Hydrogen and irradiation embrittlement, as well as various wear and ageing processes, interact with fatigue, too. The delayed fracture occurs not only under cyclic, but also under constant or slowly varying loading. Typical examples are the delayed fracture of polymers and crack initiation and propagation in metals under the combination of active environment and non-cyclic loads. The latter kind of damage, opposite of corrosion fatigue, is called stress corrosion cracking. All these phenomena taken together form a class of damage frequently called static fatigue. Later on, when no special comment is given, we consider cyclic fatigue, calling it just fatigue.

Fatigue is a gradual process of damage accumulation that proceeds on various levels beginning from the scale of the crystal lattice, dislocations and other objects of solid state physics up to the scales of structural components. Three or four stages of fatigue damage are usually distinguishable. In the first stage, the damage accumulation occurs on the level of microstructure. Where a polycrystalline alloy is concerned, it is the level of grains and intergranular layers. The damage is dispersed over the volume of a specimen or structural component or, at least, over the most stressed parts. At the end of this stage, nuclei of macroscopic cracks originate, i.e., such aggregates of microcracks that are strong stress concentrators and, under the following loading, have a tendency to grow. Surface nuclei usually can be observed visually (at least with proper magnification). The second stage is the growth of cracks whose depth is small compared with the size of the cross section. At the same time, the sizes of these cracks are equal to a few characteristic scales of microstructure, say, to several grain sizes. Such cracks are called small cracks (in the literature the term short crack is also used). The number of small cracks in a body may be large. The pattern of their propagation is different from that of completely developed macroscopic cracks. Small cracks find their way through the nonhomogeneous material. Most of them stop growing upon meeting some obstacles, but one or several cracks transform into macroscopic, "long" fatigue cracks that propagate in a direct way as strong stress concentrators. This process forms the third stage of fatigue damage. The fourth stage is rapid final fracture due to the sharp stress concentration at the crack front and/or the expenditure of the material's resistance to fracture.

The initiation and following growth of a macroscopic crack are schematically shown in Figure 1.1a for the case of a polycrystalline material under uniaxial cyclic tension. Nuclei appear near the surface of the specimen, in particular, in the local stress concentration domains as well as near the damaged or weakest grains. The initial slip planes and microcracks in grains are oriented mostly along the planes with maximal shear stresses. Small cracks are inclined, at least approximately, in the same directions. These cracks go through the grains, intergranular boundaries or in a mixed way. When one of the small cracks becomes sufficiently long, the direction of its growth changes: the crack propagates into the cross section of the specimen, in the so-called opening mode. Such a "long," macroscopic crack intersects in its growth a large number of grains. Therefore, this growth is determined mainly with averaged properties. The border between small and "long" cracks is rather conditional. In particular, it depends on the ratio of the current crack size a and the characteristic size of grains. If the grain size is of the order of 0.1 mm, a crack may be considered as "long" when its reaches the magnitude $a_0 = 0.5$ mm or 1 mm.

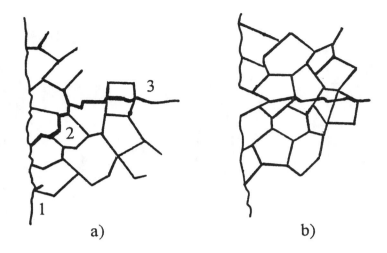

FIGURE 1.1
Fatigue crack initiation in polycrystalline material: (a) near the regular surface;
(b) near a strong stress concentrator.

The ratio of durations of these stages varies to a large extent depending on material properties, type of loading and environmental conditions. The first two stages are absent if a crack propagates from an initial macroscopic crack, sharp crack-like defect or another strong stress concentrator (Figure 1.1). In this case the position of the macroscopic crack is conditioned beforehand, and the crack begins to propagate after a comparatively small number of cycles. On the other hand, for very brittle materials the final fracture may occur suddenly, without the formation of any stable macroscopic cracks. For example, it may be a result that microcrack density attains a certain critical level.

A typical cyclic loading process is presented in Figure 1.2. The notation $s(t)$ is used here for any stress-related variable characterizing the loading process. It may be interpreted, for example, as a remotely applied normal stress relating to the whole cross section of a specimen subjected to cyclic tension/compression. If $s(t)$ is varying as a sinusoidal function, the duration of a cycle coincides with the period of the function $s(t)$. Each cycle contains maximal s_{max} and minimal s_{min} magnitudes of the applied stresses. A cycle is usually considered as a segment of the loading process limited with two neighboring up-crossings of the average stress $s_m = (s_{max}+s_{min})/2$. The cyclic loading is usually described by the amplitude stress $s_a = (s_{max} - s_{min})/2$ or the stress range $\Delta s = s_{max} - s_{min}$. Stress ratio $R = s_{min}/s_{max}$ is also a significant characteristic of the cyclic loading. For symmetrical cycles $R = -1$; when a cycle contains non-negative stresses only, then $R > 0$. The cycle number N is an integer number, although it is usually treated as a continuous variable. Actually, the cycle number corresponding to significant damage, or moreover to final failure, as a rule is very large compared with unity. Time t is useful as an independent variable when the interactions between fatigue and other time-dependent damage phenomena are considered.

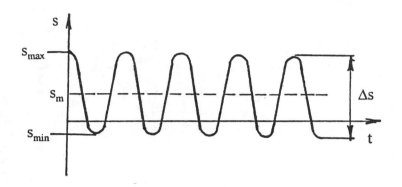

FIGURE 1.2
Cyclic loading and its characterictic stresses.

Some types of cyclic loading are shown schematically in Figure 1.3. Among them are biharmonic (a), chaotic (pseudo-stochastic) (b), and piece-wise constant (c) processes. In practice stochastic processes are often met. These processes can be narrow-band or broad-band, stationary or nonstationary. In the general case the concept of loading cycle becomes ambiguous. For example, a cycle may be determined as a segment of loading process that contains one maximum and one minimum; but often a cycle is treated as a segment limited with two neighboring

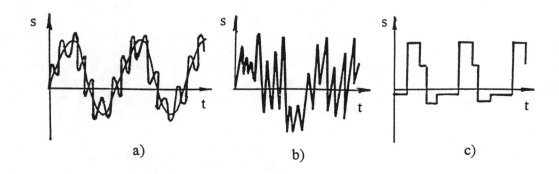

FIGURE 1.3
Cyclic loading processes: (a) biharmonic; (b) chaotic; (c) piece-wise constant.

up-crossings of a certain, generally nonstationary level. Relative freedom in choosing this level makes the concept of cycle even less determinate. For example, in Figure 1.4a so-called compound cycle is shown limited with two time moments t_{N-1} and t_N. This cycle contains interior cycles that correspond to the high-frequency components of the loading process. The situation becomes more complicated when we deal with multiple-parameter loading, in particular in the case when the components of the stress tensor vary in time in a different way. In all such cases the questions of how to divide a process into cycles and how to identify cycle parameters are to be considered separately.

A generalization of cycle loading is block loading. Each block is also called a duty cycle, a mission, etc. The block corresponds to one of multiply repeated stages in the service of a structure or machine. A block of loading corresponding to one standard (specified) flight of an aircraft can be mentioned as an example. This block includes loads during ground motions, take-off and landing, climb, cruise, and descent flights. Each block contains a large number of cycles, e.g., those due to air turbulence, impacts, and vibration at take-off and landing. If the number of blocks in the service life of a structure or machine is sufficiently large, each block can be treated as a compound cycle.

The literature on fatigue covering all its aspects is enormous. It is relevant to mention that the American Society for Testing and Materials issued more than 25 volumes of papers dedicated to fracture, and almost half of the papers deals with fatigue and related phenomena. The history of earlier studies in fatigue of materials, components and structures can be found in [99, 146]. Theoretical, experimental, and engineering aspects of fatigue are discussed in books [5, 19, 46, 59, 79, 83, 84, 113, 140] and survey papers [43, 129]. Practically every textbook in fracture mechanics contains a concise introduction in fatigue [1, 32, 41, 50, 62, 66, 68, 75, 82, 92, 114, 145]. A number of experimental data can be found in handbooks [40,

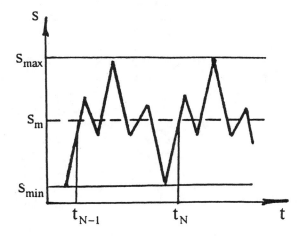

FIGURE 1.4
Compound cycle containing interior cycles.

47, 96, 112, 141, 144, 147].

The subject of this book is the analysis of fatigue and related phenomena from the standpoint of solid mechanics. Most of the book content is dedicated to mathematical modelling of fatigue crack initiation and growth. However, it seems expedient to present here a short survey of basic experimental facts concerning fatigue as well as a concise introduction to fracture mechanics.

1.2   Fatigue Tests

Two group of fatigue tests are to be distinguished. The aim of the classic tests which are descending from the Wöhler times is the estimation of the fatigue life (usually measured in cycle numbers) in function of loading parameters. This group of tests covers mostly the early stage of fatigue damage. The second group of tests is oriented to the assessment of crack growth rates in the function of loading parameters. Being based on the concepts of fracture mechanics, this group of tests will be discussed in the later sections of this Chapter.

Initial information for the design of engineering systems against fatigue is based on standard fatigue tests. Among them are tests in cyclic tension, tension-compression, rotating, bending, etc. Shapes and sizes of specimens, requirements for their manufacturing and finishing, test techniques and methods of treatment of results are specified by national and international codes. As examples, the documentation of the American Society of Testing Materials (ASTM), American Society of Mechanical Engineers (ASME), DIN, and GOST (the former USSR Committee of Standartization) can be mentioned.

Both load-controlled and displacement-controlled types of loading are used in fatigue tests. In load-controlled tests, the characteristic magnitudes of loads or of applied stresses are maintained constant throughout the duration of testing. In displacement-controlled tests, the displacements, e.g., the maximal and minimal deflections of a rotating cantilever specimen, are kept constant. In high-cycle fatigue, the type of loading affects the first stage of damage accumulation insignificantly. Its influence becomes more substantial during the stage of growth of macroscopic cracks when the specimen stiffness begins to decrease rather rapidly. The role of the type of loading in low-cycle fatigue tests may be essential.

Results of standard fatigue tests are presented in the form of fatigue curves that show the relationship between one of the characteristic cycle stresses s (the amplitude, the maximal stress, the stress range) and the limit cycle number N. Other characteristic cycle stresses, as well as parameters of the environmental conditions, including temperature, are maintained constant. The number N is counted either until the complete fracture of a specimen or until the observation of the first macroscopic crack of a given size. In particular, it can be just a visible surface crack.

Fatigue curves are plotted in the $\log - \log$ or log-uniform scales. In Figure 1.5 typical fatigue curves are shown for symmetrical ($s_{min} = -s_{max}$) and pulsating (at $s_{min} = 0$) tensile cycle loading $s(t)$. Stress ratios are $R = -1$ and $R = 0$, respectively. Figure 1.5 represents the standard test procedure. At each stress level

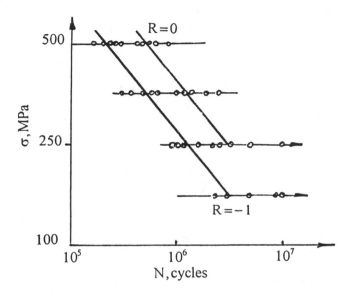

FIGURE 1.5
Fatigue curves in log − log scale for two stress ratios.

s_{max} a group of specimens is tested, and for each group the limit cycle number N is estimated. This number is subjected to significant scatter, over a half of decimal order and even more. Therefore, the fatigue curves in fact are either regression curves or fractile curves corresponding to a certain confidence level.

Fatigue tests are terminated, as a rule, when a beforehand stated cycle number N_b is achieved. In high-cycle fatigue tests, it is usually taken as $N_b = 10^6 \ldots 10^7$. If a fatigue curve in the log − log scale is close to a straight line, the analytical presentation

$$N = N_b \left(\frac{s_b}{s} \right)^m \tag{1.1}$$

is used. Here m is the fatigue curve power exponent, and s_b is a constant with the dimension of stress. These parameters, in general, depend on the mean stress (or on the stress ratio R), the loading frequency, test temperature, etc. Typical magnitudes of power exponents in high-cycle fatigue of ferrous and aluminum alloys are $m = 6 \ldots 12$.

There is evidence that, for some structural materials such as common low-carbon steels, and for polished specimens, macrocrack initiation does not take place if stress amplitudes are sufficiently small. One says in these cases that an endurance limit exists, e.g., a magnitude of stress amplitude (at the given mean stress of the cycle) such that fatigue fracture does not occur even at an arbitrary large cycle number. Fatigue curves expose, in these cases, a tendency to approach horizontal asymptotes with ordinates equal to the endurance limit. In engineering practice, however, the endurance limit is usually associated with the maximal stress or stress range corresponding to the required fatigue life or to the specified base of standard tests. In the Russian literature, the endurance limit in tension is denoted σ_R, in shear τ_R, etc., where R is the stress ratio. For example, the endurance limit for the symmetrical tension-compression cycle is denoted σ_{-1}.

If existence of the endurance limit is accepted, the analytic approximation

$$N = N_b \left(\frac{s_b}{s - s_R} \right)^m, \qquad s > s_R \tag{1.2}$$

can be used instead of (1.1). Another relation between characteristic stress s, cycle number N and fatigue failure is

$$N = \begin{cases} N_b(s_R/s)^m, & s \geq s_R \\ \\ \infty, & s < s_R \end{cases} \tag{1.3}$$

Fatigue curves corresponding to (1.2) and (1.3) are shown in Figures 1.6a and 1.6b.

The influence of the mean stress on the magnitude of the endurance limit is significant. Test results are usually plotted either on the s_m, s_a plane or on the s_m, s_{max} plane. The cycle number N is treated as a parameter. The first type of diagram is attributed to Haigh (1915) and the second to Smith (1919). An example

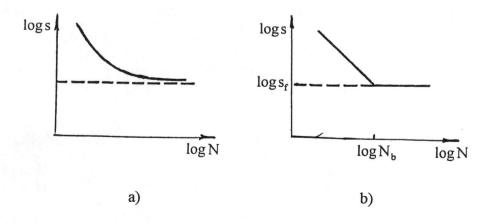

a)   b)

FIGURE 1.6
Schematization of fatigue curves in the presence of endurance limit by: (a) equation (1.2); (b) equation (1.3).

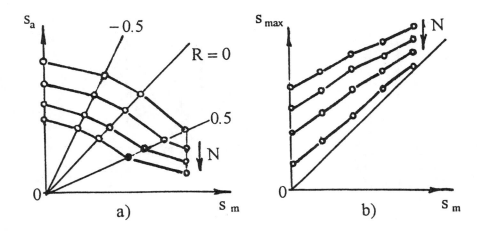

FIGURE 1.7
Schematic presentation of fatigue data in Haigh's (a) and Smith's (b) diagrams.

of such diagrams is shown in Figure 1.7. Points corresponding to the averaged cycle numbers until fatigue failure, say $N = 5 \cdot 10^4, 10^5, 10^6$, and $5 \cdot 10^6$, are connected with lines.

1.3 Low-Cycle Fatigue

The phenomenon of low-cycle fatigue can be in principle described in terms of the theory of plasticity. In fact, low-cycle fatigue is a cyclic elastoplastic deformation occurring until the expenditure of plasticity reserves. The material behavior at unloading and reversed loading, in particular the shape and size of hysteresis loops, is of essential significance in low-cycle fatigue. The relationship between maximal stresses and strains within a cycle generally differs from that in monotonic quasistatic loading. The cycle-deformation relations depend on the type of loading process. They change whether this process is load- or displacement-controlled. Some materials reveal a tendency to cyclic hardening: under the displacement-controlled loading the maximal cycle stresses grow with the cycle number. Other materials reveal a tendency to cyclic softening. An intermediate place is occupied by the so-called plastically stabilizing materials. Depending on the microstructural state and temperature, the same material may behave in various ways. Typical diagrams of uniaxial tensile/compression deformation $\sigma(\varepsilon)$ are shown in Figure 1.8. They correspond to symmetrical cycle loading with the given strain amplitude ε_a. Figure 1.8a shows the behavior of a cycle-dependent hardening material; Figure 1.8b that of a softening material.

When the strain level is high, the cycle number at fatigue failure is comparatively small, and significant one-sided residual deformations accumulate in the

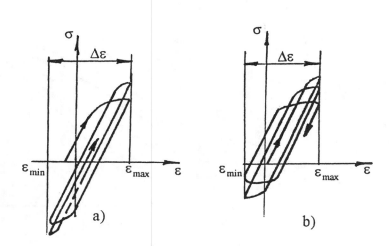

FIGURE 1.8
Stress-strain diagrams: (a) cycle-dependent hardening material, (b) cycle-dependent softening material.

specimen. At a moderate strain level, test results are convenient to represent with fatigue curves. Compared with high-cycle fatigue curves, low-cycle fatigue curves are usually plotted on the plane of characteristic strain versus cycle number at failure. Standard tests in tension or tension/compression are usually performed maintaining a constant range $\Delta\varepsilon$ of the nominal (averaged upon all the working parts of a specimen) strain ε.

The simplest equation of low-cycle fatigue was suggested by Coffin (1959):

$$\Delta\varepsilon_p N^\mu = C \qquad (1.4)$$

Here $\Delta\varepsilon_p$ is the range of plastic strain within a cycle, and μ and C are empirical constants. For carbon steels $\mu = 0.4\ldots0.6$. This corresponds to a power exponent of $\Delta\varepsilon - N$ curves equal approximately to two. The constant C is frequently connected with the ultimate strain ε_U in standard quasistatic tension tests. Assuming that equation (1.4) is applicable in monotonic loading and that fracture occurs at the end of the first quarter of the cycle, one obtains $C = \varepsilon_U/4$ at $\mu = 1/2$. On the other hand, the ultimate strain ε_U is connected with the relative area decrease ψ in the neck cross section: $\varepsilon_U = \ln(1 - \psi)^{-1}$. For symmetric cycles $\Delta\varepsilon_p = 2\varepsilon_p$. This results in the Coffin-Manson equation in the form often used by engineers:

$$\varepsilon_p N^\mu = \frac{1}{2}\ln(1 - \psi)^{-1}$$

A number of analytical relationships have been suggested that combine low- and high-cycle fatigue curves. Among them is a fatigue failure criterion connecting the range of complete (summed) strain $\Delta\varepsilon$ with the cycle number N:

$$\Delta\varepsilon = CN^{-\mu} + (\sigma_c/E)N^{-\nu} \qquad (1.5)$$

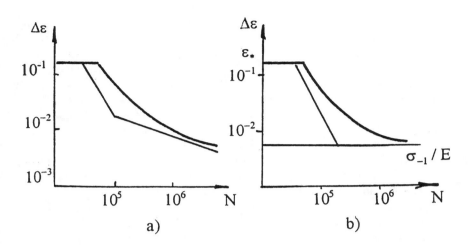

FIGURE 1.9
Wöhler's curves for low-cycle fatigue: (a) equation (1.5); (b) equation (1.6).

Here E is Young's modulus and σ_c is a material constant, e.g., $\sigma_c = 3.5\sigma_U$ when σ_U is the ultimate tensile stress. The power exponent ν for low-carbon steels is frequently assumed $\nu = 0.12$. That corresponds approximately to the exponent $m = 8$ in (1.1). Another approximation

$$\Delta\varepsilon = CN^{-\mu} + 2\sigma_R/E \qquad (1.6)$$

provides the interpolation between equation (1.4) and the asymptotic relationship $\sigma_{max} \rightarrow \sigma_R$ at $N \rightarrow \infty$. Equations (1.5) and (1.6) are illustrated in Figure 1.9. The left-hand branch of the fatigue curves corresponds to fracture after several cycles with a strain range of the order $2\varepsilon_U$.

1.4   Factors Influencing Fatigue Life

The resistance against fatigue depends essentially on a number of factors. Among them are: stress concentration, surface roughness, frequency of loading, loading history, residual stress-strain fields, temperature, environmental conditions, etc. Manufacturing process features such as heat treatment and cold deformation also affect fatigue life. Each group of factors requires special study, and hundreds of papers are published every year dedicated to their experimental analysis. The survey of these results and practical recomendations can be found in many manuals for engineers.

Much attention has been paid by experimenters to the influence of loading history. Now most experiments are performed with modern fatigue testing equipment that provides arbitrary loading programs, and are not limited to one-directional stress-strain conditions. Already, in the earlier stage of experimental studies of fatigue, a paradoxal effect of short overloadings was observed, for example, the crack growth retardation after a single overloading. This phenomenon is partially attributed to blunting of the crack tip due to plastic deformations and, mostly, to the so-called crack closure (Elber, 1970). During unloading the faces of cracks may close before the cyclic stresses reach the minimum. In this case only a part of the range of tensile stresses are active during the next cycle of loading.

Another factor is loading frequency. At sufficiently high frequencies, thermal effects due to cyclic deformation become significant, and the resulting change of mechanical properties must be taken into account. This phenomenon is important for polymers as well as for metallic materials at elevated temperatures when fatigue and creep enter in combination. However, to observe a significant cyclic heating effect, sufficiently high loading frequencies are needed.

One of the most specific features of fatigue is the statistical scatter of experimental results. Generally, the scatter is present in all experiments, in particular, due to the random microstructure of real materials. In terms of ultimate stresses, the variance coefficient usually has the order of 0.05. The fatigue life, being measured in cycle numbers until failure, is much more sensitive to material nonhomogeneities. That is directly obvious, e.g., from equation (1.1), since the power exponent m is essentially larger than unity. In addition, the fatigue ultimate stress and, moreover, the fatigue life are under the influence of many factors that interact in a complex way and are difficult to control, even in laboratory conditions. The statistical scatter of fatigue data is accompanied by a related phenomenon, a distinct size effect, i.e., by the influence of absolute sizes of geometrically similar and similarly loaded specimens on fatigue life. The presence of the size effect indicates, generally, that additional parameters of the dimension of length must be taken into account to describe the phenomenon adequately. Among such parameters are certainly the characteristic dimensions of microstructural nonhomogeneities. The latter are distributed both in the volume of a body and upon its surface. On the other hand, the nuclei of fatigue cracks are usually situated near the surface of a body, and that makes the surface roughness an important factor in fatigue.

1.5 Micro- and Macromechanics of Fracture

Deformation and fracture develop on very diverse levels characterized by length scales from the step of the crystal lattice up to the dimensions of structural components. The level of the crystal lattice has a scale of the order $10^{-10} \dots 10^{-9}$ m. The next range is occupied by the molecular structure of polymers. The characteristic scale of dislocations and so-called submicrocracks is of the order $10^{-9} \dots 10^{-7}$ m. The average distance between dislocations is situated in the range $10^{-8} \dots 10^{-4}$ m. These elements of microstructure are the subject of solid-state physics. The following levels can be attributed both to solid-state physics and to material science. These levels correspond to slips and bands of slips, microvoids and microcracks, micro-inclusions, grains, fibers, etc. The elements of surface microroughness such as characteristic heights and lengths of the surface relief occupy the scale

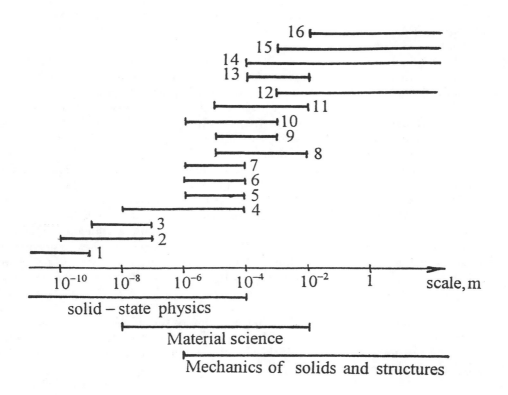

FIGURE 1.10
Size scales in deformation, damage and fracture (explanation in text).

$10^{-6} \ldots 10^{-2}$ m. These elements of microstructure can be treated from the standpoint of continuum mechanics. The larger scales, such as macroscopic cracks, crack tip plastic zones, the manufacturing flaws that are not included in the former categories, macroscopic stress concentrators and characteristic scales of stress-strain fields, are the subjects of the mechanics of solids and structures. However, two scales are to be distinguished when we stay in the framework of continuum mechanics. The smaller scale is the characteristic size of grains, fibers, microvoids, etc. The larger scale corresponds to the sizes of structural members. Respectively, we can call them micromechanics and macromechanics of deformable solids (Figure 1.10).

Macromechanics of fracture (in the conventional, narrow interpretation) is usually understood as the mechanics of deformable bodies containing macroscopic cracks. These cracks, as a rule, are modeled as mathematical cuts. The principal objective of the macromechanics of fracture is to establish stability conditions for the systems cracked body-loading with respect to crack growth, i.e., such conditions under which small perturbations do not produce an accelerated crack propagation resulting in final failure.

The primary contribution to the macromechanics of fracture was done by Griffith in the 20's. However, real interest in this domain of mechanics appeared only in the 50s. At that time, due to the development of experimental techniques and non-destructive inspection methods, it was recognized that practically every large-scale structure contains cracks and crack-like defects. Moreover, the presence of defects does not necessarily mean that a structure is unsafe or unreliable. New trends in engineering, such as fail-safe and damage-tolerance approaches, stimulated the analysis of the conditions and features of macroscopic crack growth. The objective of fracture mechanics in its current stage is to find parameters that characterize the fracture toughness of structural materials, to propose testing methods, to choose the appropriate structural materials properties, and to recommend how to maintain the safe life of engineering systems.

Linear fracture mechanics is based on a few assumptions: cracks are mathematical cuts; material is linear elastic until final failure; and deformations are small in the meaning of elasticity theory. The term linear fracture mechanics is widely accepted although it is not correct: fracture as a whole is a strongly nonlinear phenomenon even for a material supposed to be linear elastic and deformations that are small. The term mechanics of brittle fracture is frequently used as a synonym for linear fracture mechanics. Since small yield zones near the tips are present in most cases, the term mechanics of the quasi-brittle fracture is, maybe, the most appropriate one.

The fundamentals of fracture mechanics are presented in a number of textbooks and monographs. In particular, the introduction to the subject is given by Anderson [1], Broek [41], Hellan [66] and Knott [82]. Advanced study is presented in the 6-volume monograph edited by Liebowitz [92] and in the books by Cherepanov [50], Kanninen and Popelar [75].

Several approaches have been suggested to build up the foundation of fracture mechanics. Correspondingly, several variables have been introduced to describe the state of the system cracked body-loading. Similarly, various material parameters are used to characterize fracture toughness of materials under given conditions. In the pioneering works by Griffith of 1920-1924 the energy balance approach was used to state the stability condition of the system. The amount of work spent on formation of the new unit crack surface was used as a fracture toughness parameter. About three decades later Irwin introduced the concept of the stress intensity factor, characterizing the stress field in the vicinity of the crack tip. Irwin also found the relationship between the stress intensity factor and strain energy release rate for linear elastic bodies. Several years later Rice and Cherepanov introduced the concept of J-integral equal to the energy flux through a contour enveloping the crack tip. An approach based on the strain energy density criterion was suggested by Sih.

In application to brittle fracture, when a material remains linear elastic until final fracture, all of the listed approaches give identical or close results. The difference is mostly in the choice of fracture toughness parameters. These approaches were extrapolated upon the quasi-brittle fracture in the presence of plastic zones at the crack tips. Then the theory was extended, with more or less proper modifications, upon cracks in nonlinearly elastic, elasto-plastic and visco-elasto-plastic materials. Cracks in solids with such properties are the subject of non-linear fracture mechanics [71, 114, 118].

The origin of micromechanics of fracture can be traced from engineering approaches to the design of parts of machines against fatigue; in particular, from the Palmgren-Miner (1924, 1945) rule of fatigue damage summation. The simplest phenomenological models of dispersed damage accumulation suggested in the 60's by Rabotnov and Kachanov are, in fact, a continuous extension of the summation rule when a damage measure is introduced at each point of a body [73, 120]. Related stochastic models of brittle fracture originated from Weibull (1940) also may be referred to the micromechanics of fracture. Another topic of micromechanics of fracture is the theory of microvoid nucleation and coalescence. This topic is discussed in detail by Hutchinson [72] and Mura [105].

Since damage and fracture of materials develop on every scale level, micro- and macromechanics are in close connection. In the last decades the term mesomechanics began to be used also. The definition of the subject of mesomechanics is unstabilized. This subject covers many areas, from the events on the grain, microcracks and microvoid level to the prediction of bulk material properties by averaging of local properties. Sometimes, trying to avoid intersections with the terminology used by physicists, the term "mesomechanics" is applied where the scales of grain and fibers are involved, and the term "nanomechanics" in the treatment of solids on the molecular level. However, staying in the framework of continuum mechanics, one does not need such a caution in terminology. Later on, we refer to the micro- and macromechanics of fracture as well as of their synthesis without the use of the term "mesomechanics".

The damage of materials begins on the level of vacancy diffusion, dislocation glides, grain boundary sliding, etc. phenomena, covered by solid-state physics and material science. In particular, dislocation theory describes the motion of dislocation through a crystal lattice and provides the understanding of fundamental events of plastic deformation on the macroscopic level. However, due to a large number of intermediate levels, it cannot be used, say, to predict the inelastic behavior of machinery parts or structural members under actual loading conditions. In many cases it is possible to replace the analysis of damage on the microlevel with appropriate models similar to the models of continuum mechanics. Continuum damage mechanics is in fact a branch of continuum mechanics. The summary of this topic is given by Bažant [6], Kachanov [73], Krajcinovic [85], and Lemaitre [91]. Although the primary subjects under considerations are microcracks, microvoids, etc., continuum damage mechanics deals with continuum variables. They represent (in a statistical sense) size, shape and the space distribution and local properties of discrete flaws. In fact, continuum damage mechanics differs from conventional mechanics of deformable solids through additional, internal field variables. These variables enter into constitutive equations along with the common field variables such as stress and strain tensors.

1.6   Linear Fracture Mechanics

Consider the simplest problem of linear fracture mechanics, i.e., Griffith's problem of a plane opening mode crack in an unbounded medium. The crack with length $2a$ is considered as a mathematical cut, and the uniformly applied stresses σ_∞ are normal to the plane of the crack (Figure 1.11). The body is in the plane-strain

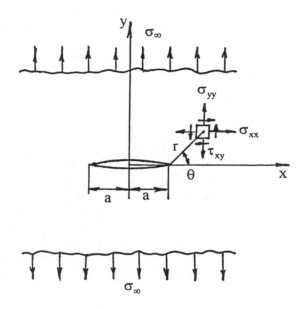

FIGURE 1.11
Griffith's problem of plane crack under uniform tension.

state, and we refer all variables to unity size in the z-direction. Material is linear elastic with Young's modulus E and Poisson's ratio ν.

The work required to advance the crack tip from a to $a+da$ is proportional to da. Griffith (1920) attributed this work to the energy of surface forces. In real structural materials the main part of this work is spent in plastic deformation and other irreversible phenomena. All these factors may be included in the specific fracture work γ. We relate this work to the unit area of the newborn crack (opposite to the tradition that this work is related to the unit area of the newborn free surface). Assume that the crack does not propagate if the increment of the potential energy of the system Π related to the crack tip advancement is less than the required fracture work, i.e., $-d\Pi < \gamma da$. When $-d\Pi > \gamma da$ the released energy exceeds the required fracture work. Then due to the energy surplus, dynamic crack growth becomes possible. For half of the body $x \geq 0$ the potential energy is

$$\Pi = \text{const} - \frac{\pi\sigma_\infty^2 a^2(1 - \nu^2)}{2E} \tag{1.7}$$

The constant in the right-hand side of (1.7) is equal to the total potential energy in the absence of the crack. Substituting (1.7) in the condition $-d\Pi = \gamma da$ we obtain the equation for the critical stress σ_c:

$$\sigma_c = \left[\frac{\gamma E}{\pi a(1 - \nu^2)}\right]^{1/2} \tag{1.8}$$

The approach by Irwin (1954) is based on the analysis of singularities of the stress field in the vicinity of the crack tip. In a linear elastic body there is a square-root singularity. If the fracture is a local phenomenon, it depends on the stress distribution close to the tip. The singular term in formulas for the stresses looks as follows:

$$\sigma_{jk}(r,\theta) = \frac{K_I}{(2\pi r)^{1/2}} f_{jk}(r,\theta) \qquad (1.9)$$

Here r is the polar radius, θ is the polar angle; subscripts j and k run for the coordinate notations x, y, z (see Figure 1.11).

Parameter

$$K_I = \sigma_\infty (\pi a)^{1/2} \qquad (1.10)$$

is called the stress intensity factor. Index I shows that the stress intensity factor relates to the opening mode of cracking (mode I). Equation (1.9) remains valid in the case of more general loading and geometry under the condition that the stress intensity factor K_I is properly calculated. A typical relation has the form

$$K_I = Y s_\infty (\pi a)^{1/2} \qquad (1.11)$$

where s_∞ is the stress applied in the remote cross section. The correction factor Y is usually of the order of unity. This factor depends on geometry and sometimes on the normalized elasticity parameters (e.g., on Poisson's ratio). For example, for a plate of width $2b$ with a central crack of length $2a$, approximate equations may be used, such as

$$K_I \approx \sigma_\infty (\pi a)^{1/2} [\sec(\pi a/2b)]^{1/2}$$

This equation yields satisfactory results for $b > 0.8a$. Many reference data can be found in hand- and textbooks, for example, in those edited by Sih [135], Murakami [106] and Panasyuk [112].

Other reference stresses may be used in (1.11) instead of the remotely applied stress; for example, the average stress in the cracked cross section. Sometimes it is convenient to take the reference stress equal to the maximal stress in the cracked section estimated by an elementary beam formula.

In terms of stress intensity factors, a crack does not grow at $K_I < K_{IC}$. The condition of the crack growth initiation takes the form

$$K_I = K_{IC} \qquad (1.12)$$

where K_{IC} is the critical stress intensity factor. The latter is frequently also called the fracture toughness. Equations (1.8) and (1.12) become equivalent if we put

$$K_{IC} = \left(\frac{\gamma E}{1 - \nu^2} \right)^{1/2} \qquad (1.13)$$

The fundamental relationship takes place between approaches based on the stress intensity factors and the energy conservation law. The amount of potential energy released when a crack advances in a unit of length is defined as follows:

$$G = -\frac{\partial \Pi}{\partial a} \tag{1.14}$$

This value, called energy release rate, has the dimension of force. Another name for G is the force driving the crack tip or, briefly, the driving force. Taking into account (1.7) and (1.14), the energy release rate in Griffith's problem is

$$G = \frac{\pi \sigma_\infty^2 a(1 - \nu^2)}{E} \tag{1.15}$$

The energy balance condition corresponding to crack growth initiation is

$$G = G_{IC} \tag{1.16}$$

where $G = K_I^2(1 - \nu^2)/E, G_{IC} \equiv \gamma$. Any of three parameters, γ, K_{IC} and G_{IC}, may be used to characterize the material fracture toughness.

Three special cases are distinguished for a plane crack in an unbounded body (Figure 1.12). They are: mode I (opening mode), mode II (transverse shear mode), and mode III (longitudinal shear or antiplane mode). Stress intensity factors for these modes are

$$K_I = \sigma_\infty(\pi a)^{1/2}, \quad K_{II} = K_{III} = \tau_\infty(\pi a)^{1/2}$$

Here σ_∞ and τ_∞ are applied (remote) stresses in tension and shear stress, respectively. Singular terms in equation (1.9) for stresses in the vicinity of crack tips can be found in any handbook or textbook on fracture mechanics.

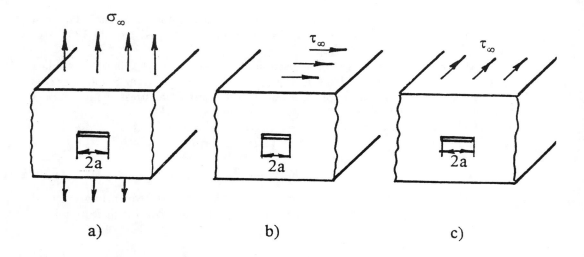

a)   b)   c)

FIGURE 1.12
Three principal modes of cracking: (a) opening (tearing); (b) transverse shear mode; (c) longitudinal shear (antiplane) mode.

Equations (1.7), (1.8), (1.13), and (1.15) correspond to the plane-strain state. In the case of the plane-stress state (for example, for an infinite thin plate), one has to replace $1-\nu^2$ with unity. The critical stress intensity factor K_{IC} corresponding to the plane-stress state depends on the plate thickness because of rather complicated patterns of crack propagation, including the shear fracture with significant plastic deformations. This is the reason that K_{IC} is used in applications as the measure of fracture toughness.

When the condition given in (1.12) or (1.16) is attained, the crack becomes unstable. It begins to propagate dynamically with the velocity limited by the Rayleigh wave velocity. Generally, the attainment of equalities in (1.12) and (1.16) does not necessarily signify instability. A typical example is an infinite plate loaded with two normal forces that are applied to the crack faces. The crack size in this problem increases with the applied forces when the equality condition takes place between the energy release rate and its critical magnitude. When $G_C = \text{const}$, the criterion of stability is

$$\frac{\partial G}{\partial a} < 0$$

1.7   Nonlinear Fracture Mechanics

Linear fracture mechanics, strictly speaking, becomes invalid when a plastic zone at the crack tip occurs. In fact, though, such zones are present in all real situations. To apply results of linear fracture mechanics to engineering design, the characteristic size λ of the plastic zone is to be small compared with the size of the crack. The size λ may be estimated by equalizing the elastic stresses as a function of the distance from the tip to the yield stress σ_Y. In particular, for the mode I crack the condition $\sigma_{yy}(x,0) \approx \sigma_Y$ gives

$$\lambda \sim K_I^2/(2\pi\sigma_Y^2)$$

A more accurate calculation with the account for stress distribution around the crack tip results in the following equations:

$$\lambda = \begin{cases} K_I^2/(3\pi\sigma_Y^2), & \varepsilon_{zz} = 0; \\[2mm] K_I^2/(\pi\sigma_Y^2), & \sigma_{zz} = 0 \end{cases} \tag{1.17}$$

The first formula relates to the plane-strain state; the second to the plane-stress one. The difference between the plastic zone sizes is explained by restrained conditions in the plane-strain state. In that case, the effective yield stress, according to the Mises criterion, is $\sqrt{3}\sigma_Y$. The approximate shape of the plastic zones is shown in Figure 1.13.

An elementary approach to include plastic deformations using the concepts of linear fracture mechanics was suggested by Irwin. He proposed to add the size of the plastic zone to the actual crack size. The modified stress intensity factor is introduced as follows:

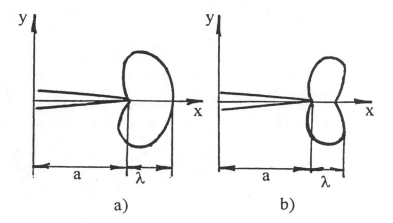

FIGURE 1.13
Plastic tip zones: (a) plane-strain state; (b) plane-stress state.

$$K_I = \sigma_\infty [\pi(a + \lambda)]^{1/2}$$

The driving force G, taking into account (1.17), becomes

$$G \approx \frac{\pi \sigma_\infty^2 a}{E} \left[1 + \frac{1}{2}\left(\frac{\sigma_\infty}{\sigma_Y}\right)^2\right] \qquad (1.18)$$

Such an approach is applicable only when the correction for plastic effects is comparatively small. Compared with (1.15), formula (1.18) refers to the plane-stress state.

The nonlinear fracture mechanics may be considered now as a special branch of fracture mechanics [71, 114, 118, 153]. In principle, nonlinear fracture mechanics ought to be based on the theory of plasticity. This is necessary when the load level is high, and plastic deformations develop in large scale occupying, say, all of the cross section of a specimen. Limit states of elasto-plastic bodies with cracks can be studied with the use of limit-state analysis in the framework of various plastic flow theories. In the framework of the theory of plasticity, fatigue is considered as cyclic plastic deformation taking into account hysteretic phenomena, strain hardening (softening), etc.

A simplified approach is frequently used that is based on the replacement of an elastoplastic material with an appropriate nonlinear elastic material. This means that effects of unloading, residual stresses and strains are neglected. Results obtained in this way may be interpreted in terms of the deformation theory of plasticity. The simplest nonlinear elastic model of fracture with a power constitutive law has been suggested by Rice and Rosengren (1968), and Hutchinson (1968).

Another simplified approach is the model of a thin plastic zone suggested by Leonov and Panasyuk (1960), and Dugdale (1960). All the plastic effects are concentrated in this zone, while the basic part of the material remains elastic. Hence,

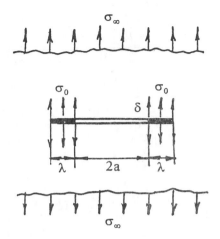

FIGURE 1.14
Model of thin plastic zone.

the evalution of the stress-strain field is reduced to a problem of linear elasticity with stresses given on the crack faces. The boundaries of the plastic zone are assumed to be infinitely thin. This model is illustrated in Figure 1.14. Within the tip zone with length λ, the opening stress $\sigma_Y(x, 0)$ is assumed to be constant. This stress, denoted later σ_0, is analogous to the yield stress σ_Y, and it may be sometimes identified with the latter. Outside the tip zone the material is supposed to remain elastic. A crack begins to grow when its tip opening displacement δ attains the critical magnitude δ_C. This magnitude is treated as a fracture toughness parameter, and the following limit condition is used instead of (1.12) and (1.16):

$$\delta = \delta_C \tag{1.19}$$

The length of the tip zone is given by equation

$$\lambda = a\left[\sec\left(\frac{\pi\sigma_\infty}{2\sigma_0}\right) - 1\right] \tag{1.20}$$

The crack tip opening displacement is

$$\delta = \frac{8\sigma_0 a}{\pi E}\log\sec\left(\frac{\pi\sigma_\infty}{2\sigma_0}\right) \tag{1.21}$$

The development of the right-hand side of (1.20) into a power series results in the approximate formula, valid at $\lambda \ll a$:

$$\lambda = \frac{\pi}{8}\left(\frac{K_I}{\sigma_0}\right)^2 \tag{1.22}$$

In the same aproximation the crack tip opening displacement is given by formula

$$\delta = \frac{K_I^2}{E\sigma_0} \tag{1.23}$$

At $\sigma_\infty \ll \sigma_0$ both Griffith's and Irwin's conditions follow from (1.19) and (1.21) under the assumptions that $\gamma = \sigma_0 \delta_C$ and $K_C = (E\sigma_0\delta_C)^{1/2}$ (the only difference is that $1 - \nu^2$ is replaced by unity). For short cracks the critical stress is close to σ_0. It means that the thin plastic zone model may be considered as a kind of interpolation between the linear fracture mechanics approach and the elementary (local) strength criterion (Figure 1.15).

Linear and nonlinear fracture mechanics may be connected through the use of the so-called path-independent integral technique. A number of integrals along trajectories around the crack tip have been suggested whose magnitudes are invariant with respect to the choice of trajectories. Some of them take into account large deformation, kinetic energy, etc. [2, 124]. However, all the consistent path-independent integrals are based on conservation laws of continuum mechanics, and their area of applicability is limited to elastic bodies, including those with nonlinear properties. There have been a lot of attempts to generalize the concept of path-independent integrals beyond this domain, e.g., to creep damage. These attempts are purely heuristic, though some of them work more or less successfully in the treatment and interpretation of experimental data. The first place among path-independent integrals is occupied by the J-integral introduced primarily in another context by Eshelby (1951). In terms of fracture mechanics the path-independent technique was suggested by Rice (1968) and Cherepanov (1968).

Consider a two-dimensional problem of a stress-strain field in a crack vicinity. Introduce at every point of a body the density of strain energy

$$w = \int \sigma_{\alpha\beta} d\varepsilon_{\alpha\beta}$$

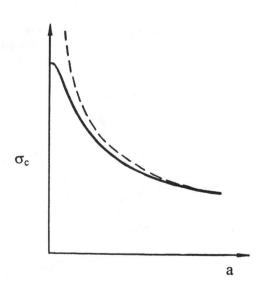

FIGURE 1.15
Critical stress versus crack size following the thin plastic zone model.

Here $\sigma_{\alpha\beta}$ is the stress tensor; $\varepsilon_{\alpha\beta}$ is the strain tensor. Hereafter, the subscript summation rule is used. The integration is performed upon all the loading history from the initial (natural) state up to the considered state. In the linear elastic case and small deformations

$$w = \frac{1}{2}\sigma_{\alpha\beta}\varepsilon_{\alpha\beta},$$

where $\sigma_{\alpha\beta}$ and $\varepsilon_{\alpha\beta}$ correspond to the actual state. The J-integral is introduced as follows:

$$J = \int_{\Gamma}\left(wdy - p_{\alpha}\frac{\partial u_{\alpha}}{\partial x}ds\right) \tag{1.24}$$

Here Γ is a contour that encloses the crack tip, ds is an element of this contour, p_{α} is the stress vector on the contour, and u_{α} is the displacement vector. The axis x is directed along the crack; axis y is orthogonal to the crack plan. The direction along Γ is counter-clockwise (see Figure 1.16). If the external forces are applied to the border of the crack, they must be included in vector p_{α}, and the contour Γ goes through the body without coming to the crack faces. The J-integral is usually interpreted as an energy flux through the surface enclosing the crack tip. When a material is linear elastic and isotropic, equation (1.24) results in Irwin's equation for the energy release rate. The concept of the J-integral is often applied to cracks in elasto-plastic materials under the assumption that the crack growth is not accompanied by unloading. When the energy release rate approach is applicable, equation (1.16) can be rewritten in the form

$$J = J_C \tag{1.25}$$

with the fracture toughness parameters denoted J_C. It is a measure of energy

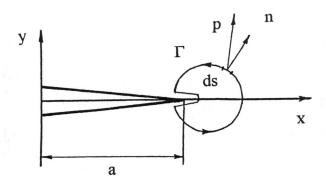

FIGURE 1.16
To the definition of *J*-integral.

spent in irreversible processes of damage and fracture. We prefer to use the more universal term "driving force" both in elastic and inelastic cases noting it by G. The corresponding resistance force we denote Γ. The condition of the crack growth initiation is

$$G = \Gamma \tag{1.26}$$

Equations (1.25) and (1.26) do not at all imply crack instability (it is, generally, not true even in the case of elastic materials). The complexity of nonlinear fracture mechanics originates from the fact that both generalized forces G and Γ depend on the loading and crack growth history. Therefore, Γ is not a constant if a material exposes ductile behavior. Several experimental techniques have been suggested for the evaluation of Γ. The general idea is based on using equation (1.14) where Π is interpreted as the amount of work performed by the applied force P through the corresponding displacement Δ. Many experimental techniques deal with the $P - \Delta$ diagram. A number of identical and identically loaded specimens with different crack sizes are required to assess generalized forces as a function of crack size. Some approximate approaches allow estimation of G using a single specimen only; however, they are based on rather far-fetched additional assumptions. Since the resistance force Γ is equal to G for a continuously advancing crack, the same techniques allow assessment of the magnitudes of resistance force.

Usually the resistance force Γ varies during crack growth. This is illustrated in Figure 1.17, where so-called "R-curves" are depicted. (R is just the common notation for the resistance force Γ, but the notation R is needed in fatigue for the stress ratio within a cycle.) In the initial stage of loading, when blunting of the fixed crack tip occurs, the R-curve goes almost vertically. The crack begins to propagate in a ductile mode. The driving force G grows in the loading process, too. The G-curves are shown schematically in Figure 1.17a. When the load level is sufficiently small, the crack size increases from the initial size a_0 in a stable way. Then the driving G force curve becomes tangent to the R-curve, and the crack growth becomes unstable. The limit condition to find the critical magnitudes $a = a_C, G = G_C$ is as follows:

$$\frac{\partial G}{\partial a} = \frac{\partial \Gamma}{\partial a} \tag{1.27}$$

It is usually assumed, at least implicitly, that R-curves depend on the difference $a - a_0$, and that in other aspects they are invariant for a given material. This assumption predicts different critical magnitudes of G and a for cracks of different initial sizes. It is illustrated in Figure 1.17b where two R-curves are shown for initial crack sizes a_0 and $a_0' > a_0$, resulting in the fracture toughness parameters $G_c' > G_c$ and critical crack sizes $a_c' > a_c$.

A useful concept of tearing moduli was introduced by Paris (1979). The tearing moduli are equal to the normalized magnitudes of derivatives entering equation (1.27):

$$T = \frac{E}{\sigma_Y^2} \frac{\partial G}{\partial a}, \qquad T_R = \frac{E}{\sigma_Y^2} \frac{\partial R}{\partial a}$$

Then equation (1.27) takes the form $T = T_R$, and the stability condition the form $T < T_R$.

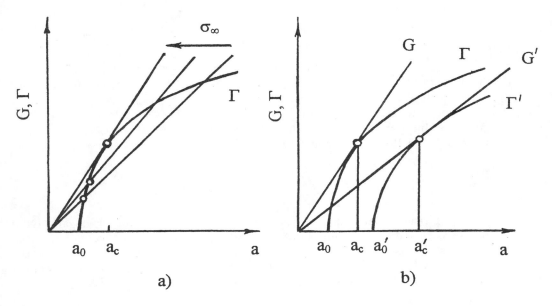

FIGURE 1.17
So-called R-curves of nonlinear fracture mechanics: (a) evaluation of critical crack size; (b) effect of initial crack size.

Many aspects of nonlinear fracture mechanics, such as transition from brittle fracture to ductile one, the effect of specimen thickness, thermal effects, etc., are beyond the framework of this short survey. The monographs by Hutchinson [71], Parton and Morozov [114], Pluvinage [118] as well as the volume edited by Wnuk [153] are almost completely dedicated to nonlinear fracture mechanics.

1.8   Fatigue Crack Growth

The growth of macroscopic cracks takes a significant part of the fatigue life of structures. Moreover, there are many situations where macroscopic cracks are assumed to be present in a structure and tolerated during the service life. Therefore the methods of prediction of fatigue crack growth are of essential practical interest.

Consider a one-parameter crack under cyclic loading. Evidently, the primary parameters controlling crack size increment during a cycle are extremal applied stresses within a cycle, s_{max} and s_{min}, and the current size of the crack a_N. Therefore,

$$a_N - a_{N-1} = f(s_{max}, s_{min}, a_{N-1})$$

Fatigue crack growth can be, as a rule, treated as a continuous function of the continuous variable N. The equivalent differential equation takes the form

$$\frac{da}{dN} = f(s_{max}, s_{min}, a) \qquad (1.28)$$

Here s_{max} and s_{min} are treated as continuous or piece-wise continuous functions of N. Instead of the couple s_{max}, s_{min}, another couple of variables can be used, the stress range $\Delta s = s_{max} - s_{min}$ and the stress ratio $R = s_{min}/s_{max}$.

Many attempts have been made to find the form of the right-hand side of (1.28). As the leading parameters controlling the fatigue crack growth are the applied stress range Δs and the current crack size a, the early proposals were presented in the form

$$\frac{da}{dN} = \text{const} \cdot (\Delta s)^{\alpha} (a)^{\beta}$$

with empirical power exponents $\alpha > 0$ and $\beta > 0$. Paris and Erdogan (1963) proposed to reduce the number of control parameters to one, the range ΔK of the stress intensity factor range, i.e., $\Delta K = Y \Delta s (\pi a)^{1/2}$. Here, and later on, the subscript indicating the fracture mode is omitted (where no special comments are present, mode I is assumed). The Paris equation

$$\frac{da}{dN} = c(\Delta K)^m \qquad (1.29)$$

remains the most popular among engineers. For the majority of metallic materials the exponent m in (1.29) takes values from $m = 2$ to $m = 6$. The magnitude $m = 4$ is frequently recommended for carbon steels under moderate stresses. Another empirical parameter, c in (1.29), varies over a wide range. For carbon steels at $m = 4$, such magnitudes as $c = 10^{-16} \ldots 10^{-12}$ (in mm^7/N^4) can be found in the literature. Using physical considerations, one might propose a connection between c and other material properties. For example, relationships were proposed such as $c \approx (\sigma_U K_{IC})^{-2}$ and $c \approx (E \varepsilon_U K_{IC})^{-2}$ for the case $m = 4$. Here σ_U and ε_U are ultimate tensile stress and strain. In general, both c and m depend on the stress ratio $R = K_{min}/K_{max}$.

Plotted in the $\log - \log$ scale, equation (1.29) presents the linear relationship between the crack growth rate da/dN and the stress intensity factor range ΔK. Many experimental data plotted in this scale show a good agreement with (1.29) under a proper choice of the constants c and m. However, this observation is valid only at moderate crack growth rate, such as $da/dN \approx 10^{-8} \ldots 10^{-4}$ m/cycle.

Experimental data show that the threshold stress intensity factor ΔK_{th} exists such that at $\Delta K < \Delta K_{th}$ the crack practically does not grow. Usually the threshold of fatigue crack growth is associated with rates of order 10^{-10} m/cycle and less. Existence of the threshold ΔK_{th} may be connected with the existence of the endurance limit s_R. Estimates like $\Delta K_{th} \approx s_R (\pi \lambda_0)^{1/2}$ were suggested, with the length parameter λ_0 of the order of the crack nucleus size. As a rule, ΔK_{th} is of the order of $(0.05 \ldots 0.1) K_C$, where K_C is the fracture toughness for considered loading and environmental conditions. On the other hand, at K_{max} of the order of fracture toughness K_C, the crack growth accelerates significantly. The general picture in $\log - \log$ scale is schematically shown in Figure 1.18. To describe all parts of the crack growth rate diagram, generalizations of equation (1.29) are necessary. A sufficiently general form is

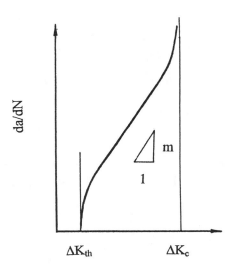

FIGURE 1.18
Fatigue crack growth rate diagram.

$$\frac{da}{dN} = c_1 (\Delta K - \Delta K_{th})^{m_1} (K_{max} - K_{th})^{m_2} (K_{fC} - K_{max})^{-m_3} \qquad (1.30)$$

Here c_1, m_1, m_2 and m_3 are positive empirical constants; ΔK_{th} and K_{th} are threshold magnitudes of the range ΔK of the stress intensity factor and of its maximal value K_{max}, respectively; and K_{fC} is the fracture toughness parameter in fatigue. The last parameter, in general, does not necessarily coincide with the fracture toughness parameter under monotonic quasistatic loading. It is a result of the far field damage, the changing conditions at the tip of fatigue cracks, etc. The right-hand side of (1.30) must be set to zero at $\Delta K < \Delta K_{th}$ or $K_{max} < K_{th}$.

All the parameters entering in equation (1.30) and similar equations of fatigue crack growth depend on a number of factors. They were mentioned above mostly in a purely qualitative manner. In particular, the threshold parameters ΔK_{th} and K_{th} significantly depend on the ratio $R = K_{min}/K_{max}$. Another way is to replace ΔK and K_{max} with their "effective" magnitudes that take into account the so-called crack closure and shielding effects. Several dozen semi-empirical equations of fatigue crack growth have been suggested. Most of them, however, are stillborn. A list of these equations can be found, for example, in the handbook of Panasyuk [112] and the survey paper by Liu [94].

Other fracture mechanics parameters sometimes are used to describe fatigue crack growth instead of the stress intensity factors. Among those parameters are the J-integral and its generalizations, the energy release rate, etc. In particular, one could assume that the driving force G is responsible for the fatigue crack growth, too. A rather general class of equations may be presented in the form

$$\frac{da}{dN} = f(G_{max}, G_{min})$$

Here G_{max} and G_{min} are extremal magnitudes of the driving force responsible for crack propagation. In conditions where this equation is applicable, it can be written as follows:

$$\frac{da}{dN} = c_1(G_{max}^{1/2} - G_{min}^{1/2})^m$$

1.9   Traditional Engineering Design against Fatigue

The aim of fatigue analysis is to provide load-carrying capacity and/or structural integrity of cyclically loaded structural components during their service life. Engineering approaches to this analysis are very diverse. Their choice depends on many factors; among them are:
  - mode of damage assumed as a structural failure;
  - loads and actions, and the type of their variability in time;
  - time segment during which the safe fatigue life is to be provided;
  - accepted (or recommended) concept of the structural reliability analysis.

The number of factors can be expanded. In addition, different approaches are used in different branches of industry, e.g., in civil and mechanical engineering, in aircraft and shipbuilding industries. The subject has been discussed in many hand- and textbooks, and most of them are centered in the area of activity of their authors.

Historically, for a long time the formation even of a single macroscopic (e.g., visible) crack has been considered as a failure of a structural component. This conclusion was based on the observation of fatigue tests of smooth laboratory specimens. The stage of dispersed damage accumulation takes a major part of the life of such specimens, and the stage of crack growth up to final rupture only a minor part. Apart from that, the opinion was that no macroscopic cracks could be tolerated in stressed structural components. This viewpoint partially remains up to this time.

If loading is stationary, i.e., if characteristic stresses of the cycle are constant in time, and the endurance limit exists, the reliability condition is

$$s \le s_R/n \tag{1.31}$$

where $n > 1$ is the safety factor. This condition is purely deterministic. Nevertheless, probabilistic considerations are hidden here, either in choosing the magnitude of n or in choosing the limit stress s_R. In the latter case, s_R is a percentile endurance limit corresponding to an acceptable reliability level. This means that the statistical scatter is partially taken into account in the treatment of fatigue data.

Although the condition (1.31) corresponds to the simplest situation, and it is of an extremely simple form, its practical use is not simple. A large number of factors must be accounted for the application of test results of polished standard laboratory specimens to real full-scale structures. The endurance limits of machinery parts and structural components differ, as a rule, more significantly than for those of laboratory specimens. Stress concentration, scale factor and surface roughness are only a part of the factors to be considered.

For example, the influence of stress concentration is usually characterized by the so-called effective stress concentration factor. It is equal to the ratio of two endurance limits, one for smooth specimens, and the second for specimens with concentrators of a given shape. Even in the linear elastic case these endurance limits do not coincide. The typical semi-empirical equation is

$$\kappa_{eff} = 1 + q(\kappa - 1)$$

Here κ and κ_{eff} are theoretical and effective stress concentration factors, respectively, and q is a coefficient characterizing the material sensitivity to stress concentration in fatigue. The magnitude of q varies in the range $0 \dots 1$ depending on material properties. In particular, it depends on microstructure and on surface roughness which is, in fact, a set of microscopic stress concentrators.

Another example presents the influence of the stress ratio R. As a rule, reference data relate to the fatigue endurance at $R = -1$ (symmetrical cycle) and $R = 0$ (pulsating cycle). To estimate the limit stress for an intermediate R, a linear interpolation is used. It is usually based on Haigh's or Smith's diagrams [46]. A typical approach is to replace the applied stress in (1.31) with an effective stress such as

$$s_{eff} = s_a + \psi s_m.$$

Here s_a is the stress amplitude, and s_m is the mean stress of the cycle. Empirical value, ψ, is equal to the tangent factor of the straight line in Haigh's diagram (see Figure 1.6a). For carbon steels it is frequently assumed that $\psi = 0.1 \dots 0.2$ at $-1 < R < 0$.

The type of stress state also, as a rule, enters into engineering fatigue analysis with some effective, reduced stresses. For instance, when bending and torsion are combined, as in machinery shafts, the reduced stresses are used such as $(\sigma^2 + 4\tau^2)^{1/2}$ or $(\sigma^2 + 3\tau^2)^{1/2}$. Here σ and τ are characteristic tensile and shear stresses in the cross section of a shaft. Such an approach results in the popular equation for the resulting safety factor

$$\frac{1}{n^2} = \frac{1}{n_\sigma^2} + \frac{1}{n_\tau^2}$$

The partial safety factors n_σ and n_τ are calculated from (1.31) for bending and tension separately.

Frequently there is no reason to assume the existence of the endurance limit, but the cycle number N during the specified life time or the time between inspections is known. Then formula (1.31) remains applicable under the condition that the endurance limit s_R is replaced with the limit stress $s(N)$ corresponding to the cycle number N. Fatigue curves $s = s(N)$ are used for this purpose. The magnitude of the safety factor n depends on the probabilistic level of the fatigue curve. This curve is either a regression line of fatigue tests, or a precentile line corresponding to a certain survival probability. The safety factor can be introduced with respect to the limit cycle number $N(s)$ under the given stress level. Then instead of (1.31) the condition

$$N < \frac{N(s)}{n_N}$$

is to be satisfied. The scatter of fatigue life is significantly larger than that of the limit stress. Therefore, always $n_N > n_s$ where the safety factors n_s and n_N relate to the stress and to the fatigue life, respectively.

In practice, however, loading processes seldom are stationary. In some cases fatigue curves can be obtained from specially programmed tests. For example, if the loading is a narrow-band normal random process, its mean square stress can be used as a characteristic stress. The number of maxima up-crossings of the mean stress level takes the place of the cycle number N. In some branches of industry, the loading history can be specified for certain types of structures and certain in-service conditions. As an example, a standard flight or a standard mission may be mentioned for civil and military airplanes, respectively. The simplest approach to take into account nonstationarities of loading is based on the so-called linear damage summation rule. This rule, suggested primarily by Palmgren (1924) for the design of the life of bearings, was later widely used in engineering design against fatigue. Let $N(s_k)$ be the limit cycle number under stationary loading with the characteristic stress s_k (other parameters of loading are assumed fixed). Introduce the damage measure D, equal to zero when damage is absent, and to unity for a completely damaged specimen, a structural component, a part of a machine, etc. Then the magnitude $\Delta D_k = 1/N(s_k)$ may be interpreted as an elementary damage associated with one loading cycle and the characteristic stress s_k. The summation rule is based on the assumption that elementary damage measures are additive, and the result does not depend on the order of sequential cycles. Under the above assumptions, the complete damage or fracture occurs when the sum of elementary damages ΔD_k attains the limit magnitude, i.e., unity. The equation for evaluating of the limit cycle number N_* is as follows:

$$\sum_{k=1}^{N_*} \frac{1}{N(s_k)} = 1 \qquad (1.32)$$

Each term in the left-hand side is to be repeated as many times as cycles with equal s_k entered into the loading history. Equation (1.32) is called usually Miner's damage summation rule.

Evidently, equation (1.32) does not take into account the order in the sequence of applied stresses. For example, if the loading process contains two steps with stresses s_1 and s_2, equation (1.32) results into

$$\frac{n_1}{N(s_1)} + \frac{n_2}{N(s_2)} = 1$$

Here n_1 and n_2 are the cycle numbers with stresses s_1 and s_2. If this equation is valid, the relationship between n_1 and n_2 is a straight line. Then the total fatigue life does not depend on the order of application of stresses s_1 and s_2. In reality, step-wise fatigue tests show a rather strong influence of the loading sequence (see the dashed lines in Figure 1.19). Sometimes preliminary overloading prolongs the summed fatigue life, sometimes opposite. To take into account these phenomena, a number of equations have been suggested. Most of them can be presented in the form of a differential equation with respect to the damage measure:

$$\frac{dD}{dN} = f(D, s) \tag{1.33}$$

Under certain assumptions equation (1.33) reduces to the summation rule

$$\int_0^{N_*} \frac{dn}{N(s)} = 1 \tag{1.34}$$

i.e., to a continuous equivalent of (1.32). In other cases effects of loading history can be described with a correspondingly chosen right-hand side of (1.33). However, as was shown by Bolotin [19], the natural scatter of fatigue data measured in cycle numbers until final failure is so significant that it covers the deviations from the summation rule. Equation (1.34) is useful when a loading process is of stochastic or pseudo-stochastic nature, being described with the density function $p(s)$ of the characteristic cycle stress s. Then one may put in (1.34) that $dn = p(s)\,ds$. The limit cycle number is determined as follows:

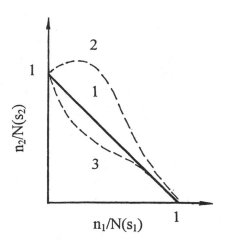

FIGURE 1.19
Application of linear summation rule to two-stage loading process.

$$N_* = \left[\int_0^\infty \frac{p(s)ds}{N(s)}\right]^{-1}$$

This equation, evidently, is semi-deterministic, and N_* is of the order of the expected total cycle number up to the final failure.

1.10 Current State and Trends in Design against Fatigue

Most contemporary design codes in power engineering, shipbuilding, aircraft industry, etc. allow the existence of macroscopic cracks and crack-like defects. These codes contain a number of requirements that provide the specified level of structural reliability and safety in the presence of crack-like defects.

The following modes of failures in the presence of macroscopic cracks may be listed:

- the attainment by a macroscopic crack to a stated, specified size (e.g., 1/4, 1/3, etc. of the wall thickness of piping or pressure vessel);

- the formation of a through crack in a pipe, pressure vessel, etc., i.e., the loss of structural integrity of a pressurized component;

- the approach of a crack to critical size after which unstable, dynamic crack propagation begins;

- the approach of a crack to the size that corresponds to the loss of load-carrying capacity of a structural component with respect to other loadings (e.g., buckling of a compressed component due to the stiffness drop as a result of crack growth).

All the listed cases require the integration of fatigue crack growth equations for the given initial conditions and non-failure requirements. In particular, the fracture toughness parameters of material must be known to estimate the level of safety factors with respect to crack instability.

The simplest equation of fatigue crack growth is (1.29). Let us assume that $K = Y s_\infty (\pi a)^{1/2}$, where the correction factor Y does not depend on the crack size. Integrating (1.29) for the initial condition $a(N_0) = a_0$, it is easy to obtain that

$$c(N - N_0)(Y\pi^{1/2}\Delta s_\infty)^m = \begin{cases} \log(a/a_0), & \text{if } m = 2; \\ m_1^{-1}(a_0^{-m_1} - a^{-m_1}), & \text{if } m \neq 2 \end{cases}$$

Here $m_1 = (m/2) - 1$. At the given a_0 and N_0, the size $a(N)$ attained in the N-th cycle can be calculated.

The situations met in practice are not so simple. First, initial crack size, crack shape and its position are not known with certainty. Not only the stochastic nature of cracking, but also uncertainties, the lack of sufficient information, and the limited reliability of crack detection procedures are to be taken into account, and that requires the use of probabilistic and related models of reliability theory. Second, loading and environmental conditions are usually far from deterministic. Therefore, the critical crack sizes are subject to at least two sources of variability: the randomness of material properties that results in the scatter of the fracture toughness, and the temporal variability of current maximal stresses. The latter source also affects, in a crucial way, the critical crack size. For example, in situations when equations (1.10) and (1.12) are applicable, the critical crack size $a_c(N)$ is

$$a_C(N) = \frac{K_c^2}{\pi s_{max}^2(N)}$$

where $s_{max}(N)$ is the maximal stress randomly varying in time. Hence, both nominator and denominator are of stochastic nature.

General philosophy of the reliability approach in the presence of fatigue cracks has been discussed by many authors [3, 7, 43, 97, 137]. This philosophy is illustrated schematically in Figure 1.20 where sample functions $a(t)$ are shown. Three probability density functions are depicted there: that of the initial crack size $p(a_0)$; that of the size attained at the time T; and that of the critical size which, in general, varies in time, too. It is natural to use, in this case, a probabilistic safety index, say the probability that the critical size a_c will not be attained during the specified time (the planned lifetime of a structure, or an interval between inspections and maintenance procedures). Such approaches have been widely used in various branches of industry, among them aircraft and nuclear engineering.

However, the real problems are much more complicated than may be concluded from Figure 1.20. Apart from the complexities of structural models adequate for real engineering systems, and those of loading, actions and environmental conditions, two difficulties will be met which are of purely probabilistic nature. The first is the necessity to estimate very small probabilities (e.g., of the order 10^{-6} and even less) in the conditions of uncertain and insufficient information. The second difficulty

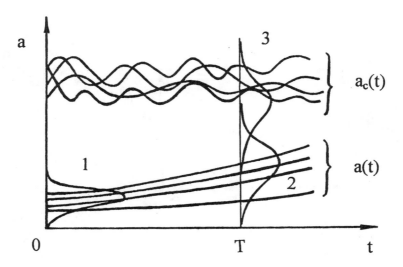

FIGURE 1.20
Transition from the stable crack growth to instability in the process of loading.

is connected to the character of processes leading to structural failures. Many of them are to be considered as excursions of random processes from certain admissible regions in the parameter space. This curcumstance requires the application of the theory of random crossings [7, 19, 97, 137]. The situation can be understood from Figure 1.20, where the critical crack size $a_c(N)$ varies in time nonmonotonously due to variability of $s_{max}(N)$. In complex structures subjected to a combination of loads and actions varying in time, the situation becomes more complicated.

There is an aspect of the engineering analysis of fatigue almost not touched in the present discussion. It is the counting of cycles and the identification of characteristic stresses in the case of irregular loading. More precisely, it is the problem of how to use the data relating to stationary sinusoidal loading in reliability analysis when the loading process consists of cycles with a complex structure.

Two typical examples are presented in Figure 1.21. Samples of stochastic or, maybe, pseudo-stochastic processes of loading are depicted there. In Figure 1.21a the loading is a narrow-band process. Without appealing to the concept of power spectral density, a narrow-band process can be defined as that for which the ratio of the mean number of maxima to the mean number of up-crossings of the mean level is close to unity. Oppositely, for broad band processes this ratio is large compared to unity. It is illustrated in Figure 1.21b. If a loading cycle is understood as a segment between two neighboring up-crossings of the mean stress s_m, it will contain internal cycles. Although the stress ranges of internal cycles are smaller than those of low-frequency components, they carry an input both into the total damage and the crack growth rate. It is rather clear how to count cycles and their characteristics if a process is narrow-band. In particular, the cycle number may be equalized by the number of maxima, and the stress ranges by the differences of sequential maxima and minima. The first thought concerning a broad-band process (Figure 1.21b) is to consider each segment as a cycle containing one maximum and

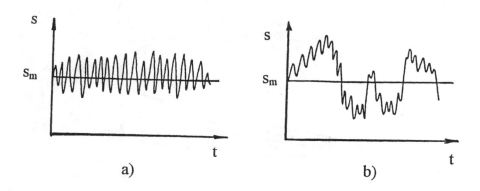

FIGURE 1.21
Stochastic loading processes: (a) narrow-band; (b) broad-band.

one minimum. But the resulting schematized loading process may be less damaging than the original process. The cause is that some of the high-range cycles occur hidden among the low-range cycles. To avoid that, special methods have been suggested, e.g., rain-flow counting. The prevalent idea is to begin counting from the internal high-frequency cycles in such a way that the residue contains the large-range, low-frequency cycles only. This special field of engineering design against fatigue is discussed in a number of handbooks and manuals [3, 46, 47, 84, 116].

1.11 Synthesis of Micro- and Macromechanics

Fatigue fracture and related phenomena such as the delayed fracture under sustained loading, the fatigue associated with creep, corrosion, irradiation, hydrogenization and other deterioration processes develop on two levels: on the level of material microstructure and on the macroscopic level. The dispersed damage accumulation is certainly a subject of micromechanics, while macrocrack formation and short crack growth are positioned on the boundary between micro- and macromechanics. It is less obvious that the propagation of "long," macroscopic cracks depends on the microdamage accumulation process, too.

The interaction between the microdamage near the crack tips and the energy balance in the system cracked body-loading is, in fact, the main mechanism that governs fatigue crack growth. A crack does not propagate until the force driving the crack is less than the corresponding resistance force. The magnitude of the resistance force depends on the microdamage in the domains close to the crack tip. A part of this microdamage is accumulated in the far field, before material particles approach the tip; but the larger part of microdamage is a result of stress-strain conditions in the near field, i.e., in the process zone. In any case, the crack advances a certain step when the resistance force decreases, because of microdamage, until a level where the system cracked body-loading becomes unstable with respect to dimensions of the crack. This approach may be considered as an adequate and, by the way, a pure mechanical model of the fatigue crack growth process.

The model described above has been suggested by Bolotin (1983) and developed later in a series of publications [14, 15, 18, 21]. This model allows inclusion of a wide variety of material properties, loads, actions and environmental conditions, as well as most of the damage and deterioration that are accompanied with cracking: high-cycle and low cycle fatigue, cracking of polymers, fatigue of metals associated with creep, corrosion fatigue, stress corrosion cracking, different modes of cracking and fracture of composite materials under cyclic or sustained loading.

To develop a theory that covers such a broad range of phenomena, a generalization of standard fracture mechanics appears to be necessary. Many approaches have been developed during the last decades to analyze behavior of cracked bodies. Most of them may be differentiated with respect to the choice of the generalized forces describing loading conditions, and of the corresponding resistance forces that characterize material toughness against crack propagation. From this viewpoint, one can distinguish the approaches based on the concepts of stress intensity factors, energy release rate, J-integrals, and crack tip opening displacements (K-, G-, J-

and δ-approaches, respectively). The limitations put on each approach are more or less evident. For example, the K-approach is of a local nature while the G-approach is based on global energy balance. The J-approach occupies an intermediate place. In simple problems, such as the classical Griffith's problem of linear fracture mechanics, all approaches are valid and produce identical or close results; but in more complicated situations one can see a number of limitations, controversies and other disadvantages of classical approaches. It is enough to refer to three-dimenional problems, e.g., to a plane crack of an arbitrary shape in a three-dimensional body. A lot of similar examples may be found when we turn to kinking, branching and interaction of cracks, not to mention problems concerning small (short, shallow) cracks.

All these problems have in common that the cracks are described not with a single parameter (say, like the half-length of a crack in Griffith's problem), but with two and more parameters. In some problems we meet a continuous set of parameters. To describe such problems adequately, a multiparameter extension of fracture mechanics is necessary. Since this extension covers various types of material behavior including microdamage accumulation, the theory can serve as a foundation for development of an adequate theory of fatigue crack growth.

In this book, using the synthesis of micromechanics and the extended version of fracture mechanics (referred here as analytical fracture mechanics), we study the process of crack nucleation and crack propagation until final fracture. Of course, no theory could predict quantitatively all the features of fatigue and related phenomena. It is impossible to describe in an analytical form a large number of factors affecting the safe life of structural components under loads and actions varying in time in an arbitrary way. The situation in fatigue, as in corrosion and wear, is in some sense paradoxical: all these phenomena are very complicated to treat analytically; meantime the experimentation (at least when it is interpreted on the phenomenological level) is rather simple. In such circumstances, the aim of the theory is rather modest: to build up a unified, consequent and transparent approach to the development of models of fatigue, remaining within the framework of rational mechanics and mechanics of solids.

1.12 Summary

A brief survey of mechanical aspects of fracture and fatigue was presented in this chapter. Both conventional approaches to design against fatigue and techniques based on the concepts of fracture mechanics were discussed. The need for future development was substantiated with up-to-date trends in various branches of engineering, such as the damage-tolerance approach. The possibilities were discussed to build up a consistent mechanical theory of fatigue based on the synthesis of the mechanics of dispersed damage accumulation (in the form of micromechanics or continuum damage mechanics) and the mechanics of cracked bodies, i.e., fracture mechanics in its standard sense.

Chapter 2

FATIGUE CRACK NUCLEATION AND EARLY GROWTH

2.1 Introductory Remarks

The aim of this chapter is to discuss analytical models for the prediction of fatigue crack nucleation and early growth. The preliminary stage frequently takes a significant portion of the total fatigue life. Moreover, even now, when the fail-safe and damage-tolerant concepts are widely spread in engineering applications, the formation of a macroscopic crack, e.g., a visible one, is sometimes considered as a kind of failure after which the structural component must be withdrawn out of operation. Putting apart structural safety considerations, one must state that the theory of crack nucleation and early growth enters, as an essential part, into the theory of fatigue. In particular, if a crack is born in a "natural" way, final conditions of the first stage become initial conditions for the stage of regular fatigue crack growth. We do not mention that some models used in the theory of fatigue crack nucleation intrinsically enter into the theory of fatigue crack propagation.

The problem of fatigue crack initiation and early growth may be referred to both micro- and macromechanics of fracture (see Section 1.5). In general, the border between micro- and macrocracks is rather vague. From the practical viewpoint, it is expedient to draw such a border, taking into account detectability of cracks with common nondestructive inspection techniques. Then the border depends on the current state of these techniques. It depends also on the scale of the material's microstructure. For example, the detectable size of cracks in aircraft engineering is of the order 1 mm, and those in large concrete structures may be much larger, say, of the order 10 mm. In addition, the distinction between micro- and macrocracks is complicated by the problem of so-called "short," "physically small" or just "small" cracks. Such cracks cover the domain between the cracks which may be surely referred to as micro- and macromechanics [90, 102]. To make the later presentation clearer, we name a crack localized within a grain a microscopic one, or briefly, microcrack. When a crack covers several grains, we call it mesoscopic crack, or mesocrack. Such cracks propagate both in the depth of a body and along the surface. In the latter case the crack is shallow but long, and the terms "short" and even "small" may seem misleading. Last, a crack that includes a large number of ruptured grains is referred to as a macroscopic crack, or macrocrack. From the viewpoint of analytical study, the most essential difference between meso- and macrocracks is that the growth of the former is controlled by local grain properties, and that of the latter by averaged properties. This difference is of a conceptual nature and has a strong influence on the choice of analytical models. Crack nucleation and early growth are to be described by probabilistic models of material properties while the propagation of macrocracks usually fits in purely deterministic models (if no other stochastic factors are involved).

Actually, one of the main features of fatigue crack nucleation is its stochastic character. For example, before a fatigue test of a smooth specimen neither the time (or cycle number) until initiation nor the location of a crack can be predicted deterministically. The scatter of fatigue test data is of common knowledge. Any consistent analytical model of fatigue crack nucleation must be probabilistic. Other phenomena related to fatigue, e.g., crack growth, kinking and branching, as well as final failure considered as a loss of integrity, instability, or both, are subjected to the influence of stochastic factors. Most of these factors can be assessed, at least in principle, from the statistical analysis of experimental data. So-called "uncertainties" also may be included in probabilistic models.

Random factors that enter true models of fatigue can be divided in three groups: randomness of material properties; random defects and imperfections of structural components; random loads, actions and environmental conditions. The two latter groups are of special significance for structural reliability and safety. These factors and their contribution to fatigue damage are the subject of special studies [19, 97, 137]. In this book only the randomness of material properties is considered. Three kinds of material randomness may be distinguished: within-specimen, specimen-to-specimen, and batch-to-batch. The within-specimen randomness is inherent to the microstructure of materials, i.e., to nonhomogeneities, flaws and imperfections on the level of grains, fibers, etc. This kind of randomness is inevitable due to the nature of materials and methods of their manufacturing. The scale of this randomness varies within a wide range. We do not discuss here the randomness that is a subject of solid-state physics. When we refer to microstructure, we mean the scales beginning from $1\,\mu$m and up to larger scales that are nevertheless small compared with characteristic dimensions of structural components. They may be of the order of 1 mm, and even more, e.g., of the scale of particles in large concrete structures. Not only properties distributed upon the volume of a body, but also properties of the surface, namely, its small-scale roughness, can be included into this group of random factors. Random fields with various properties are used to describe the randomness of microstructure on this level. The fields are continuous for amorphous polymers, and piece-continuous for multi-phase materials or materials with multiple microcracking.

The specimen-to-specimen randomness of mechanical properties is well known to experimenters who observe the scatter of bulk mechanical properties. This scatter is significant even if all specimens are made from the same sheet and carefully controlled before testing. As the characteristic size of microstructure is small compared with the size of specimens, and most of the properties expose themselves in an averaged form, only a part of the inherent material randomness is responsible for the specimen-to-specimen scatter. Another group of randomness is born from instabilities and imperfections of manufacturing. In fact, mechanical properties of commercial materials vary in a rather large range even when they are specified and controlled. Moreover, on the design stage of a structure, an engineer cannot be sure that the specified properties will be realized in the future. This uncertainty also can be described with a probabilistic model. Opposite to the within-specimen randomness described with random fields, the specimen-to-specimen and batch-to-batch randomness are described with a set of random variables. Among them, in general, may be parameters of random fields describing local material properties. For example, the fracture toughness may be a random function of coordinates within a specimen. Parameters of this function (mean fracture toughness, its variance, etc.) may be random variables characterizing the specimen-to-specimen and/or batch-to-batch scatter of material properties.

2.2   Phenomenological Models of Fatigue Crack Nucleation

The simplest models of fatigue crack nucleation are based on phenomenological scalar damage measures. Such a measure, denoted D, was discussed briefly in Section 1.9. This measure is related to a specimen as a whole or to its certain part, for example, to a unit of length, area or volume. The magnitude $D = 0$ means that there is no damage in the considered domain, and $D = 1$ corresponds to the state when this domain is interpreted to be damaged completely. Selection of states corresponding to complete damage is rather arbitrary. In the context of fatigue, the magnitude $D = 1$ might correspond both to the first macroscopic crack initiation as well as to the final failure due to fatigue crack growth up to the critical size.

In this section, we interpret "damage" as the initiation in the considered domain of at least one macroscopic crack, i.e., such a crack which, being detectable with conventional inspection procedures, can grow under loading in a more or less steady way up to the final failure. To describe the macrocrack initiation under cyclic loading we use the generalized Miner rule (Section 1.9). Let the equation of damage accumulation be

$$\frac{dD}{dN} = f[s(N), D(N)] \tag{2.1}$$

Here N is the running cycle number treated as a continuous variable. The right-hand side is a non-negative function that depends on a load parameter (or a set of such parameters) treated as a continuous or piece-continuous vector function $s(n)$ of the cycle number N. It is known that under certain conditions equation (2.1) is equivalent to the common Miner rule. It means that its solution satisfying the boundary conditions $D(0) = 0$, $D(N_*) = 1$ results in

$$\int_0^{N_*} \frac{dn(s)}{N(s)} = 1 \qquad (2.2)$$

Here N_* is the cycle number to complete damage, and $N(s)$ is to be taken from the standard $S - N$ curve.

Generally, the question remains open, what is the meaning of the notion "complete damage." In the pre-fracture-mechanics times, when engineering analysis had been based mostly on standard fatigue tests of specially fabricated specimens, the failure of a specimen was usually associated with the observation of the first macroscopic crack. Actually, for small, smooth and initially "flawless" specimens the crack nucleation phase takes $80 \ldots 90\%$ of the total fatigue life. Taking into account the large statistical scatter of the fatigue life, the duration of the second phase might be considered nonsignificant compared with that of the first phase. In any case, we treat later $N(s)$ in (2.1) and (2.2) as the cycle number to macrocrack initiation.

Probabilistic models based on equations (2.1) and (2.2) were studied by many authors. A survey of different approaches to the randomization of these equations can be found in [137]. Later on, we discuss in detail a special class of models leading to $S - N$ distributions [7, 19].

In the beginning, let us consider a body or its part subjected to steady cyclic loading. In particular, it can be the loading with the constant stress range at the fixed stress ratio. Take a small part of the body with a measure M_0. This measure may be, for example, the unit of the surface area if surface flaws are responsible for fatigue crack initiation. Assume that equation (2.1) is given in the form of power-threshold law

$$\frac{dD}{dN} = \begin{cases} 0, & N \leq N_0 \text{ or } s \leq s_{th}; \\ \dfrac{1}{N_c} \left(\dfrac{s - s_{th}}{r - s_{th}} \right)^m, & N > N_0 \text{ or } s > s_{th} \end{cases} \qquad (2.3)$$

Here $N_0 \geq 0$ is the cycle number at which the damage accumulation begins; N_c is a characteristic cycle number; s is the characteristic stress, for example, the applied stress range within a cycle; s_{th} is the threshold stress parameter analogous to endurance limit; m is the power exponent similar to the fatigue curve exponents in standard fatigue test presentation. The only random variable in (2.3) is r which can be interpreted as a characteristic resistance stress against fatigue crack initiation. The distribution of the random value r is supposed to be known.

Assume that for a part of the body with the measure M_0 the cumulative probability distribution function $F_0(r)$ of the random variable r is presented in the form:

$$F_0(r) \approx c(r - s_{th})^\alpha, \qquad r \geq s_{th} \qquad (2.4)$$

Here c and α are empirical deterministic constants (both positive). Equation (2.4) is valid only in the small vicinity of the threshold resistance stress s_{th} (dashed line in Figure 2.1). Let the measure for the whole body be M. Each of M/M_0 parts of the body may be responsible for macrocrack initiation. Under macroscopically homogeneous distribution of both stresses and mechanical properties upon all the

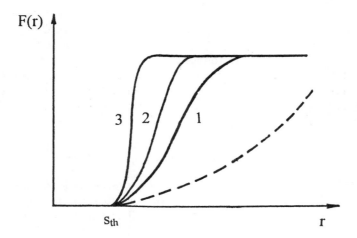

FIGURE 2.1
Cumulative distribution function for the resistance stress for various body measures $M_1 < M_2 < M_3$.

body, the weakest part will be responsible for damage. Hence, the probability distribution function for the ultimate stress is

$$F(r) = 1 - [1 - F_0(r)]^{M/M_0} \tag{2.5}$$

At $M/M_0 \gg 1$ equations (2.4) and (2.5) result into the asymptotic distribution

$$F_r(r) = 1 - \exp\left[-\frac{M}{M_0}\left(\frac{r - s_{th}}{s_c}\right)^{\alpha}\right] \tag{2.6}$$

with the notation $s_c = c^{-1/\alpha}$. Equation (2.6) is Weibull's distribution for the representative resistance stress r. At $r < s_{th}$ we must put $F_r(r) = 0$.

Solution of equation (2.3) at $D(0) = 0, s = \text{const} > s_{th}$ has the form:

$$D(N|s) = \left(\frac{s - s_{th}}{r - s_{th}}\right)^m \left(\frac{N - N_0}{N_c}\right) \tag{2.7}$$

If parameter r is deterministic, then, assuming in (2.7) the equality $D(N|s) = 1$, we obtain the equation of fatigue curve:

$$N = N_0 + N_c \left(\frac{r - s_{th}}{s - s_{th}}\right)^m, \qquad s > s_{th} \tag{2.8}$$

To obtain a probabilistic equivalent of (2.8), let us return to (2.6) characterizing the random property of r. The probability distribution function of the cycle number at the macrocrack initiation under a given loading level s is determined as follows:

$$F(N|s) = \Pr\left\{D(N|s) > 1\right\} = F_r\left[s_{th} + (s - s_{th})\left(\frac{N - N_0}{N_c}\right)^{1/m}\right]$$

Here $\Pr\{\cdot\}$ is the probability of the event enclosed in the braces. Taking into account (2.6) we obtain

$$F(N|s) = 1 - \exp\left[-\frac{M}{M_0}\left(\frac{s-s_{th}}{s_c}\right)^\alpha\left(\frac{N-N_0}{N_c}\right)^\beta\right] \qquad (2.9)$$

where notation $\beta = \alpha/m$ is introduced.

Equation (2.9) is in fact Weibull's distribution for the cycle number N to macrocrack initiation under the condition that the stress level s is given. When the cycle number N is given, we obtain from (2.9) Weibull's distribution for the critical stress level s corresponding to the given cycle number N

$$F(s|N) = 1 - \exp\left[-\frac{M}{M_0}\left(\frac{N-N_0}{N_c}\right)^\beta\left(\frac{s-s_{th}}{s_c}\right)^\alpha\right] \qquad (2.10)$$

Such a reciprocity reflects the fact that the weakest links of material resistance are responsible both for the minimal cycle number to crack initiation, and for the minimal critical stress level. The Weibull distribution, being an asymptotic distribution of extremes, is quite natural to describe the resistance against fatigue crack initiation. Namely, in connection with fatigue this distribution was introduced by Weibull (1939); further analysis can be found in [7].

Fractiles of the distributions $F(N|s)$, i.e., the roots of the equation $F(N|s) = \gamma$

$$N = N_0 + N_c\left(\frac{M_0}{M}\right)^{1/\beta}\left(\frac{s_c}{s-s_{th}}\right)^m\left(\ln\frac{1}{1-\gamma}\right)^{1/\beta} \qquad (2.11)$$

are shown schematically in Figure 2.2. The mean life to macrocrack initiation at given stress s is

$$\mathbf{E}[N(s)] = N_0 + N_c\left(\frac{M_0}{M}\right)^{1/\beta}\left(\frac{s_c}{s-s_{th}}\right)^m\Gamma\left(1+\frac{1}{\beta}\right) \qquad (2.12)$$

where $\mathbf{E}[\cdot]$ is the operator of averaging, and $\Gamma(\cdot)$ is the gamma function. The standard deviation of fatigue life is

$$\sigma_N(s) = N_c\left(\frac{M_0}{M}\right)^{1/\beta}\left(\frac{s_c}{s-s_{th}}\right)^m\left[\Gamma\left(1+\frac{2}{\beta}\right)-\Gamma^2\left(1+\frac{1}{\beta}\right)\right]^{1/2} \qquad (2.13)$$

Similarly, the mean critical stress corresponding to equation (2.10) is

$$\mathbf{E}[s(N)] = s_{th} + s_c\left(\frac{M_0}{M}\right)^{1/\alpha}\left(\frac{N_c}{N-N_0}\right)^{1/m}\Gamma\left(1+\frac{1}{\alpha}\right) \qquad (2.14)$$

and the standard deviation

$$\sigma_s(N) = s_c\left(\frac{M_0}{M}\right)^{1/\alpha}\left(\frac{N_c}{N-N_0}\right)^{1/m}\left[\Gamma\left(1+\frac{2}{\alpha}\right)-\Gamma^2\left(1+\frac{1}{\alpha}\right)\right]^{1/2} \qquad (2.15)$$

The considered model (as well as those discussed later) take into account, along with the scatter of fatigue data, another important phenomenon — the scale factor in fatigue. Sensitivity of the critical cycle number with respect to the dimension of a body enters into (2.11)-(2.13) with the multiplier $(M_0/M)^{1/\beta}$. Sensitivity of the stress level enters into (2.14)-(2.15) with the multiplier $(M_0/M)^{1/\alpha}$. Equations (2.9)-(2.15) correspond to homogeneous macroscopic stress fields in the considered

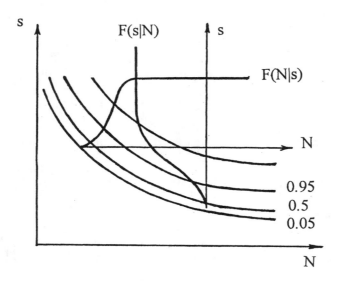

FIGURE 2.2
Fatigue curves corresponding to various probabilities of fatigue failure.

domain. When stresses and/or mechanical properties are heterogeneous, and/or there are several groups of flaws responsible for crack initiation, more sophisticated probabilistic models are needed. In particular, a more general form of probability distribution is

$$F_N(N|s) = 1 - \exp\left[-\left(\frac{N - N_0}{N_c}\right)^\beta \int\limits_M \left(\frac{sg(x) - s_{th}}{s_c}\right)^\alpha \frac{dM}{M_0}\right] \quad (2.16)$$

with the stress field characterized with function $g(x)$ and the loading parameter s. In the book by Bolotin [7] other suggestions can be found, e.g., the fatigue life distribution for the case when both surface and volume flaws are responsible for crack initiation.

If a structure consists of subsystems with different mechanical properties and they are differently loaded, obvious generalizations of (2.16) are to be used such as

$$F_N(N|s) = 1 - \exp\left[-\sum_i \left(\frac{N_i - N_{0,i}}{N_{c,i}}\right)^{\beta_i} \int_{M_i} \left(\frac{sg_i(x) - s_{th,i}}{s_{c,i}}\right)^{\alpha_i} \frac{dM_i}{M_{0,i}}\right] \quad (2.17)$$

Here M_i is the measure of i-th component or a part of a component, and the subscript i at other variables entering (2.17) relates to i-th component or its part. The manifolds $M_1, M_2, \ldots$ may be one-, two- and three-dimensional. For example, beams and columns are usually treated as one-dimensional manifolds, plates and shells as two-dimensional ones.

Note that Weibull's distribution is not the only family of distributions used in phenomenological modelling of fatigue. Other distributions of extremes may be used. For example, assuming instead of (2.4) the probability distribution in the form

$$F_0(r) \approx c \exp(r - s_{th}), \qquad r \geq s_{th} \tag{2.18}$$

we come to the double-exponential (Gumbel's) distribution

$$F_r(r) = 1 - \exp\left[-\frac{M}{M_0} \exp\left(\frac{r - s_{th}}{s_c}\right)\right]$$

and to the conditional distribution for the cycle number to macrocrack initiation

$$F(N|s) = 1 - \exp\left\{-\frac{M}{M_0} \exp\left[\left(\frac{s - s_{th}}{s_c}\right)\left(\frac{N - N_0}{N_c}\right)^{1/m}\right]\right\} \tag{2.19}$$

Fractiles of this distribution are

$$N(s) = N_0 + N_c \left(\frac{s_c}{s - s_{th}}\right)^m \log\left[\frac{M_0}{M} \log\left(\frac{1}{1 - \gamma}\right)\right]^m \tag{2.20}$$

Compared with equations (2.11) and (2.13) corresponding to power law fatigue curves, equation (2.20) describes an exponential relationship between the stress level s and the cycle number N. Note that the parameters entering equations (2.18)-(2.20), in general, differ from those in (2.4)-(2.6). In engineering practice other distributions are used, in particular, log-normal ones. There is a powerful argument in favor of Weibull's and Gumbel's distributions: they are asymptotic distributions of extremes.

2.3   Models of Continuum Damage Mechanics

Continuum damage mechanics is in fact an extension of classical mechanics of deformable media for modelling of various types of damage on the mesoscopic level. Compared with the theories of elasticity, plasticity, etc., additional field variables are introduced in continuum damage mechanics characterizing in some averaged, statistical or homogenized way the distribution of microcracks, micropores and other defects. From another viewpoint, continuum damage mechanics is a phenomenological theory with internal variables which are governed with additional kinetic equations. These variables enter into the constitutive equations along with common stresses, strains, and other field variables of continuum mechanics. However, these internal variables can be directly connected with microdamage distributions being interpreted in terms of microcrack density, porosity, etc. In the ideal case, equations of continuum damage mechanics are to be obtained from micromechanical models with the use of any kind of homogenization such as statistical, self-consistent, or energy smoothing approaches. In reality, as the previous experience has shown, many models of continuum damage mechanics were introduced in a purely phenomenological way with or without their interpretation in terms of micromechanics.

Continuum damage mechanics was initiated by Kachanov (1958) and Rabotnov (1963). They introduced, mainly with applications to creep of materials, a scalar microdamage measure ω that characterizes intensity of microcracking. This measure takes magnitudes from the segment [0,1]. It is assumed equal to zero in the case when there is no damage, and to unity for completely damaged material. Damage is assumed to be local: ω is a function of coordinates and of time, cycle number, or other temporial variable. In any case, $\omega(x)$ is a scalar field.

In terms of uniaxial tension, ω can be interpreted as the ratio of the damaged area in the cross section of a specimen to the total cross section area. Noting the applied (nominal) stress σ, we estimate the "true" stress as $\sigma/(1-\omega)$. The simplest equation of damage accumulation connects the damage accumulation rate $\partial\omega/\partial t$ with a certain power of the "true" stress:

$$\frac{\partial\omega}{\partial t} = \frac{1}{t_c}\left[\frac{\sigma}{\sigma_d(1-\omega)}\right]^{m_d} \qquad (2.21)$$

Here σ_d and m_d are material parameters that, generally, depend on temperature. The time constant t_c can be chosen arbitrarily, say, equal to the unity of time. Complementing (2.21) with the analogous equation for the creep strain ε

$$\frac{\partial\varepsilon}{\partial t} = \frac{\varepsilon_0}{t_c}\left[\frac{\sigma}{\sigma_c(1-\omega)}\right]^{m_c} \qquad (2.22)$$

we arrive to a set of constitutive equations for the considered simplified case. Here σ_c and m_c are material parameters, too. The same parameter, σ_d, may be used both in (2.21) and (2.22) if the additional parameter ε_0 of the dimension of strain is introduced in (2.22). There is a wide freedom in how to choose the right-hand sides in (2.21) and (2.22).

The methods of generalization of the above equations are obvious. Various three-dimensional models taking into account nonhomogeneity of stress, strain and temperature fields are discussed in [86, 91, 95]. The general structure of equations with respect to damage tensor ω and strain tensor ε is of the form

$$\frac{\partial\omega}{\partial t} = \Phi_d(\sigma,\omega,T), \qquad \frac{\partial\varepsilon}{\partial t} = \Phi_\varepsilon(\sigma,\omega,T) \qquad (2.23)$$

The right-hand sides in (2.23) are functions of stress tensor σ, damage tensor ω, and temperature T. The damage accumulation is usually assumed being governed with a certain characteristic stress: mean normal stress (for micropore formation), maximal normal stress (for opening mode microcracking), the deviatoric stress intensity (for plastic slips formation), etc.

A group of mostly consistent models was developed in connection with creep of metals. They, as a rule, take the origin in mechanical models of micropores nucleation and growth [72, 105]. The coalescence of micropores is sometimes also discussed. Until micropores can be treated as spherical, microdamage remains isotropic both in the micro- and macroscale. This allows to develop constitutive equations of continuum damage accumulation and to describe the influence of damage on bulk material properties such as elastic and creep compliances.

To take into account microcrack orientation and, in general, three-dimensionality of microdamage, tensorial measures are needed. A typical tensor model is proposed in the paper by Vakulenko and Kachanov [150]. Each microcrack is considered as a discontinuity in a body. It is characterized with the crack area S and the jump of the displacement vector $\mathbf{u}$ on the crack surfaces: $\mathbf{b} = \mathbf{u}_+ - \mathbf{u}_-$. Evidently that vector $\mathbf{b}$ is similar to Burger's vector in the theory of dislocations (it is stressed in the notation). Let $\mathbf{n}$ be the unity normal vector to S. Introduce the direct product of vectors $\mathbf{b} \otimes \mathbf{n}$, i.e., a second rank tensor. After symmetrization and integration upon the crack surface, we obtain

$$\omega_s = \frac{1}{2} \int_S (\mathbf{b} \otimes \mathbf{n} + \mathbf{n} \otimes \mathbf{b}) dS \qquad (2.24)$$

Equation (2.24) is valid for opening, shear and mixed mode cracks, though fails in the case of crack closure. Assume the additivity of the tensor measure ω_s. To obtain a smoothed measure, let us draw a small sphere around the considered point of the media and summarize all magnitudes ω_s for the cracks located within this sphere. The resulting value is

$$\omega = \frac{1}{2} \sum_k \frac{1}{V} \int_{S_k} (\mathbf{b} \otimes \mathbf{n} + \mathbf{n} \otimes \mathbf{b}) dS_k$$

where V is the sphere volume. The right-hand side is also a second rank tensor. Considered as a function of coordinates, this tensor could be used as an additional field variable to build up a continuum damage theory. It is to be noted that the first invariant of ω is in fact the specific porosity, i.e., the characteristic which is equivalent to Kachanov-Rabotnov's scalar measure ω.

Most of tensorial models of continuum damage mechanics are rather vulnerable. For example, it is vague how to make distinction between spheroidal pores and sharp cracks using equation (2.24). There are a lot of other problems: how to account for the microcracks interaction, the presence of several different mechanisms of damage, local stress and strain fields, etc. A survey of various approaches to development of continuum damage mechanics as well as a comprehensive list of references up to the first half of the 80's can be found in the books by Kachanov [73], Krajcinovic [86] and Lemaitre [91] and the survey paper by Bažant [6]. Problems of homogenization are discussed in [74].

Apart from tensorial field variables, the sets of scalar variables may be used for the quantative description of dispersed damage. Two and more scalar fields are needed when there are different sources of damage. In particular, we have to distinguish the damage under cyclic loading, under sustained loading, and the damage produced by single short-time overloadings. Various types of damage of a nonmechanical nature are met in engineering systems: chemical, electro-chemical, biological, radiational. Generally, there is an interaction between various types of damage, and the typical equations governing the damage accumulation are:

$$\frac{\partial \omega_j}{\partial t} = \Phi_j(\mathbf{s}, \omega), \qquad (j = 1, \ldots, n) \qquad (2.25)$$

Here $\mathbf{s}$ is the set of parameters describing loads and actions, and ω is the set of damage parameters. Both are functions of time and coordinates. In classical fatigue time t is to be replaced with the cycle number N. However, in more general cases as environmentally assisted fatigue, the use of the calendar or operating time t is more appropriate.

2.4 Continuum Approach to Fatigue Crack Nucleation

Continuum damage mechanics is a useful tool to generalize Miner's rule and related phenomenological approaches in terms of continuum mechanics. The main idea is to connect the damage field with the probability of formation of even a single nucleus of the macroscopic fatigue crack in the considered domain of a body. This idea realized by Bolotin [8] is presented briefly below.

Introduce a scalar function $\|\omega(\mathbf{x}, N)\|$ of the local damage measure $\omega(\mathbf{x}, N)$. Here $\mathbf{x}$ is a spatial independent variable, say the reference vector, and the cycle number N is a temporal variable treated as a continuous parameter. In the simplest case when ω is a scalar variable with magnitudes from $[0,1]$, we have $\|\omega\| \equiv \omega$. In other cases it may be a norm of vector, an invariant of tensor or a function of such invariants. The limitations are that $\|\omega\| = 0$ for nondamaged material, $\|\omega\| = 1$ for completely damaged material, and $\|\omega(\mathbf{x}, N)\|$ grows in time during damage accumulation. We call this variable the local norm of the field $\omega(\mathbf{x}, N)$, though the functional space $\omega(\mathbf{x}, N)$ and the scalar field $\|\omega(\mathbf{x}, N)\|$, generally, do not satisfy all the conditions of normalized spaces.

In the beginning, consider macroscopically homogeneous deterministic stress, strain and microdamage fields in a certain domain M_0 of the body. It may be, for example, the volume or surface area of a standard specimen. Experimenters usually consider a specimen damaged when they detect the first macroscopic crack. Therefore, if $\|\omega\| = 1$ in the domain with measure M_0, we may expect that in average one macroscopic crack appears in this domain. Moreover, we assume that the mean number of macrocracks initiated in domain M_0 on time segment $[0, N]$ is a function of the norm $\|\omega(t)\|$ of measure $\omega(t)$. Notate with $k(N)$ the point-wise process of macrocrack initiation in M_0, and with $\mu(N)$ its mathematical expectation (mean value), i.e.,

$$\mu(N) = \mathbf{E}[k(N)]$$

Postulate that the process $\mu(N)$ is a function of $\|\omega(N)\|$:

$$\mu(N) = f(\|\omega(N)\|) \tag{2.26}$$

The function $f(\cdot)$ is differentiable and satisfies the conditions $f'(\cdot) > 0, f(0) = 0,$ $f(1) = 1$. A typical example is

$$f(\omega) = \omega^\beta \tag{2.27}$$

where ω is a scalar measure, and $\beta > 1$.

Formula (2.26) corresponds to a homogeneous distribution of microdamage in the domain with measure M_0. The natural generalization of (2.26) upon nonhomogeneous fields is

$$\mu(N) = \int_M f(\|\omega(\mathbf{x}, N)\|) \frac{dM}{M_0} \tag{2.28}$$

If the domain M consists of subdomains $M_1, M_2, \ldots$ of dimensions which may be different, we obtain, similarly to (2.17), the equation

$$\mu(N) = \sum_i \int_{M_i} f_i(\|\omega_i(\mathbf{x}, N)\|) \frac{dM_i}{M_{0i}} \tag{2.29}$$

As a rule, in the further discussion we will refer to equation (2.26) with the scalar measure ω, and $\|\omega\| \equiv \omega$. Generalizations based on (2.28), (2.29) and tensorial measures ω are easy to perform.

To connect the probability of formation of macrocracks with the expected value of their number, a Poisson model was used in [19]. In fact, the formation of nuclei of macroscopic cracks is a rare event. The characteristic size of macrocracks a_* is small compared with the sizes of the body, and interaction between neighboring nuclei may be considered as negligible. Then the probability of formation of k macrocracks in the domain M_0 is determined as

$$Q_k = \frac{\mu^k}{k!} \exp(-\mu); \qquad (k = 0, 1, \ldots) \tag{2.30}$$

Substituting (2.26) in (2.30) at $\|\omega(x, N)\| \equiv \omega(N)$ results in

$$Q_k(N) = \frac{f^k[\omega(N)]}{k!} \exp\{-f[\omega(N)]\}; \qquad (k = 0, 1, \ldots)$$

The probability that at least one macrocrack initates at cycle number N is

$$Q(N) = 1 - \exp\{-f[\omega(N)]\} \tag{2.31}$$

The right-hand side of (2.31) coincides with the distribution function $F_*(N_*)$ of cycle number N_* at the appearance of the first macrocrack. The discussed model is illustrated in Figure 2.3a. There the damage measure $\omega(N)$ and the expected number of macrocracks $\mu(N)$ are shown varying in time. A realization of the stepwise process $k(N)$ is presented there also. It is assumed that in (2.26) the equality $\mu = \omega^2$ holds. Figure 2.3b shows the growth of the sizes of macroscopic cracks from the initial magnitude a_0 considered as the size of a macrocrack nucleus.

If the damage field $\omega(\mathbf{x}, N)$ is nonhomogeneous in the domain M but varies there sufficiently slowly, we obtain with use of (2.28) that

$$F_*(N_*) = 1 - \exp\left\{-\int_M f[\mathbf{x}, \omega(N)] \frac{dM}{M_0}\right\} \tag{2.32}$$

Here the notation of the left-hand side is changed putting $Q(N_*) \equiv F_*(N_*)$.

The distribution given in (2.31) is related to Weibull's distribution. In special cases it coincides with the latter, in particular with the distribution given in (2.9). In fact, let the equation related to (2.3) be in the power-threshold form

$$\frac{\partial \omega}{\partial N} = \begin{cases} 0, & s \leq s_0 \text{ or } N \leq N_0; \\[2mm] \dfrac{1}{N_c}\left(\dfrac{s - s_0}{s_c}\right)^m, & s > s_0, N > N_0 \end{cases} \tag{2.33}$$

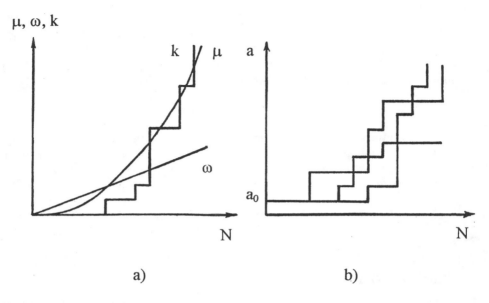

FIGURE 2.3
Model of crack initiation: (a) damage measure and expected number of macro-cracks; (b) samples of early crack initiation.

Here $N_c > 0$ is a constant, and $N_0 \geq 0$ is a certain threshold cycle number necessary to initiate the damage accumulation process. Other parameters are of the same meaning as in (2.3). At $s = \text{const} > s_0$, $\omega(0) = 0$, equation (2.33) gives

$$\omega(N) = \left(\frac{s - s_0}{s_c}\right)^m \left(\frac{N - N_0}{N_c}\right) \tag{2.34}$$

Substituting (2.27) and (2.34) into (2.31) we obtain

$$Q(N) = 1 - \exp\left[-\left(\frac{s - s_0}{s_c}\right)^\alpha \left(\frac{N - N_0}{N_c}\right)^\beta\right] \tag{2.35}$$

where similarly to (2.9) notation $\alpha = m\beta$ is used. The only difference compared with (2.9) is that the multiplier M/M_0 is absent. It is obvious that replacing $\omega^\beta(N)$ with $(M/M_0)\omega^\beta(N)$ or using (2.28), we reconstruct equation (2.9) completely.

Equation (2.35) is not the only consistent relationship between the number of macroscopic cracks nucleated due to microdamage accumulation. Moreover, if the microdamage $\omega(x, N)$ is a random function of space and/or time (for example, if loading is stochastic), equation (2.26) becomes invalid since $\mu(N)$ is by definition a deterministic function. At least three ways to overcome this obstacle may be listed under which the basic relationship (2.26) remains valid for deterministic functions $\omega(N)$. First, $\omega(N)$ can be replaced with its expected value. Second, the averaging procedure can be applied to the right-hand side of (2.35). Third, the most consistent way, one can interpret probabilities $Q_k(N), Q(N)$ and $F_*(N)$ in (2.30)-(2.32) as conditional ones. For example, in the latter approach instead of (2.31) the resulting probability is

$$Q(N) = 1 - \int_0^\infty \exp[-f(\omega)] \, dF_\omega(\omega; N) \tag{2.36}$$

Here $F_\omega(\omega; N)$ is the cumulative distribution function of the measure ω at the cycle number N. A more general equation for the probability $Q(N)$ is

$$Q(N) = \Pr\left\{\sup \|\omega(x, N)\| \geq \omega_\star; x \in M\right\}$$

that takes into account both randomness of the damage field, and its nonhomogeneity in the domain M. The special forms of equation (2.36), etc. are given here as examples only to demonstrate possibilities to obtain the final equations similar to those commonly used in the statistical treatment of fatigue data.

2.5 Micromechanical Models of Fatigue Crack Nucleation

In the preceding analysis, macrocrack initiation was treated in the framework of continuum mechanics. Now we are going to show that similar results can be obtained considering damage accumulation and macrocrack formation on the microstructural level.

From the viewpoint of micromechanics, fatigue fracture is a result of a long chain of events developing on various scales. It is practically impossible to build up a unified theory beginning, say, from diffusion of atomic vacancies. It is a long way from the concepts of solid-state physics to the reliable quantitative prediction of plastic deformation and fracture on the macroscopic scale. At least it is expedient to build up a practical and inherently consistent theory of fatigue going from the level of grains, micropores, microcracks, and microinclusions. Multiple microcracking and microvoid formation are usually treated on the mesomechanical level, in the framework of mechanics of continua. Microcracks and macrovoids are assumed to form regular deterministic arrays or, at least, homogeneously distributed random sets. The same approach is used when the interaction takes place between a macroscopic crack and a set of microdefects on the vicinity of this crack. The microstructure of real materials with their essential nonhomogeneities cannot be covered with such models. Moreover, the initiation of small fatigue cracks is complicated with the microcracks-microvoids-grains-boundaries interaction. Most of the fatigue cracks are initiated near the surface. They develop primarily, before their transformation into macrocracks, as a kind of shear cracks that find the easiest way to propagate among randomly nonhomogeneous materials.

One of the patterns of fatigue failure is the total loss of integrity due to microdamage accumulation dispersed upon the total volume of a body or its critical part. This type of failure may occur in some composite materials, for example those with brittle fibers and comparatively ductile matrix.

Consider a body with a standard measure M_0 that consists of a large number n_0 of microstructural elements. Let in the beginning the nominal stress be distributed in the body uniformly. Introduce the damage measure ψ as the ratio of the number of ruptured elements n to their total number n_0, i.e.,

$$\psi = n/n_0 \tag{2.37}$$

The measure ψ, generally, does not coincide with the measure ω in (2.27). To achieve the correspondence, ψ has to be properly normalized. For example, we may put $\omega = \psi/\psi_{**}$ where ψ_{**} is the critical magnitude of ψ which is not necessarily equal to unity.

Further discussion is performed assuming that damage density is sufficiently small, so that it is possible to neglect the interaction between fractured elements. At the same time, the number of fractured elements n is supposed to be large. For example, a standard specimen of carbon steel contains $10^6 \dots 10^7$ grains. Then even at $\omega = 10^{-2}$ the number $n = \psi n_0$ is very large. Thus, we assume that

$$n_0 \gg 1, \qquad \psi n_0 \gg 1, \qquad \psi \ll 1 \tag{2.38}$$

Considering the damage accumulation measure as a function of cycle number N, and denoting the life of elements with N_k, equation (2.37) results into

$$\psi(N) = \frac{1}{n_0} \sum_{k=1}^{n_0} H(N - N_k) \tag{2.39}$$

where $H(\cdot)$ is the Heaviside function. Due to the assumption of mutual independence of fractured elements, and taking into account (2.38), we use the Bernoulli scheme and Moivre-Laplace limit theorem. Then the asymptotic probability distribution function for the damage measure ψ at the cycle number N takes the the form

$$F_\psi(\psi; N) \approx \Phi\left\{ \frac{[\psi - F_0(N)]n_0^{1/2}}{(F_0(N)[1 - F_0(N)])^{1/2}} \right\} \tag{2.40}$$

Here $\Phi(z)$ is the normalized Gaussian distribution function, i.e.,

$$\Phi(z) = \frac{1}{(2\pi)^{1/2}} \int_{-\infty}^{z} \exp\left(-\frac{u^2}{2}\right) du$$

The probability distribution function $F_0(N)$ of the cycle number at the fracture of an arbitrarily chosen element enters into equation (2.40). The step-wise function given in (2.39) is replaced with an appropriately smoothed function. It follows from (2.39) and (2.40) that the variance of ψ is very small. Ignoring the scatter of ψ, we come to a semi-deterministic approximation

$$\psi(N) \approx F_0(N) \tag{2.41}$$

which will be used in the discussion below.

Estimate the probability that in the domain M at least one nucleus (a set of n_* neighboring fractured elements) will be born at cycle number N. Generally, it is a rather complicated problem. Various combinations of fractured elements, stress redistribution and stress concentration around these elements are to be taken into account to build up a complete space of independent events. These difficulties are wellknown in composite mechanics. Even in the case of simplified single-row fiber models, the total number of combinations of single ruptures and load redistribution grows tremendously even at a moderate number of elements. An heuristic approach was discussed in [11] in connection with the fracture of fiber composite materials.

The cycle number N_* to macrocrack initiation has been evaluated based on the following scheme. Initiation occurs when, in the considered domain, at least one fractured element appears with $n_* - 1$ fractured neighbors. Assuming that all these events take place at $N = N_*$, and neglecting the load redistribution, the probability distribution function $F_*(N_*)$ is determined as follows

$$F_*(N_*) = 1 - \left\{ 1 - [\psi(N_*)]^{n_*-1} \right\}^{\psi(N_*)n_0}$$

The asymptotic equivalent of this equation is

$$F_*(N_*) = 1 - \exp[-n_0 \psi^{n_*}(N_*)] \qquad (2.42)$$

Generalizing this equation upon nohomogeneous fields and using (2.41), we obtain

$$F_*(N_*) = 1 - \exp\left(-n_0 \int_M F_0^{n_*}(N_*) \frac{dM}{M_0} \right) \qquad (2.43)$$

Equations (2.42) and (2.43) are similar to (2.31) and (2.32). This analogy is a natural consequence because both models connect macrocrack initiation with the rupture of weakest elements of microstructure.

To apply equations (2.42) and (2.43), the distribution function $F_0(N)$ of the fatigue life of individual elements must be known. That requires the introduction of a special model of damage accumulation in a single randomly chosen element. If the microdamage measure $\chi(N)$ is a scalar variable, equations similar to (2.33) are applicable. For example, assume the equation

$$\frac{d\chi}{dN} = \begin{cases} \dfrac{1}{N_c} \left(\dfrac{s - s_{th}}{s_c} \right)^m, & s > s_{th}, \ N > N_0 \\ 0, & s \leq s_{th} \text{ or } N \leq N_0 \end{cases} \qquad (2.44)$$

where, opposite to (2.33), parameters m, N_c, s_c and s_{th} relate to an element of microstructure, e.g., to a grain or a fiber segment with attached matrix. Let the characteristic stress be applied to the element, for example, the range of the tensile stress during a cycle $s = $ const. Then, repeating the calculations resulting in (2.35), we obtain

$$F_0(N) = 1 - \exp\left[-\left(\frac{s - s_{th}}{s_c} \right)^\alpha \left(\frac{N - N_0}{N_c} \right)^\beta \right] \qquad (2.45)$$

The right-hand side of (2.45) according to (2.41) is the damage measure $\psi(t)$ attained at cycle number N. Substitution of (2.45) into (2.43) at $\psi \ll 1$ results:

$$F_*(N_*) = 1 - \exp\left[-n_0 \int_M \left(\frac{s - s_{th}}{s_c} \right)^{n_*\alpha} \left(\frac{N - N_0}{N_c} \right)^{n_*\beta} \frac{dM}{M_0} \right] \qquad (2.46)$$

Compared with (2.45), formula (2.46) contains power exponents $n_*\alpha, n_*\beta$ instead of α, β, respectively. One of the reasons is that the parameters entering in equations (2.45) and (2.46) in fact relate to different levels of consideration and, therefore, they could have quite different magnitudes. By the way, for surface initial cracks in granular alloys the power exponent n_* takes magnitudes $n_* = 3$ or maybe 4. For internal cracks n_* is much larger, $n_* = 4 \ldots 6$ and even more. For example, if the packing of grains is cubic-centered, the number of neighbors $n_* - 1 = 8$, and that results in $n_* = 9$. Combining results of direct observations of the fracture process with the statistical treatment of fatigue tests and using relationships similar to (2.46), one can estimate parameters entering (2.44). The discussed model is useful for the prediction of mechanical properties of composite materials. The properties of components (fibers, particles, interface) usually may be treated as known. But composites very often appear as newborn materials with properties that must be predicted beforehand. Some more advanced models of fatigue crack nucleation in composite materials will be discussed in Chapter 9.

2.6 Formation and Early Growth of Fatigue Cracks

Two principal patterns of formation of macroscopic fatigue cracks are to be distinguished. First is the coalescence of microdefects into a connected damaged domain. This is possible as a result of clustering of defects randomly distributed in a body or at its surface. Damaged grains may be separated with a number of nondamaged ones, but the bridges between them are to be ruptured due to stress concentration around the interacting flaws. The second mechanism consists of the formation of mesocracks, their growth along the surface and the beginning of their in-depth penetration. The first pattern is met in composite materials such as unidirectional fiber composites. The second pattern is more typical for polycrystalline materials. This case is discussed in the present section.

It is well known that there is a significant difference between the behavior of mesoscopic and macroscopic cracks. This difference is illustrated in Figure 2.4 where the crack growth rate da/dN is plotted versus the range ΔK of the stress intensity factor K. The mode I fatigue cracks are discussed here; however, we omit the subscripts I such as in ΔK_I. The branches 1 and 2 correspond to mesocracks, the branch 3 to macrocracks. When a crack is a small enough covering, say, one or two grains in the body's depth, diminishing of the rate da/dN is frequently observed. Most micro- and mesocracks, being the candidates to develop in macroscopic cracks, stop growing at all, while some of them begin to penetrate into depth; then we come to branch 3.

Another peculiarity of mesoscopic fatigue cracks is significant scatter of their dimensions and, respectively, of their growth rates. This behavior is the result of random structure of real materials, polycrystalline ones in particular. There is a scatter of mechanical properties of grains (elastic moduli, yield limits, ultimate stresses) as well as of the grains' orientation with respect to the directions of applied stresses and crack growth. When the number of ruptured grains entering the cracked domain is small, the effect of local material randomness is essential. The scatter decreases when the number of grains crossed by the crack front becomes large, i.e., when the crack becomes a macroscopic one.

Consider a body whose surface contains a number of microcracks. The aim is

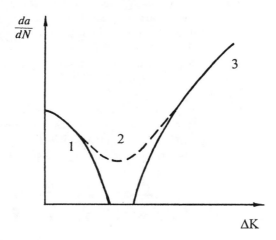

FIGURE 2.4
Early crack growth diagrams (lines 1 and 2 correspond to mesocracks, line 3 to macrocracks)

to follow the process of damage accumulation in the vicinity of a crack consisting of a moderate number of ruptured grains, and the further transformation of this mesocrack into a regular macroscopic fatigue crack. It is obvious that such a problem, stated in the whole scale, is very complicated. A number of simplifications should be used to make the study more transparent.

Let a mode I crack be initiated from the surface of a body subjected to cyclic tension with remotely applied stresses σ_∞. The nucleus of a crack is modelled as an initial surface notch with the characteristic size h. Due to damage accumulation in the neighboring grains some of them become ruptured. This means the propagation of the cracked area from the nucleus in a body and the formation of a mesocrack. For simplification we assume that only the grains situated in the cross-section plane are subjected to rupture. Hence, the crack is modelled as a slit of constant thickness h and of arbitrary shape in the plane (Figure 2.5). It means that we neglect kinking, branching, crack tip blunting, and other effects accompanying the fatigue crack propagation in real polycrystalline materials. With the intention of applying the finite element technique and eight-knot brick-work mesh, we present grains as elementary cubical bricks with dimensions h. In real materials, it may be $h = 10~\mu$m, or $100~\mu$m, etc. Mechanical properties are randomly distributed among the grains. The first-row grains are assumed weaker than those situated in the depth. This takes into account the presence of surface flaws and their significance for fatigue cracks initiation. However, the properties are assumed distributed independently among the neighboring grains. Among material parameters that are of stochastic nature, there are a lot of variables such as elastic moduli, yield stresses, ultimate stresses. When we consider grains as crystals, we must take into

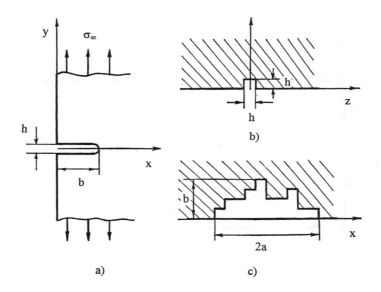

FIGURE 2.5
Modeling of mesocrack growth from a nucleus.

account their anisotropy and random distribution of their orientation with respect to applied stresses and crack growth directions. Involvement of these factors makes the problem too cumbersome. To minimize the number of parameters subjected to randomization, we assume that all the grain properties are given by a random value s interpreted as the resistance stress against damage accumulation. Compared with this resistance stress, the scatter of compliance characteristics such as elastic moduli seems to be less significant. In principle, following the idea of stochastic finite element techniques, one can take into account this kind of randomness, too.

As in Section 2.5, the grain damage is discussed in terms of continuum damage mechanics using a scalar damage measure $\chi(N)$. Here $0 \leq \chi \leq 1$, and the limits correspond to virgin and ruptured grains, respectively. For each grain a separate equation of damage accumulation is used. In this study we assume this equation is in the form similar to equation (2.44)

$$\frac{d\chi}{dN} = \begin{cases} \left(\dfrac{\Delta\sigma - \Delta\sigma_{th}}{r} \right)^m (1-\chi)^{-n}, & \Delta\sigma > \Delta\sigma_{th} \\ 0, & \Delta\sigma \leq \Delta\sigma_{th} \end{cases} \qquad (2.47)$$

Here $\Delta\sigma$ is the range of the tensile stress in the considered grain; $r, \Delta\sigma_{th}, m$, and n are material parameters. The resistance stress r is a random value; the power exponents m and n are given deterministically. The threshold resistance stress $\Delta\sigma_{th}$ may be either random or deterministic. In the first case, not to multiply the number of random variables, we may assume $\Delta\sigma_{th}$ proportional to r.

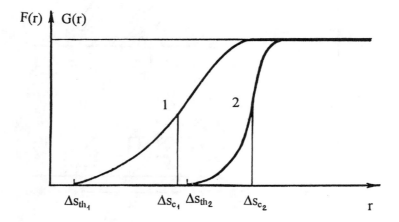

FIGURE 2.6
Cumulative distribution functions of resistance stresses to damage accumulation in grains.

It is natural to use the Weibull distributions for the resistance stress r from equation (2.47). For the first-row grains let the cumulative distribution be similar to that given in (2.6):

$$F(r) = 1 - \exp\left[-\left(\frac{r - \Delta s_{th_1}}{s_{c_1}} \right)^{\alpha_1} \right] \tag{2.48}$$

For other grains we assume

$$G(r) = 1 - \exp\left[-\left(\frac{r - \Delta s_{th_2}}{s_{c_2}} \right)^{\alpha_2} \right] \tag{2.49}$$

Here $s_{c_1}, s_{c_2}, \Delta s_{th_1}, \Delta s_{th_2}, \alpha_1$, and α_2 are material parameters, and $r \geq s_{th_1}$ in equation (2.48), $r \geq s_{th_2}$ in equation (2.49). To take into account the comparable weakness of the first-row grains, we assume $s_{c_1} < s_{c_2}$. As higher scattering of resistance is expected for the first-row grains, we assume $\alpha_1 < \alpha_2$ (Figure 2.6). When the threshold stress $\Delta\sigma_{th}$ in equation (2.47) is deterministic, it is expedient to put $\Delta s_{th_1} = \Delta\sigma_{th}$ for the first-row grains and $\Delta s_{th_2} = \Delta\sigma_{th}$ for other grains.

The procedure of numerical simulation is illustrated by the flow-chart given in Figure 2.7. This procedure includes computation of the stress ranges $\Delta\sigma_1(N), \ldots, \Delta\sigma_n(N)$ in each grain entering into the fine mesh domain; generation of samples of resistance stresses $r_1, \ldots, r_n$ in these grains; solution of equation (2.47) for these grains and comparison of damage measures $\chi_1, \ldots, \chi_n$ with the critical level; chang-

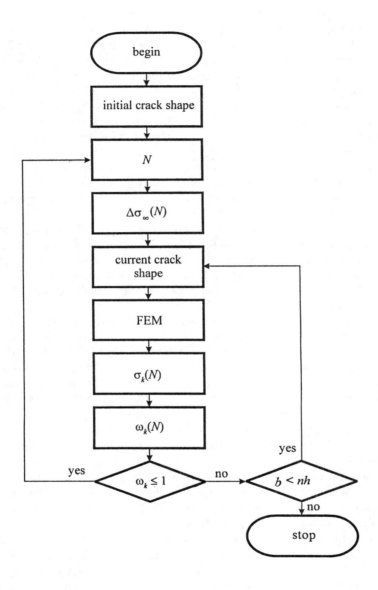

FIGURE 2.7
Flow-chart of numerical simulation of early crack growth.

ing of the crack shape when some of grains are ruptured, or the prolongation of the cyclic loading process $\Delta\sigma_\infty(N)$. In addition, statistical treatment of results is included such as the estimation of means and standard deviations for crack dimensions and their rates. The fine mesh domain counts in half of the body's $62,500$ elements. The mesh step $h = 10$ μm is assumed coinciding with the characteristic size of grains.

The following numerical data for parameters entering equations (2.47)-(2.49) were used: $s_{c_1} = 2.2$ GPa, $s_{c_2} = 3.0$ GPa, $\Delta s_{th_1} = \Delta s_{th_2} = 300$ MPa, $\alpha_1 = 2, \alpha_2 = 4, m_1 = m_2 = 4, n = 1, \Delta\sigma_\infty = 150$ MPa, $R = \sigma_\infty^{min}/\sigma_\infty^{max} = 0$. The initial conditions were stated corresponding to a single completely ruptured grain which becomes the nucleus of a crack. Other grains were assumed to be initially nondamaged.

The process of early crack growth is illustrated in Figure 2.8. A single nucleus near the surface is present in state (a). State (b) is attained at $N \approx 70 \cdot 10^3$ when the start of in-depth propagation occurs. In addition, two grains are damaged, one at the surface and another in the depth. At $N \approx 131 \cdot 10^3$ state (c) is attained. Then the process of crack growth proceeds more intensively: the crack grows both in depth and along the surface. States (d), (e), and (f) are separated only with several thousands of cycles. The additional nuclei do not propagate, though they could merge with the main crack in the future.

The shape of mesocracks is close to semi-elliptical with in-depth semi-axes which is usually less than a half of the along-the-surface size. Therefore, it is expedient to notate the characteristic sizes of cracks in such a way that, for a semi-ellipsoidal cavity, the conventional relationship holds between semi-axes, namely $a \geq b \geq c$. Then the in-depth size of a crack is notated with b, and the size measured along the surface size with $2a$.

Several samples of along-the-surface size $a(N)$ and in-depth size $b(N)$ are shown in Figure 2.9. The scatter of samples is significant: the cycle number at the in-depth advancement in several grains varies from $50 \cdot 10^3$ to $150 \cdot 10^3$. The scatter of samples for crack propagation along the body surface is of the same order of magnitude. To propagate a crack up to the size containing 40 grains, the cycle number varies from $70 \cdot 10^3$ to $140 \cdot 10^3$. The processes $a(N)$ and $b(N)$ are certainly step-wise. To make the picture clearer, these step-wise processes are replaced with continuous ones by connecting the points at the middle of each step with straight lines. Hence, in Figure 2.9 (as later also) the "smoothed" samples are plotted.

Samples of a/b are presented in Figure 2.10. In the beginning of crack propagation, the ratio varies in a large scale, from 0.5 to 2.0 and more. Later this variation becomes more moderate. Toward the end of the numerical simulation, the distribution of a/b becomes comparatively compact. The coarse estimation gives $a/b = 1.5$ that is in agreement with experimental data.

To estimate the crack growth rates da/dN and db/dN, we apply the "smoothing" procedure again: the middle points of each step of samples da/dN and db/dN are connected with straight lines. The results are presented in Figure 2.11a and b. The scatter of db/dN covers one order of magnitude and is in agreement with experimental results on short crack behavior [90, 102]. At the initial stage of crack growth, the rates are decreasing, and the minima of db/dN are of the order of 10^{-10} m/cycle. Then the steady increase of db/dN is observed. When a crack has penetrated 6-8 grains, one may say that the stage of stable crack growth begins. Several samples of da/dN are shown in Figure 2.11a also. Two items ought to be mentioned. Compared with db/dN, the scatter of da/dN is much higher and the samples of da/dN do not exhibit an initial stage of decrease before steady growth. One might explain this phenomena by the initial assumption: the scatter of the properties of near-surface grains was assumed larger than that of internal grains.

Results of statistical treatment of 50 samples are shown in Figures 2.12 and 2.13.

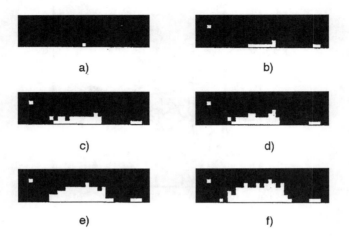

FIGURE 2.8
Stages of mesocrack development: (a) crack nucleus; (b) beginning of in-depth propagation; (c), (d), (e), (f) in subsequent states.

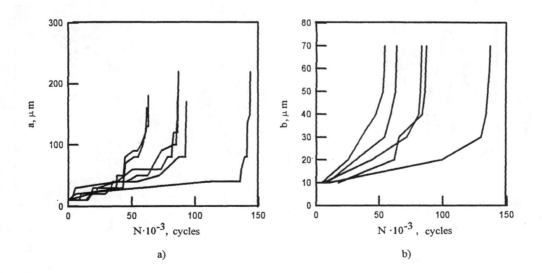

FIGURE 2.9
Samples of in-depth and along-the-surface early crack growth.

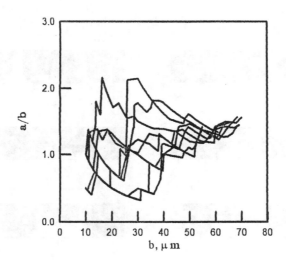

FIGURE 2.10
Samples of the ratio of crack dimensions.

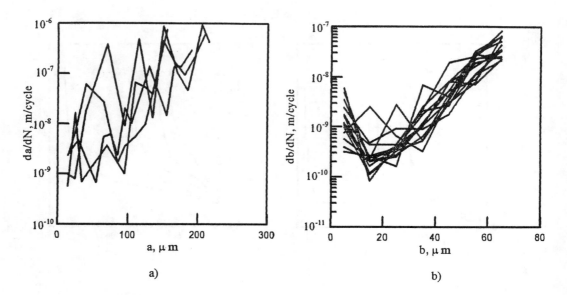

FIGURE 2.11
Samples of along-the-surface (a) and in-depth (b) crack growth rates.

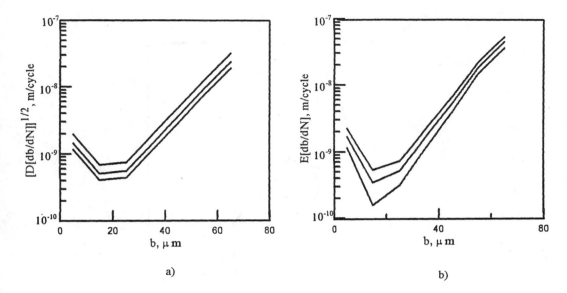

a)   b)

FIGURE 2.12
Estimates of mean along-the-surface crack growth rate (a) and its standard deviation (b).

The common approaches were applied to estimate the mean value of the in-depth crack growth rate as well as its standard deviation. The estimates of the mean crack growth rate db/dN (Figure 2.12a) and its standard deviation (Figure 2.12b) are plotted by the middle line. Two other lines correspond to the confidence level 0.9. The coefficient of variation of db/dN is high at the beginning of crack propagation (Figure 2.13). However, a distinct decrease takes place later until it reaches the magnitude of the order 0.5.

Additional information is given in Figure 2.14. There the estimates are presented for the cumulative distribution function $F(N; b)$ of cycle number when a crack propagates in the given depth b. The numbers at the lines indicate the crack depth measured in numbers of ruptured grains. Line 1, for example, corresponds to the first expansion of cracks along the surface at $b = 1$. Lines 2, 3, etc., correspond to crack propagation in the depth equal to $2h$, $3h$, etc.

When the number of ruptured grains in the cracked area becomes sufficiently large, the process is further controlled, besides loading parameters, by the averaged material characteristics such as mean resistance stress. For the case when the distributions (2.48) and (2.49) are valid, the mean resistance stresses are

$$s_{m_1} = \Delta s_{th_1} + s_{c_1} \Gamma \left(1 + \frac{1}{\alpha_1}\right), \quad s_{m_2} = \Delta s_{th_2} + s_{c_2} \Gamma \left(1 + \frac{1}{\alpha_2}\right) \quad (2.50)$$

Let us assume that a crack advances in a step h when a corresponding grain on the crack boundary is ruptured. Then at $m_1 = m_2 = m, n = 0$ we obtain

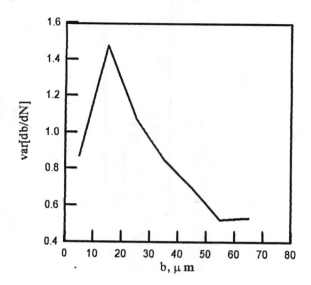

FIGURE 2.13
Coefficient of variation of the crack growth rate.

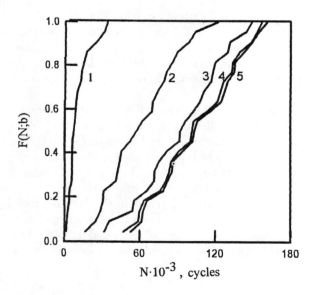

FIGURE 2.14
Estimates for cumulative distribution function of cycle numbers for growing mesocracks; numbers 1, 2, 3, ...correspond to crack depths $b = h, 2h, 3h, \ldots$.

$$\frac{da}{dN} = h\left(\frac{\kappa_A \Delta\sigma_\infty - \Delta s_{th_1}}{s_{m_1}}\right)^m, \quad \frac{db}{dN} = h\left(\frac{\kappa_B \Delta\sigma_\infty - \Delta s_{th_2}}{s_{m_2}}\right)^m \quad (2.51)$$

Here κ_A and κ_B are stress concentration factors at points A and B (Figure 2.15); s_{m_1} and s_{m_2} are mean stresses given in equation (2.50). Let the crack have a semi-elliptical shape with semi-axes a and b. To find an approximate value of stress concentration factors, the analogy between stress concentration and stress intensity factors may be used. The stress intensity factor in mode I for near-surface semi-elliptical cracks is defined as

$$K(\beta) = \frac{Y\sigma_\infty(\pi b)^{1/2}}{\mathbf{E}(k)}\left[\sin^2\beta + \left(\frac{b}{a}\right)^2\cos^2\beta\right]^{1/4} \quad (2.52)$$

where β is the elliptical angle, Y is the correction factor, $\mathbf{E}(k)$ is the elliptical integral of the second kind

$$\mathbf{E}(k) = \int_0^{\pi/2}\left(1 - k^2\sin^2\beta\right)^{1/2}d\beta, \quad k^2 = 1 - \frac{b^2}{a^2} \quad (2.53)$$

Following Mura [106] who has considered internal elliptical hollows, we use an heuristic formula for stress concentration factors based on the analogy with (2.52). This formula is

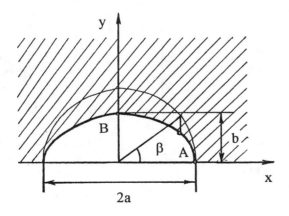

FIGURE 2.15
Model of semi-elliptical surface crack.

$$\kappa(\beta) = \frac{2Y}{\mathbf{E}(k)} \left(\frac{b}{\rho}\right)^{1/2} \left[\sin^2\beta + \left(\frac{b}{a}\right)^2 \cos^2\beta\right]^{1/4} \tag{2.54}$$

where $\rho(\beta)$ is the curvature radius of the crack tip. This radius actually characterizes stress concentration near the crack tip with complex fractography, as well as increasing of the material compliance due to the near field damage.

The finite element analysis shows satisfactory agreement with results given by equation (2.54) when a hollow is sufficiently flattened. It is demonstrated in Figure 2.16 where the stress concentration factors κ_A and κ_B are plotted against the ratio a/b. Solid lines are drawn according to equation (2.54) at $\beta = 0$ and $\beta = \pi/2$, respectively. Separate points are obtained by the finite element method. Though the mesh is coarse, and maximal stresses refer to stresses in the center of the element situated, near the crack tip, the results are rather close. However, for our purposes, even coarser estimates are enough. As $Y \sim 1, E(k) \sim 1, \rho \sim h$, one may state that $\kappa_A = b/(\rho a)^{1/2}, \kappa_B \approx (b/\rho)^{1/2}$. This means that $\kappa_A/\kappa_B \approx (a/b)^{1/2}$. At $\kappa_A \sigma_\infty \gg \Delta\sigma_{th_2}, \kappa_B \sigma_\infty \gg \Delta\sigma_{th_1}$, the estimate follows from equation (2.51)

$$\frac{\Delta b}{\Delta a} \approx \left(\frac{a}{b}\right)^{m/2} \left(\frac{s_{m_1}}{s_{m_2}}\right)^m$$

When a crack propagates in the depth of the body, remaining however shallow compared with the body thickness, the ratio of semi-axes approaches to a constant estimated as

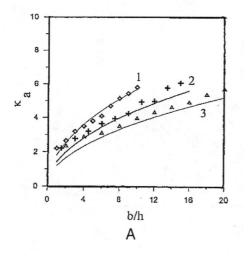
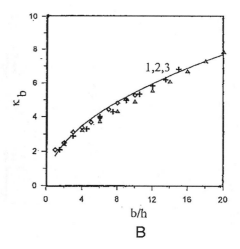

FIGURE 2.16
Stress concentration factors at points A and B.

$$\frac{a}{b} \approx \left(\frac{s_{m_2}}{s_{m_1}}\right)^{2m/(m+2)} \qquad (2.55)$$

We have $a \geq b$ at all $s_{m_1} \leq s_{m_2}, m > 0$. In particular, at $m = 2$, equation (2.55) gives $a/b = s_{m_2}/s_{m_1}$; at $m = 4$ we obtain $a/b = (s_{m_2}/s_{m_1})^{4/3}$. When $m \gg 1$, the ratio a/b approaches $(s_{m_2}/s_{m_1})^2$. We see that most parameters of the micromechanical model enter equations (2.51)-(2.53) which are valid, at least approximately, at the macroscopic level. This means that these parameters may be estimated from common fatigue tests.

We can see that, despite a number of simplifications, the model covers all main features of early fatigue crack growth. The model predicts high scattering of crack growth rates in the early stage when crack is to be considered mesoscopic. The temporary retardation of crack growth on this stage and the transition to a more regular growth when the depth of a crack becomes significant is also covered by the model. In addition, the model predicts quite a reasonable ratio of the crack in-depth and along-the-surface dimensions. Most of the material parameters entering the micromechanical model can be estimated by measuring the crack sizes during the early growth stage.

2.7   Summary

Three approaches to the problem of crack nucleation were discussed in this chapter: a purely phenomenological approach based on a generalized version of the damage summation rule; an approach within the framework of continuum damage mechanics; the modeling of real material as a media with the random microscale structure, i.e., as a subject of micromechanics. It was shown that, under a proper choice of constitutive equations, similar equations with respect to the cycle number until fatigue crack nucleation can be obtained with the use of all the three approaches. The size effect is included in all the discussed models, both with respect to life until macroscopic crack nucleation and with respect to the stress level corresponding to crack nucleation at a given cycle number.

To predict the early stage of fatigue crack propagation, a simple model was used treating a body as a set of microscale elements with randomly distributed mechanical properties. Numerical simulation was performed to study the evolution of the shape of short cracks, and a qualitative comparison with experimental data was presented.

Chapter **3**

MECHANICS OF FATIGUE CRACK GROWTH

3.1 General Outlook

This chapter is dedicated to the theory of fatigue crack growth based on the synthesis of fracture mechanics and mechanics of microdamage accumulation (used mostly in the form of continuum damage mechanics). The first sketch of this theory was suggested by Bolotin (1983) and later developed in a number of publications including the book [19]. A brief survey of the theory was given in Section 1.11.

The interaction between the dispersed damage and the balance of forces and energy in the system cracked body-loads or loading device is very complex because this interaction contains a number of feedbacks (Figure 3.1). The input of this flow-chart is loading in general, i.e., a set of external forces, prescribed displacements, thermal, chemical and radiation actions, etc. These loads and actions produce in the body the fields of stresses, strains, and temperature. These fields may be treated on the macroscopic level, in terms of continuum mechanics.

Microdamage changes a number of material properties. Among them are material's compliance characterized by elastic moduli, yield limit and related parameters of plasticity. As a rule, microdamage accumulation results in decreasing of stiffness fracture toughness. Sometimes a reverse effect occurs such as the growth of fracture toughness due to shielding with an array of microscopic cracks and voids. In addition, microdamage produces residual stresses, both on the micro- and macrolevel.

The change of a material's compliance is the origin of additional complications of the theory of fatigue. Actually, because of damage distributed nonuniformly over the bulk of a body, the material becomes nonhomogeneous and, frequently, anisotropic even if initial properties have been isotropic and homogeneous. That affects the stress and strain distribution in a body, especially near the crack tips. In its turn, the stress-strain distribution influences the microdamage accumulation,

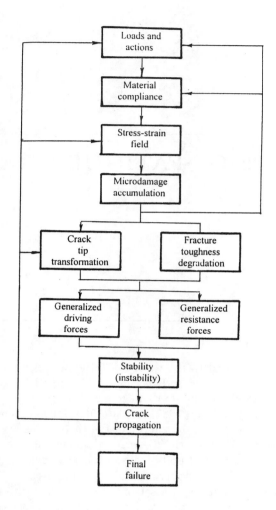

FIGURE 3.1
Flow-chart illustrating fatigue damage accumulation and crack propagation.

and so on.

Another factor is the residual stresses and strains occurring during the cyclic loading. This factor is especially important in low-cycle fatigue and in fatigue phenomena that are accompanied with overloadings and, as a result, with plastic deformations in the vicinity of crack tips. Due to residual stress-strain phenomena, one has to distinguish the plastic zone, where plastic strains are developed, and the cyclic process zone with residual plastic deformations.

The third and most important factor, from the viewpoint of the theory of fatigue crack growth, is the influence of microdamage on fracture toughness, i.e., resistance to crack propagation. In contrast to the factors formerly listed some of which may be neglected in simplified modeling, the effect of microdamage on fracture toughness is of primary importance.

The general pattern to predict fatigue crack propagation is the following. After the assessment of microdamage fields and its effect on material properties, we return to the macroscopic level. It is the level of macromechanics of fracture. In terms of analytical fracture mechanics, we evaluate the generalized driving and resistance forces with accounts of microdamage. Microdamage close to the crack (the near field damage) is of primary significance here. It affects essentially on the resistance against crack propagation, although sometimes the influence of the far field damage also must be taken into account. Comparing the driving and resistance forces, we judge on stability (instability) of the system cracked body-loading (Figure 3.1).

The character of propagation may be continuous and discontinuous (jump-wise). It depends mostly on the distribution of microdamage in the near field. The final failure as a result of fatigue crack growth is included in this scheme, too.

The presented scheme of fatigue crack growth is rather complicated; but it is the complexity of the phenomenon, not that of the chosen approach to its study. Some simplifications are available. Among them are the above mentioned neglect of the influence of microdamage on material compliance and the influence of the latter on the stress-strain field. Another simplification can be achieved by neglect of the far field damage.

The comparative contribution of micro- and macromechanical phenomena depends on the stage of crack growth which is considered. Let the crack depth be small, e.g., equal to several grain sizes, as well as small compared with cross-section dimensions of the body. Then stability considerations play a negligible part in the mechanism of crack growth. In that case the process of crack growth is governed with microdamage only. For such "short" cracks we must evaluate microdamage near the crack tips and compare the level of microdamage with a certain initially stated critical level. To the contrary, close to the final failure, the microdamage ahead of tips is comparatively small. Crack growth is governed mainly by the global energy balance in the system. On the middle stage both factors are important.

In linear fracture mechanics, as a rule, cracks are schematized as mathematical cuts. This schematization introduces singularities when elastic or linear visco-elastic bodies are considered; however it is not treated as an essential drawback of the theory. Studying fatigue, we must take into account the microdamage accumulation near the crack tips and, therefore, the schematization of cracks as mathematical cuts is not acceptable. We will schematize cracks and crack-like defects as narrow slits with finite radii of curvature. However the fractographic picture near the crack tips in real materials is usually rather complex. Cracks pass through the strongly damaged material, and the state at the crack tips depends on the type of damage. For example, in a polycrystalline material crack tips propagate both through grains and intergranular boundaries. In amorphous polymers crack tips pass through ruptured ligaments and voids. In both cases the damaged material almost does not carry stresses. Therefore, the curvature on the tips is a kind of measure for stress concentration. Another way is to model cracks as mathematical cuts but to estimate the characteristic damage not at the tips, but at certain points situated a little ahead. Such an approach is used in the application of damage mechanics to creep problems [119, 120, 126]. In both approaches additional variables with the dimension of length enter in the modeling of crack growth. It even follows from dimensional considerations. Crack tip radii and/or related length parameters are, generally, the variables intrinsic for any fatigue theory. A typical example where these variables play an important role is the crack growth accompanied by tip bluntings and sharpenings.

3.2   Cracked Body as a Mechanical System

In this section we discuss a generalized version of fracture mechanics, more precisely, the mechanics of cracked bodies. This generalization is performed in several aspects, and one of them is a multiparameter approach. The theory is applicable to bodies containing multiparameter cracks and sets of cracks; to bodies subjected to a variety of loadings, etc. The foundation for such a generalization is classical, analytical (rational) mechanics. In this book we refer to this extension as analytical fracture mechanics [32].

Fracture is an irreversible phenomenon. Cracks in structural materials are "non-healing" when the special repair is not provided. From the viewpoint of analytical mechanics, bodies with such flaws must be considered as systems with unilateral constraints. Together with the multiparameter approach, the account for cracks irreversibility is an essential point of the analytical fracture mechanics. The aim of the theory is to develop a sufficiently transparent analytical approach applicable to a wide class of problems of fracture and fatigue. The commonest of these problems is that they concern bodies with crack-like defects that can grow irreversibly up to the final failure.

The main variables in analytical mechanics are generalized coordinates and generalized forces. In addition to common (Lagrangian) generalized coordinates, (later on, for brevity, L-coordinates), additional variables are introduced that describe shape, size and position of cracks. In the paper by Bolotin [14] it was proposed to name these variables, in honor of Griffith, Griffithian generalized coordinates (later on, G-coordinates). Actually, Griffith (1920) was the first to introduce this kind of coordinate (such as crack half-lengths) into analytical consideration. In simple cases, the corresponding generalized forces coincide with the well-known and widely used generalized forces of fracture mechanics. Another fundamental concept is the concept of Griffith's variations (G-variations), named also in honor of Griffith. They are variations of G-coordinates calculated under certain conditions that will be stated later.

Analytical fracture mechanics is based upon the principle of virtual work for quasistatic problems. This principle has a wider field of applicability than the energy equations used in fracture mechanics. The principle of virtual work provides an adequate approach to multiparameter problems where more than one independent G-coordinate is involved. The realization of the principle of virtual work is not as trivial as the constraints put on the G-coordinates are unilateral.

Consider a body of material with arbitrary mechanical properties under arbitrary quasistatic loading. The body contains a number of cracks of varying origin, shape and position. Some of them may be modeled as mathematical slits, others as shallow cavities with finite tip curvatures.

The parameters of cracks are G-coordinates and, for simplification, we assume that the set G-coordinates is bounded. Denote them $a_1, \ldots, a_m$. Since flaws are assumed irreversible, G-coordinates can be chosen in such a way that their variations satisfy conditions

$$\delta a_j \geq 0, \quad j = 1, \ldots, m \tag{3.1}$$

The common L-coordinates usually are not subjected to such limitations if unilateral constraints of structural origin are absent.

Several examples are shown in Figure 3.2. A plane through crack in a plate is depicted in Figure 3.2a. The crack can propagate at their two tips nonuniformly. This means that two independent length parameters are necessary, a_1 and a_2 with variations $\delta a_1 \geq 0, \delta a_2 \geq 0$. A plate of two materials with a flat crack near the boundary is shown in Figure 3.2b. This crack may have four parameters if all possible cases are taken into account including propagation of the crack into the depth of the other material. All G-coordinates, $a_1, a_2, a_3,$ and a_4, satisfy condition (3.1). In Figure 3.2c a plane crack of elliptical shape in a three-dimensional body is shown. It is described with semi-axes a and b. Under the assumption that the crack remains elliptical, its growth can be described with two G-coordinates with variations $\delta a \geq 0, \delta b \geq 0$.

It is easy to present examples of cracks with a continuous set of G-coordinates. If a plane crack in a plate is subjected to kinking, and the angle φ of kinking is not known beforehand, the G-variation δa of the initial size a is a function $\delta a(\varphi)$. The angle φ takes the place of the index j in (3.1). Instead of (3.1) we have

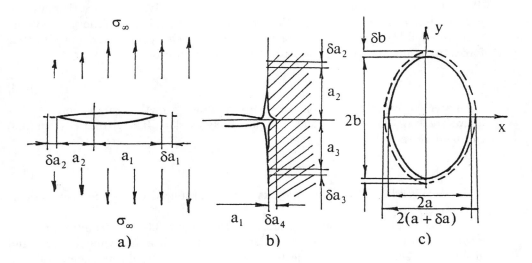

FIGURE 3.2
Griffith's coordinates and their variations for: (a) plane through crack; (b) crack at the boundary of two materials; (c) penny-shaped elliptical crack.

$$\delta a(\varphi) \geq 0, \quad \varphi \in [-\pi, \pi] \tag{3.2}$$

Apart from G-coordinates, the common Lagrangian generalized coordinates (L-coordinates) are necessary to describe any state of a system. For continuous systems, the displacement field functions $u(x)$ form a set of L-coordinates. Here u is a displacement vector, and x is a reference vector. The latter, depending on properties of the structural model, may be one-, two- or three-dimensional, and even may take various dimensions in different parts of the model. We assume that no restrictions similar to (3.1) and (3.2) are put on L-coordinates. When G-coordinates are fixed, we denote the displacement field $u(x \mid a)$. Here a means the set of G-coordinates, in particular, $a = \{a_1, \ldots, a_m\}$ if their number is finite. We will also call a the crack parameter vector. Lagrangian variations are calculated for a body with fixed cracks, i.e., $\delta_L u(x) \equiv \delta u(x \mid a)$. Variations of the displacement field because of the variation of crack parameters

$$\delta_G u(x) \equiv \sum_{j=1}^{m} \frac{\partial u(x|a)}{\partial a_j} \delta a_j$$

are later called G-coordinates of this field.

3.3   Application of Principle of Virtual Work

The principle of virtual work is the foundation of the analytical statics. For systems with unilateral ideal constraints this principle is formulated as follows: a system is in an equilibrium state if and only if the sum of elementary works of all active forces upon all small admissible displacements is non-positive

$$\delta W \leq 0 \tag{3.3}$$

Later on, we relate the forces resisting to crack growth to active forces. This allows treating the constraints in a cracked body as ideal. The virtual work in (3.3) can be divided in two parts. Then

$$\delta W = \delta_L W + \delta_G W \leq 0 \tag{3.4}$$

where $\delta_L W$ and $\delta_G W$ are the amounts of work produced on L- and G-variations, respectively.

The concept of G-variation is illustrated in Figure 3.3. By definition, the following conditions are to be satisfied: time, given surface and volume forces and given displacements are not subjected to variations; all equilibrium, compatability, heat-transfer and constitutive equations are satisfied in the volume of the body (except, maybe, the vicinities of crack tips). We assume also that both initial and perturbed states are stable in the Lyapunov sense at fixed crack parameters.

The perturbed states of the system are equilibriums from the standpoint of mechanics. Therefore, the virtual work produced on Lagrangian variations is equal to zero. Equation (3.4) takes the form

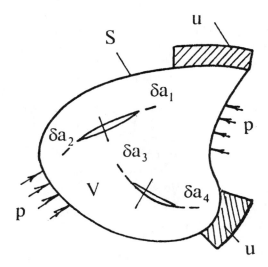

FIGURE 3.3
To the definition of Griffith's variations.

$$\delta_G W \leq 0 \qquad (3.5)$$

Represent the virtual work $\delta_G W$ as

$$\delta_G W = \delta W_e + \delta W_i + \delta W_f \qquad (3.6)$$

Here δW_e is the virtual work of external forces; δW_i is the virtual work of internal forces in the volume of the body except, maybe, tip zones. The virtual work in the tip zones is included in the last term δW_f of (3.6):

$$\delta W_f = -\sum_i \int_{S_i} \gamma_i \mid ds_i \times \delta \lambda_i \mid.$$

Here γ_i is specific fracture work, i.e., the amount of work required for formation of the unit of crack surfaces (opposite to tradition we count two newborn crack surfaces one time only). The remaining notations are: ds_i is a length element of the contour S_i of i-th crack, and $\delta \lambda_i$ is an increment of the size of i-th crack. Summation covers all the cracks in the body. Quantities γ_i, in general, may vary within the limits of each crack. They may depend also on crack dimensions and modes of fracture.

With use of the concept of virtual work produced on G-variations, the states of the system cracked body-loading can be classified in two levels: with respect to equilibrium, and with respect to stability. Such a classification was introduced primarily in papers by Bolotin [14, 15]. The states for which the virtual work is negative for all $\delta a_j > 0$ are called sub-equilibrium states. The states for which such variations $\delta a_j > 0, j = 1, \ldots, m_1$ exist that $\delta_G W = 0$, and for remaining variations

$\delta_G W < 0$, are called equilibrium states (with respect to coordinates $a_1, \ldots, a_{m_1}$). Both types of states are equilibriums from the viewpoint of classical mechanics. If even one variation exists that $\delta_G W > 0$, i.e., condition (3.5) is violated, we say that a non-equilibrium state takes place.

Conditions of stability can be also expressed in terms of virtual work. Sub-equilibrium states are, evidently, stable. An additional amount of energy is needed to transfer a system in any neighboring state, and there are no such energy sources in the system. Non-equilibrium states cannot be realized as equilibrium ones and, therefore, are unstable. Equilibrium states (in the sense of analytical fracture mechanics) may be both stable and unstable.

Let the work $\Delta_G W$ on the increments $\Delta a_1, \ldots, \Delta a_m$ be a differentiable smooth function of $a_1, \ldots, a_{m_1}$. In the vicinity of equilibrium with respect to $a_1, \ldots, a_{m_1}$ this work can be presented in the form

$$\Delta_G W = \delta_G W + \sum_{k=1}^{m} \frac{\partial(\delta_G W)}{\delta a_k} \delta a_k + \cdots$$

Therefore, in the study of stability conditions for equilibrium states, the sign of the second term in the right-hand side

$$\delta_G(\delta_G W) = \sum_{k=1}^{m} \frac{\partial(\delta_G W)}{\delta a_k} \delta a_k \tag{3.7}$$

is to be decisive. It must be noted that $\delta_G(\delta_G W)$ in (3.7) and later on is not a second variation of a certain functional but just a notation for the first variation of $\delta_G W$.

Consider an equilibrium state. When $\delta_G W \equiv 0$, the principle of virtual work requires that $\delta_G(\delta_G W) \leq 0$. The inequality $\delta_G(\delta_G W) < 0$, being fulfilled for all arbitrary non-vanishing variations $\delta a_j (j = 1, \ldots, m_1)$, means stability of the equilibrium state. If such variations exist that $\delta_G(\delta_G W) > 0$, the equilibrium state is unstable. When such variations $\delta a_j > 0$ exist that $\delta_G(\delta_G W) = 0$, and for remaining variations $\delta_G(\delta_G W) < 0$, the state of the system is neutral. It may be a critical state, e.g., corresponding to a boundary between stability and non-stability, or a questionable state. For the neutral state, a study is required of the following terms of expansion of $\delta_G W$ in a power series with respect to δa_j.

The above presented classification is illustrated in Figure 3.4. Relationships $\delta_G W = 0$, $\delta_G W < 0$, etc., are to be understood according to the above statements. The classification has two levels: the level of equilibrium and the level of stability. A state enters into one of the classes of the first level (sub-equilibrium, equilibrium and non-equilibrium) depending on the sign of $\delta_G W$. A state refers to one of the classes of the second level (stable, neutral or unstable) depending on the signs of $\delta_G W$ and $\delta_G(\delta_G W)$.

The analogy between analytical fracture mechanics and classic analytical statics was illustrated with a simple model suggested by Bolotin [32]. It is a heavy cylinder placed on a rigid geared cylindrical surface. The body is restrained against the backward motion by means of a ratchet. Another illustrative example is shown in Figure 3.5. It is a heavy body (for example, a cylinder) on the smooth rigid surface limited by a left-hand side rigid vertical wall. The constraint condition is, evidently,

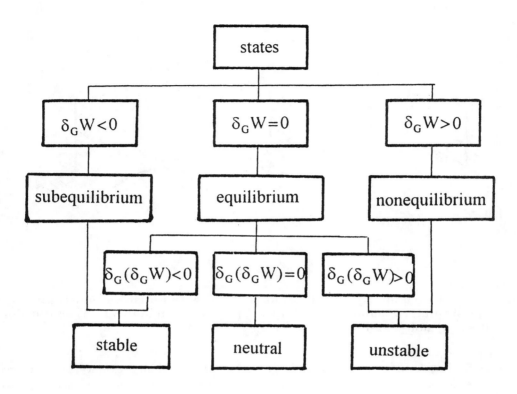

FIGURE 3.4
States of cracked bodies classified from the viewpoint of equilibrium and stability.

$\delta a \geq 0$, where a is the generalized coordinate measured from the wall. Therefore, the constraint is unilateral. State 1 of the system is sub-equilibrium, state 2 is stable equilibrium. State 3 corresponds to neutral equilibrium, state 4 to unstable equilibrium. In case 5 the state of the system is non-equilibrium and, therefore, unstable. The example in Figure 3.5 is related to the common illustration of the Lagrange-Dirichlet theorem for conservative systems. The main difference is that in the latter case variations δa of both signs are admissible.

The assumption that all cracks are irreversible is not of primary significance for analytical fracture mechanics. We may include in the consideration reversible, "healing" cracks, too. If for any $\delta a_j < 0$ inequality $\delta_G W > 0$ holds, it means that the state of the system is unstable with the tendency of corresponding cracks to closure. We call such states anti-equilibrium [32], opposite from the non-equilibrium states that result in crack propagation and final failure.

In concluding this section, it is expedient to note that the above analysis is not

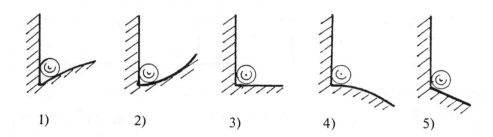

1)   2)   3)   4)   5)

FIGURE 3.5
Illustration to classification of states for systems with unilateral constraints: (1)
sub-equilibrium, (2) equilibrium stable, (3) equilibrium neutral, (4) equilibrium
unstable, (5) non-equilibrium.

rigorous from the viewpoint of the general theory of stability. However, crack propagation is a strongly dissipative, energy-consuming process. Therefore, the presented approach to stability seems to be appropriate. In any case, no one contradiction has been met with the results obtained on the basis of other approaches.

3.4   Generalized Forces in Analytical Fracture Mechanics

All the introduced concepts are easy to formulate in terms of generalized forces. As all variations are isochronic, the terms in the right-hand side of (3.6) are linear forms of δa_j. Hence one can present them as follows:

$$\delta W_e + \delta W_i \equiv \sum_{j=1}^{m} G_j \delta a_j, \quad \delta W_f \equiv -\sum_{j=1}^{m} \Gamma_j \delta a_j. \tag{3.8}$$

The multipliers G_j at variations δa_j are generalized driving forces. Multipliers Γ_j are generalized resistance forces.

The division of generalized forces into two parts might seem to be ambiguous. The virtual work of internal forces, generally, includes the work transformed into stored (potential) energy, and a part of this work is dissipated in the volume of the body. There is no distinct boundary between the tip zone and the main bulk of the body. In ductile fracture, especially when a plastic zone is not confined within a body, and the complete ligament is subjected to yielding, most of the work $\delta_G W_i$ is to be related to resistance forces. As equilibrium and stability properties of the system cracked body-loading depend on the differences

$$H_j = G_j - \Gamma_j, \quad (j = 1, \ldots, m) \tag{3.9}$$

and not on the forces G_j and Γ_j separately, the approach to division may be rather arbitrary. Later on we attribute all the processes within process zones to generalized resistance forces and those in the bulk of the body to generalized driving forces.

Consider a system cracked body-loading that is in the sub-equilibrium state. Then for all δa_j the condition $\delta_G W < 0$ is satisfied. With use of (3.8) the conditions of sub-equilibrium take the form

$$G_j < \Gamma_j, \quad (j = 1, \ldots, m) \tag{3.10}$$

Equilibrium states need special consideration. We will say that a system cracked body-loading is in the equilibrium state with respect to μ generalized coordinates $a_1, \ldots, a_\mu$ if the generalized forces satisfy conditions

$$G_j = \Gamma_j, \quad (j = 1, \ldots, \mu)$$

$$G_k < \Gamma_k, \quad (k = \mu + 1, \ldots, m) \tag{3.11}$$

However, the equilibrium of multi-parameter systems with respect to all generalized coordinates is an exception, if we do not count the problems with symmetry of higher order or with the sets of similar cracks. It very often happens that $\mu = 1$. Then a system is in the equilibrium state with respect to one of the generalized coordinates, and in the sub-equilibrium state with respect to all other coordinates. A system cracked body-loading will be in a non-equilibrium state, if only one $\delta a_j > 0$ exists for which $G_j > \Gamma_j$. This state is certainly unstable.

Following the paper by Bolotin [18], let us consider in detail the stability conditions for equilibrium states. Let a system be in an equilibrium state with respect to coordinates $a_1, \ldots, a_\mu$ and in a sub-equilibrium state with respect to the remaining $m - \mu$ coordinates. Then the latter may be treated as parameters. The variation $\delta_G(\delta_G W)$ is a quadratic form (3.7) of variations of G-coordinates. Assume that the generalized forces (3.9) are differentiable with respect to all $a_1, \ldots, a_\mu$. Taking into account (3.8) and (3.9), we replace equation (3.7) with the following

$$\delta_G(\delta_G W) = \sum_{j=1}^{\mu} \sum_{k=1}^{\mu} \frac{\partial H_j}{\partial a_k} \delta a_j \delta a_k \tag{3.12}$$

An equilibrium state is stable if the quadratic form (3.12) is definitely negative, and unstable if it is definitely positive or sign-indefinite. If the form (3.12) is non-positive, we must study variations of a higher order.

When a body is elastic and the external forces are potential, the symmetry conditions are satisfied:

$$\frac{\partial H_j}{\partial a_k} = \frac{\partial H_k}{\partial a_j}, \quad (j, k = 1, \ldots, m)$$

The quadratic form (3.12) is definitely negative if the Cauchy-Sylvester conditions

$$(-1)^n \det \left\| \frac{\partial H_j}{\partial a_k} \right\|_1^n > 0 \quad (n = 1, \ldots, \mu) \tag{3.13}$$

are satisfied. However, all cracks are irreversible. Therefore, it is sufficient to require that $\delta_G(\delta_G W) < 0$ in the first reference ortant and conditions (3.13) may be weakened. At $\mu = 2$ instead of conditions

$$\frac{\partial H_1}{\partial a_1} < 0, \qquad \frac{\partial H_1}{\partial a_1} \frac{\partial H_2}{\partial a_2} - \left(\frac{\partial H_1}{\partial a_2} \right)^2 > 0$$

we arrive to the weaker conditions:

$$\frac{\partial H_1}{\partial a_1} < 0, \qquad \frac{\partial H_2}{\partial a_2} < 0, \qquad \frac{\partial H_1}{\partial a_2} < \left(\frac{\partial H_1}{\partial a_1} \frac{\partial H_2}{\partial a_2} \right)^{1/2}$$

As has been mentioned, the most typical equilibrium state is that with respect to a single G-coordinate. Let us denote this coordinate a_1, and omit the indication on the dependence on other coordinates. Let the loading process be given with μ parameters $s_1, \ldots, s_\mu$. The equilibrium condition takes the form $H_1(a_1, s_1, \ldots, s_\mu) = 0$, and the stability condition

$$\partial H_1(a_1, s_1, \ldots, s_\mu)/\partial a_1 < 0 \tag{3.14}$$

Let the function $H_1(a_1, s_1, \ldots, s_\mu)$ satisfy the conditions of the theorem on an inexplicit function with respect to a_1. Solving the equilibrium condition $H_1 = F(a_1, s_1, \ldots, s_\mu) = 0$ with respect to a_1, denote the solution $a_1 = F_1(s_1, \ldots, s_\mu)$. If the loading parameters satisfy conditions $\partial H_1/\partial s_i > 0$ for all i, we obtain

$$\frac{\partial F_1}{\partial s_i} > 0 \quad (i = 1, \ldots, \mu) \tag{3.15}$$

Condition (3.15) states that an equilibrium crack is stable if increasing loads are needed for the further equilibrium growth of the crack. Formula (3.14) and (3.15) are transparent from the physical viewpoint. In particular, condition (3.14) coincides with the stability condition $T < T_R$ if the notation for tearing moduli is used (see Section 1.7).

To compare the developed approach with the conventional concepts of fracture mechanics, we consider the simplest problem, a one-parameter crack, and a potential system cracked body-loading. Then $\delta W_e + \delta W_i = -\delta \Pi$ where Π is the potential energy of the system. Then

$$G = -\frac{\partial \Pi}{\partial a} \tag{3.16}$$

where subscripts for one-parameter cracks are omitted. The right-hand side of (3.16) is in fact the energy release rate. Let the loading system not produce work, i.e., $\delta W_e \equiv 0$, and the body be elastic with the strain energy U; then $\delta \Pi \equiv \delta U$. Equation (3.16) takes the form

$$G = -\frac{\partial U}{\partial a} \tag{3.17}$$

and G is the strain energy release rate. In both cases the generalized resistance force Γ is the common fracture toughness measured as the critical magnitude of the energy release rate G_c. For a plane crack in isotropic linear elastic media and plane strain-state we arrive at Irwin's equation (1.15) that connects the driving force with stress intensity factors.

To establish the relation with the path-invariant integrals of fracture mechanics, consider the virtual work of external forces

$$\delta W_e = \int_S p_\alpha \delta u_\alpha dS + \int_V X_\alpha \delta u_\alpha dV \qquad (3.18)$$

Here p_α is the stress vector on the body surface S, X_α is the volume force vector in the body's volume V, and u_α is the displacement vector. The virtual work of internal forces is

$$\delta W_i = -\delta \left(\int_V w dV \right), \qquad w = \int_{[0,t]} \sigma_{\alpha\beta} d\varepsilon_{\alpha\beta} \qquad (3.19)$$

Here $\sigma_{\alpha\beta}$ and $\varepsilon_{\alpha\beta}$ are stress and strain tensors, and w is equal to the density of internal work produced from the initial time moment $t = 0$ when the state of the body is natural, up to the considered time moment t. Quantities p_α, X_α and $\sigma_{\alpha\beta}$ are to be taken for the unperturbed state, and variations are understood in Griffith's sense. Substitution of (3.18) and (3.19) into the first of equations (3.8) results in the formula

$$G_j = \int_S p_\alpha \frac{\partial u_\alpha}{\partial a_j} dS + \int_V X_\alpha \frac{\partial u_\alpha}{\partial a_j} dV - \frac{\partial}{\partial a_j} \int_V w dV. \qquad (3.20)$$

A similar equation for the case $m = 1$ can be found in many publications. There G is attributed not to virtual work but to the energy release rate due to the small actual crack tip advancement. An extension of such an approach on inelastic bodies is discussed by Atluri [2] and Rice [124].

The aim of the next study is to show that the integration domain in (3.19) and (3.20) may be stretched continuously up to domains enveloping the crack tips. Instead of the given body with volume V and surface S (Figure 3.6a), we consider a body with a piece-smooth surface S' situated inside the surface S (Figure 3.6b). The embedded volume V' contains a definite crack tip. The stress vector on S' is denoted p'_α. The part $V \backslash V'$ of the body is in equilibrium. Hence

$$\int_S p_\alpha \delta_0 u_\alpha dS + \int_{S'} p'_\alpha \delta_0 u_\alpha dS + \int_{V \backslash V'} X_\alpha \delta_0 u_\alpha dV - \delta_0 \int_{V \backslash V'} w dV = 0$$

where $\delta_0 u$ and $\delta_0 w$ are Lagrangian variations corresponding to kinematically admissible variations of the stress-strain state for the body with fixed crack sizes. Let G-coordinate variations in the area V' produce Lagrangian variations in the domain $V \backslash V'$. Then active generalized forces G_j will be invariant with respect to

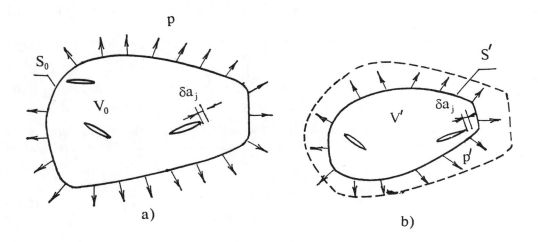

FIGURE 3.6
Transfer of the integration domain in the evaluation of generalized driving forces.

the choice of the integration domain enveloping the crack tip with the coordinate a_j. The explicit formula for generalized forces takes the form

$$G_j = \int\limits_{S'_j} p'_\alpha \frac{\partial u_\alpha}{\partial a_j} dS + \int\limits_{V'_j} X_\alpha \frac{\partial u_\alpha}{\partial a_j} dV - \frac{\partial}{\partial a_j} \int\limits_{V'_j} w dV. \qquad (3.21)$$

In particular, under certain conditions the integration area V'_j with the boundary S'_j may be stretched up to the vicinity of the crack tip with the coordinate a_j.

Thus, we come to the sufficient condition for the transfer of the integration domain within a body. This condition states that G-variations within the internal domain are to produce L-variations in the external domain. The condition is satisfied for elastic (both linear and nonlinear) bodies. It is not connected with the existence of the potential of loads. However, the condition almost surely is violated if material in the external domain is deformed plastically. Similarly, when material in the external region is subjected to dispersed damage, one cannot be sure, without an additional analysis, that variations of the displacement field associated with crack size variations can be considered as Lagrangian variations [23].

The generalized driving force given in equation (3.21) is related to the J-integral and under certain conditions coincides with the latter. Let volume forces be absent and the plane stress- or strain-state is realized approximately near the crack tip. Denote with C_j a cross section of S_j with a plane orthogonal to the crack front, and with Ω_j the area enclosed with C_j. The axis Ox_j is directed along the external normal to the front (Figure 3.6b). Then the force G_j related to the unit of the length of the crack front is

$$G_j = \int_{C_j} p'_\alpha \frac{\partial u_\alpha}{\partial a_j} dS - \int_{\Omega_j} \frac{\partial w}{\partial a_j} d\Omega \qquad (3.22)$$

Here the additional assumption is used that the integration area Ω_j does not depend on a_j. Therefore, the differentiation with respect to parameter a_j may be inserted before the sign of integral:

$$\frac{\partial}{\partial a_j} \int_{\Omega_j} w d\Omega = \int_{\Omega_j} \frac{\partial w}{\partial a_j} d\Omega \qquad (3.23)$$

Let us replace the variation with respect to a_j by moving the contour C_j on the nonvaried crack. Then

$$\frac{\partial}{\partial a_j} = -\frac{\partial}{\partial x_j} \qquad (3.24)$$

Using (3.23) and transforming the second integral in (3.22) we obtain

$$G_j = \int_{C_j} \left(w dy_j - p'_\alpha \frac{\partial u_\alpha}{\partial x_j} ds \right) \qquad (3.25)$$

The right-hand side of (3.25) is in fact the J-integral (1.24). But formula (3.23) and (3.24) are not, in general, identical transformations. For example, formula (3.23) is certainly nonvalid for beam delaminations. For elastic bodies with vanishing tip zones the error (from the viewpoint of analytical fracture mechanics) transformings (3.23) and (3.24) mutually compensate themselves, and equation (3.25) remains valid. But if the size of the tip zone is to be varied together with the size of the crack, equation (3.25) becomes invalid.

3.5 Equations of Fatigue Crack Growth

To present equations of fatigue crack growth, we have to decide which independent variable is to be used to describe the development of the process in time. In regular fatigue the cycle number N is commonly used as an independent variable. Frequently this variable may be considered as a continuous one because the total number until final failure is large, and the jumps in crack propagation, even if they are present, are small enough. In classic high-cycle fatigue the size of jumps is on the order of the characteristic microstructure scale. Assuming a mean crack growth rate of the order $10^{-9} - 10^{-7}$ m/cycle and a microstructure scale of the order $10^{-5} - 10^{-4}$ m, we obtain that one discontinuity of crack growth takes place in $10^3 - 10^4$ cycles. For a practically significant crack size, say $a = 10$ mm, we have to count more than $10^2 - 10^3$ jumps before this size will be attained. Evidently, in approximate engineering analysis such a process may be treated as a continuous one. Moreover, in real polycrystalline and other microscopically nonhomogeneous materials the jump-wise character of fatigue growth might be not displayed at all. Mechanical properties of microstructural elements have a statistical scatter. The front of the crack acts as a kind of "averager" over the array of these elements.

However in low-cycle fatigue the total cycle number up to final failure is much lower. The dimension of the plastic zone enters here as a characteristic scale. The jump-wise character of crack propagation is easy to observe in low-cycle fatigue. It is also observable during the concluding phase of fatigue damage when the steps between striations become significant. Even then the continuous approximation is justified, especially for engineering design purposes. Sometimes, when the loading process is described in terms of blocks or missions, their number N may be comparatively small. In that case, N is to be treated as a discrete variable.

In the case of damage under sustained loading such as creep cracking or stress corrosion cracking, the use of the natural time t is necessary. It is useful also when a combination of cyclic loading and other actions developing in time is considered. At last, the use of time is more appropriate when the loading process is cyclic but its parameters vary in time.

At the beginning, let us formulate the governing equations using time t as an independent variable. Then the set of G-coordinates $a_1(t), \ldots, a_m(t)$ form the vector process $\mathbf{a}(t)$, the set of loading parameters form vector process $\mathbf{s}(t)$, and the set of effective crack tip curvature radii-form vector process $\rho(t)$. Dimensions of these vectors are, in general, different.

To describe dispersed damage in a body, we apply the concepts of continuum damage mechanics. Let this damage be characterized by the field function $\omega(\mathbf{x}, t)$ where $\mathbf{x}$ is the reference vector. This function may be a tensor field of any rank, in particular, of zero rank, i.e., a scalar field (see Section 2.3). Microdamage at the crack tips is of primary significance. We describe it with the measures $\psi_1(t), \ldots, \psi_\nu(t)$ forming the microdamage vector process $\psi(t)$. Each of the components $\psi_k(t)$ is linked with one of the components of vector $\mathbf{a}(t)$ so that $\nu \geq m$ where m is the dimension of $\mathbf{a}(t)$. In the general case components $\psi_k(t)$, along with the time, depend on the angles of orientation to the direction of prior crack propagation. We will discuss this situation in Chapter 5 in the context of kinking and branching of cracks.

The connection between the damage measures $\omega(\mathbf{x}, t)$ and $\psi(t)$ may be rather complicated. For example, invariants of the fields $\omega(\mathbf{x}, t)$ may enter into the components of the process $\psi(t)$. When $\omega(x, t)$ is a scalar field, and a crack propagates in the fixed plane with the single G-coordinate $x = a(t)$, we have

$$\psi(t) = \omega[a(t), t] \tag{3.26}$$

The general relationship between $\psi(t)$ and $\omega(x, t)$ may be symbolically presented as

$$\psi(t) = \Psi\{\omega[a(t), t]\} \tag{3.27}$$

Here $\Psi\{\cdot\}$ is a certain functional, and substitution $x = a(t)$ signifies that the crack tips are considered.

To evaluate the far field damage, we have to know the crack trajectories. Sometimes these trajectories are known beforehand, e.g., from the symmetry considerations. In the general case, the evaluation of crack trajectories is a part of the problem's solution.

Equations of microdamage accumulation can be presented in the general form:

$$\varepsilon(x,t) = E\left\{\mathbf{s}(\mathbf{x},\tau), \omega(\mathbf{x},\tau)\right\}, \quad \omega(x,t) = \Omega\left\{\mathbf{s}(\mathbf{x},\tau), \varepsilon(\mathbf{x},\tau)\right\} \qquad (3.28)$$

Here σ and ε are stress and strain tensors. Functionals $E\{\ldots\}$ and $\Omega\{\ldots\}$ take into account the effects of the stress history at $\tau \in [0,t]$ on the strain and microdamage fields.

The process $\rho(t)$ with components $\rho_1(t), \ldots, \rho_\nu(t)$ characterizes the stress concentration near the crack tips. To close the statement of the problem, we need an equation with respect to $\rho(t)$. Let us present this equation in the form similar to (3.28).

$$\rho(t) = P\left\{\mathbf{a}(\tau), \psi(\tau)\right\} \qquad (3.29)$$

Here, by definition, the crack tip curvatures are functionals of the crack sizes and microdamage measures at the tips.

Both active and passive generalized forces depend on the level of microdamage accumulated near the crack tips. The system cracked body-loading can be in sub-equilibrium, equilibrium or non-equilibrium states with respect to generalized coordinates a_j depending on the signs in relations

$$G_j[\mathbf{a}(t), \mathbf{s}(t), \psi(t), \rho(t)] \lessgtr \Gamma_j[\mathbf{a}(t), \mathbf{s}(t), \psi(t), \rho(t)] \quad (j = 1, \ldots, m) \qquad (3.30)$$

Both sides in (3.30), generally, depend on the crack parameters vector $\mathbf{a}(t)$, loading vector $\mathbf{s}(t)$, tip microdamage vector $\psi(t)$, and effective tip curvature vector $\rho(t)$.

3.6 Stability of Fatigue Cracks

To study stability of fatigue cracks, let us use the general concepts of analytical fracture mechanics (Section 3.3). A fatigue crack does not grow with respect to G-coordinate $a_j(t)$ if $G_j < \Gamma_j$. In particular, if all $G_j < \Gamma_j, j = 1, \ldots, m$, none of the cracks grow. Fatigue crack propagation initiates if at a certain time moment t_* the equality $G_j = \Gamma_j$ is primarily attained. The crack growth will be continuous if the attained state is stable equilibrium, and jump-wise if this state is an unstable equilibrium. The jump-wise growth takes place also when the sign of inequality $G_j < \Gamma_j$ suddenly changes into the opposite, $G_j > \Gamma_j$. It is possible at overloading or if a crack enters into a weaker material.

Let a system be in the equilibrium state with respect to μ generalized coordinates. In terms of generalized forces the variation $\delta_G(\delta_G W)$ of the virtual work $\delta_G W$ takes the form

$$\delta_G(\delta_G W) = \sum_{j=1}^{\mu} \sum_{k=1}^{\mu} \frac{\partial H_j}{\partial a_k} \delta a_j \delta a_k + \sum_{j=1}^{\mu} \sum_{i=1}^{\nu} \frac{\partial H_j}{\partial \psi_i} \delta a_j \delta \psi_i \qquad (3.31)$$

where notation (3.9) is used, i.e.,

$$H_j = G_j - \Gamma_j. \qquad (3.32)$$

Compared with (3.12), the right-hand side of (3.31) contains the isochronic variations $\delta\psi_i$ of microdamage measures on the moving crack tips. An equilibrium state is stable if $\delta_G(\delta_G W) < 0$ for all $\delta a_j, j = 1, \ldots, \mu$, and unstable if such variations δa_j and $\delta\psi_i$ exist that $\delta_G(\delta_G W) > 0$.

Discrete argument N equal to the number of cycles or blocks is useful in the study of cyclic fatigue. For simplification we use the term "cycle" only. Conditions placed on $\delta_G W$ may be expressed through the upper bounds of differences $H_j = G_j - \Gamma_j$ attained during N-th cycle:

$$H_j(N) = \sup_{t_{N-1} \leq t < t_N} \{G[\mathbf{a}(t), \mathbf{s}(t), \psi(t), \rho(t)] - \Gamma[\mathbf{a}(t), \mathbf{s}(t), \psi(t), \rho(t)]\} \quad (3.33)$$

Here $[t_{N-1}, t_N]$ is a time segment corresponding to the N-th cycle. The system cracked body-loading is staying in a sub-equilibrium state during the N-th cycle if all $H_j(N) < 0$. It becomes unstable if even a single $H_j(N) > 0$. Equality $H_j(N) = 0$ holds for cracks which are in the equilibrium state with respect to a_j. Conditions of stability of equilibrium states remain to be $\delta_G(\delta_G W) < 0$ where $\delta_G(\delta_G W)$ is determined with (3.31) and (3.33) is used instead of (3.32).

If at $N = 0$ the system cracked body-loading is in the sub-equilibrium state, then at certain $N > 0$ we have

$$H_j(N) < 0 \quad (j = 1, \ldots, m) \quad (3.34)$$

The first violation of these inequalities corresponds to the growth initiation of at least one of the cracks. Denote with N_* the corresponding cycle number. As a rule, only one of $H_j(N)$ attains zero at $N = N_*$ while the other generalized forces satisfy equations (3.34). Assume that the crack growth begins with respect to a_k. The further propagation of the crack depends on stability of the attained equilibrium state. If $\delta_G(\delta_G W) < 0$, the crack will be growing in a stable way. If $\delta_G(\delta_G W) > 0$, a jump with respect to a_k will take place. It means either the final failure or the transition to a new sub-equilibrium state.

The condition of jump-wise growth with respect to generalized coordinate a_k during N-th cycle is $H_k(N) \geq 0$. The magnitudes $a_{k*}(N)$ and $a_k^*(N)$ of a_k and after a jump satisfy the relationship following from the energy conservation law:

$$\int\limits_{a_{k*}(N)}^{a_k^*(N)} H_k\left[\mathbf{A}(t_*), \mathbf{s}(t_*), \psi(\mathbf{A}, t_*), \rho(\mathbf{A}, t_*)\right] da_k \geq 0 \quad (3.35)$$

Here the crack propagation is considered as an instant process at $t = t_*$, where t_* is instant when the supremum in the right-hand side of (3.33) is attained. Vector $\mathbf{A}(t)$ differs from $\mathbf{a}(t)$ only with the component a_k which is considered as an independent variable. Neglecting the loss of energy during the jump due to heat, transfer, acoustic emission, radiation, etc., we obtain from (3.35) the following equation for

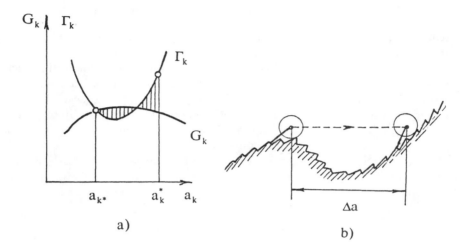

FIGURE 3.7
Illustration of jump-wise crack propagation.

the new magnitude $a_k^*(N)$ when $a_{k\cdot}(N)$ is known:

$$\int\limits_{a_{k\cdot}(N)}^{a_k^*(N)} G_k\left[\mathbf{A}(t_*), \mathbf{s}(t_*), \psi(\mathbf{A}, t_*), \rho(\mathbf{A}, t_*)\right] da_k =$$

$$= \int\limits_{a_{k\cdot}(N)}^{a_k^*(N)} \Gamma_k\left[\mathbf{A}(t_*), \mathbf{s}(t_*), \psi(\mathbf{A}, t_*), \rho(\mathbf{A}, t_*)\right] da_k \tag{3.36}$$

Equation (3.36) is illustrated schematically in Figure 3.7 where the relationship between $G_k(a_k)$ and $\Gamma_k(a_k)$ is shown for a jump with coordinate a_k. The dashed areas in Figure 3.7a correspond to the equal amounts of energy. Mechanical interpretation of this equality is presented in Figure 3.7b where the model of a system with unilateral constraint suggested in [15] is used. It is a heavy rigid cylinder with a ratchet placed on a rigid, geared cylindrical surface. The ratchet restricts the opposite motion of the cylinder. The left-hand position is an unstable equilibrium state; the right-hand position is a subequilibrium (stable) state. The energy level is the same in both cases.

The jump-wise process described in equation (3.36) is usually sufficiently slow if we disregard the final failure as well as large jumps as a result of sudden load increase or a sudden decrease of the material's resistance. In the continuous approximation, we assume that a crack grows with respect to G-coordinate a_k under the equilibrium condition

$$H_k(N) = 0 \tag{3.37}$$

or, in the notation of (3.36),

$$G_k[\mathbf{A}(N), \mathbf{s}(N), \psi(N), \rho(N)] = \Gamma_k[\mathbf{A}(N), \mathbf{s}(N), \psi(N), \rho(N)] \tag{3.38}$$

All the functions of the continuous argument N entering equation (3.38) correspond to supremum states indicated in (3.33). The interaction between two mechanisms is taken into account with equations (3.30)–(3.38). The first mechanism is microdamage accumulation at the crack tips and along the future crack trajectories, and the second one is the balance of forces and energy in the system cracked body-loading. If a crack is short or the load level is low, microdamage near the tip plays a decisive part. Opposite, near the final fracture a crack propagates at a low microdamage level, and the general balance becomes decisive.

If the crack growth is governed with microdamage only, the analytical formulation of the theory becomes much simpler. Let $\omega_k(N)$ be the microdamage measure corresponding to G-coordinate $a_k(N)$, and $\psi_k(N)$ be its magnitude at the crack tip. Then the crack does not grow if

$$\psi_k(N) < \psi_k^*$$

Here ψ_k^* is a critical magnitude of microdamage. The stable growth takes place when the condition holds:

$$\psi_k(N) = \psi_k^*$$

A more general equation of crack growth may be formulated in terms of generalized forces:

$$\Gamma_k\left[\mathbf{a}(N), \mathbf{s}(N), \psi(N), \rho(N)\right] \approx 0$$

This equation corresponds to (3.38) under the assumption that the driving force G_k is small compared with the resistance force Γ_k for nondamaged material.

3.7   Patterns of Fatigue Crack Growth

Different patterns of fatigue crack propagation can be distinguished in the framework of the developed theory. To illustrate the origin of these patterns, consider a single-parameter crack described with G-coordinate a. Similarly, let s, ω, ψ and ρ be scalar parameters, too. The crack does not propagate at

$$G < \Gamma \tag{3.39}$$

and propagates in a continuous, stable way when

$$G = \Gamma, \qquad dG/da < d\Gamma/da \tag{3.40}$$

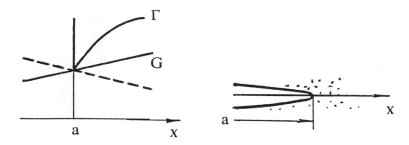

FIGURE 3.8
Relationship between generalized forces during continuous crack growth.

The crack becomes unstable in an equilibrium state at

$$G = \Gamma, \qquad dG/da > d\Gamma/da \tag{3.41}$$

When

$$G > \Gamma \tag{3.42}$$

we have a non-equilibrium, unstable state. Later on, for simplification, assume the loading level s and the effective tip radius ρ are constant, the driving force G depends on the crack size, and the resistance force Γ depends on ψ.

The first type of fatigue crack growth is illustrated in Figure 3.8 where the relationship between the generalized forces G and Γ is shown in the moment when an equilibrium state is attained. Because of the character of damage distribution in the vicinity of the tip, the inequality $dG/da < d\Gamma/da$ holds at the tip $x = a$, and the inequality $G < \Gamma$ at all points $x > a$ ahead of the tip. This means that the crack will propagate in a continuous pattern.

The second type is illustrated in Figure 3.9. A distinct plastic zone is present at the tip. When damage is distributed almost uniformly in this zone, the inequality $dG/da > d\Gamma/da$ takes place for the equilibrium state. This state will be unstable that results in a jump of crack tip at the distance $\Delta a > \lambda$. However, in the case of G decreasing with a (see the broken line in Figure 3.9), the process of crack growth will be continuous even in the presence of the plastic zone. Sometimes, due to stress and strain distribution in the process zone, the maximal damage is accumulated not at the tip but a little ahead of it (Figure 3.10). Then a jump-wise crack propagation is expected even in the case of the driving force G slightly decreasing at $x > a$.

All three patterns meet in fatigue, and they sometimes vary in the process of crack propagation. For example, due to the damage accumulated near the moving crack, cyclic softening, etc. the distinct process zone can disappear. Then the jump-wise growth will transfer in a continuous process. On the other hand, the pattern

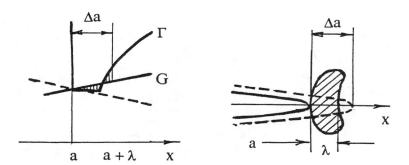

FIGURE 3.9
Relationship between generalized forces in the presence of the process zone.

can change significantly when we approach the final failure. Then the increments of the driving force become large even for a single loading cycle or a short cycle sequence. This is illustrated in Figure 3.11 where we have $G(N) < \Gamma(N + 1)$ but $G(N + 1) > \Gamma(N + 1)$. As a result, the crack advances in Δa at each cycle of loading.

A typical picture of continuous crack growth is presented in Figure 3.12. The

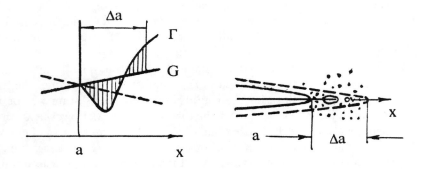

FIGURE 3.10
Relationship between generalized forces in the presence of significant damage ahead of the crack tip.

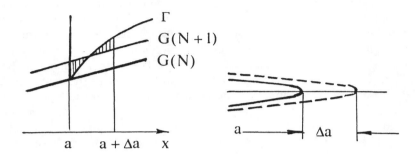

FIGURE 3.11
Jump-wise crack propagation per each loading cycle.

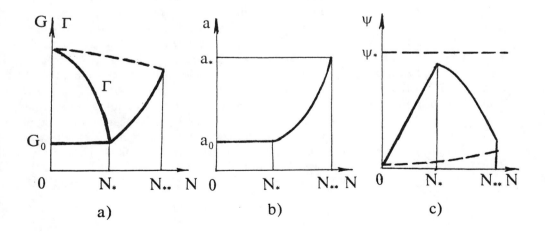

FIGURE 3.12
Parameter evolution during the continuous crack growth: (a) generalized forces,
(b) crack size, (c) damage measured at the tip and in the far field.

driving force G is a monotonically increasing function of the crack size a. The resistance force Γ (with account of microdamage) is a monotonically increasing function, at least in the vicinity of the crack tip. This case is typical for high-cycle fatigue damage resulting in quasi-brittle fracture. In particular, we have such a relation between G and Γ in the classic opening mode crack problem under the condition that no plastic zone is present at the crack tip. Let the sub-equilibrium condition $G < \Gamma_0$ hold at the beginning of the loading. Then due to microdamage accumulation, the resistance force Γ decreases with the cycle number N. The equilibrium is primarily attained at $N = N_*$, and this equilibrium is stable. If Γ increases more rapidly than G, the state of the system remains to be stable at $N_* < N < N_{**}$. At $N = N_{**}$ the loss of stability occurs as the equalities $G = \Gamma$ and $dG/da = d\Gamma/da$ are attained simultaneously. The critical crack size a_{**}, generally, is less than that for the short-time monotonically loading determined from the equation $G = \Gamma_0$. Figure 3.12a illustrates how the inequality $G < \Gamma$ changes into the equality $G = \Gamma$ at $N \geq N_*$. Figure 3.12b shows the crack size growth, and Figure 3.12c exhibits the microdamage accumulation process at the moving crack tip $x = a(N)$. The microdamage measure ψ grows monotonically at $N < N_*$ not necessary approaching too closely the critical level $\psi_* = 1$. When the crack begins to propagate, the microdamage at the tip begins to diminish. The broken line shows the microdamage measure ω_{ff} accumulated in the far field, i.e., before the tip arrives at the fixed material point. Near $N = N_{**}$ the tip damage ψ diminishes approaching the field damage ω_{ff}.

Another case is illustrated in Figure 3.13. There the generalized force G decreases as the crack propagates. Such decrease takes place, for example, when the external forces are applied to the face surfaces of the crack. Equilibrium states are stable in this case. The generalized forces as functions of the cycle number N are presented in Figure 3.13a. To make another distinction from Figure 3.12, it is assumed that the initial loading level is high enough to provide the inequality $G(a_0, s) > \Gamma_0$ at the initial moment. The instantaneous quasi-brittle crack growth results from the initial magnitude a_0 to the new one $a_0^* > a_0$ (Figure 3.13b). Then the equilibrium crack size asymptotically approaches a constant magnitude, while the microdamage measure ψ its critical level ψ_* (Figure 3.13c).

The process of jump-wise fatigue crack propagation is illustrated in Figure 3.14. It is assumed that a distinct process zone exists at the crack tip. At $N = N_*$ the first equilibrium state is attained. But the second condition (3.40) is not satisfied, and the state of the system is unstable. The crack tip advances at the distance $\Delta a > \lambda$. The new state of the system is sub-equilibrium. The crack remains to be arrested until the resistance force Γ is not diminished to the level G. Then the next jump occurs, etc. The jump sizes Δa_* may be found from equation

$$\int\limits_a^{a+\Delta a} G(a)da = \int\limits_a^{a+\Delta a} \Gamma(a, \psi)da \qquad (3.43)$$

which is a special case of equation (3.36). In the further growth, the process can take a continuous form as is shown in Figure 3.14 where G, Γ, a and ψ are plotted against the cycle number N. Close to the end of the process, a new stage of jump-wise propagation can be observed when each loading cycle produces a finite crack tip advancement (Figure 3.11).

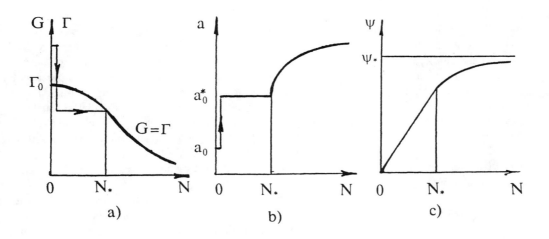

FIGURE 3.13
The case of generalized driving forces decreasing during continuous crack growth: (a) generalized forces, (b) crack size, (c) damage measured at the tip and in the far field.

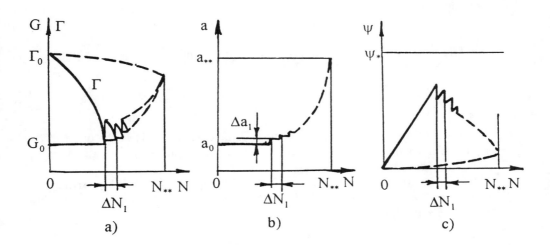

FIGURE 3.14
The case of jump-wise crack growth: (a) generalized forces, (b) crack size, (c) damage measured at the tip and in the far field.

3.8   Quasistationary Approximation in Mechanics of Fatigue

The constitutive equations of fatigue crack growth presented in Section 3.4 are of general character. As a matter of fact, they are literally not the set of equations because the key relationship given in (3.30) is an inequality. This inequality changes the sign when a crack is subjected to starts or arrests. In addition, the properties of the quadratic form in the right-hand side of (3.31) may also be important. Meantime, the empirical and semi-empirical equations of fatigue crack growth such as Paris and Forman equations are just simple ordinary differential equations solved explicitly with respect to the crack growth rate. The question arises of the possibility of reducing the set of primary relationships to differential equations with respect to G-coordinates $a_1, \ldots, a_m$. The answer is given by the so-called quasistationary approximation [14].

First, we treat the fatigue crack propagation as a continuous process considering all involved variables as functions of the cycle number. Second, we assume that cracks propagate under the condition of equilibrium with respect to corresponding G-coordinates. This means that equation (3.38) is valid during the entire crack growth process with respect to $a_1(N), , a_\mu(N)$, where $1 \le \mu \le m$. Third, we refer each component of the damage process vector $\omega(N)$ to a certain component of the crack size vector process $a(N)$. For example, without substantial limitations, we assume that ν_j components $\omega_{ij}(N)$ of the process $\omega(N)$ are referred to G-coordinate $a_j(N)$. And finally, we introduce the process zones for each G-coordinate and for each component of the damage vector. The lengths λ_{ij} of these zones along crack trajectories are assumed to be known beforehand or, at least, are governed by given equations. Damage within the process zones is the near field damage. Remote sections of crack trajectories are considered as the far field. The level of the far field damage is treated as small or at least moderate compared with the critical level.

The characteristic cycle numbers that are required for a crack tip to cross process zones are:

$$\Delta N_{ij} = \lambda_{ij} \left(\frac{da_j}{dN} \right)^{-1} \qquad \left(\begin{array}{l} i = 1, \ldots, \nu_j \\ j = 1, \ldots, \mu \end{array} \right) \tag{3.44}$$

We assume here that the change of the loading over intervals of the order of ΔN_{ij} is sufficiently small. At the same time these intervals are long enough to treat the cycle number within them as a continuous variable.

It is expedient to specialize equations (3.28) and (3.29). In particular, we present the second equation (3.28) in the form solved with respect to the rate of damage accumulation:

$$\frac{\partial \omega(N)}{\partial N} = f\left[s(N), a(N), x\right] g\left[\omega(x, N)\right] \tag{3.45}$$

The right-hand side is a product of functions, one of which depends on the damage vector ω. This separation allows to simplify the further calculations. The function $f(s, a, x)$ is supposed integrable with respect to all variables; the function $g(\omega)$

satisfies conditions $g(\omega) > 0, g'(\omega) > 0$ for all $\|\omega\| \neq 0$. In addition, we suppose that the ij-th component of (3.45) takes the form:

$$\frac{\partial \omega_{ij}}{\partial N} = f_{ij}\left[\mathbf{s}(N), \mathbf{a}(N), \mathbf{x}\right] g_{ij}\left[\omega_{ij}(\mathbf{x}, N)\right]$$

$$(i = 1, \ldots, \nu_j; j = 1, \ldots, \mu)$$

(3.46)

Represent the components ω_{ij} as a sum of the near field damage $\omega_{ij,nf}$ accumulated before the process zone approaches the considered material point, and the far field damage $\omega_{ij,ff}$. The tip damage ψ_{ij} satisfies equations

$$h_{ij}(\psi_{ij}) - h_{ij}(\omega_{ij,ff}) = \int_{N-\Delta N_{ij}}^{N} f_{ij}\left\{\mathbf{s}(N_1), \mathbf{a}(N_1), \mathbf{x}[\mathbf{a}(N_1)]\right\} dN_1 \quad (3.47)$$

Hereafter the notation used is

$$\int \frac{d\omega}{g_{ij}(\omega)} = h_{ij}(\omega) + \text{const}$$

Applying the mean value theorem to the integral in (3.47), we obtain as a result of (3.44)

$$\int_{N-\Delta N}^{N} f_{ij}\left\{\mathbf{s}(N), \mathbf{a}(N), \mathbf{x}[\mathbf{a}(N)]\right\} dN \approx \lambda_{ij} \left(\frac{da_j}{dN}\right)^{-1} f_{ij}\left\{\mathbf{s}(N), \mathbf{a}(N), \mathbf{x}[\mathbf{a}(N) + \lambda_{ij}^0]\right\}$$

where $0 \leq \lambda_{ij}^0 \leq \lambda_{ij}$. Substitution of this result in (3.47) gives

$$h_{ij}(\psi_{ij}) \approx h_{ij}(\omega_{ij,ff}) + \lambda_{ij}\left(\frac{da_j}{dN}\right)^{-1} f_{ij}\left\{\mathbf{s}(N), \mathbf{a}(N), \mathbf{x}[\mathbf{a}(N) + \lambda_{ij}^0]\right\} \quad (3.48)$$

Now we return to the equilibrium condition (3.38):

$$G[\mathbf{s}(N), \mathbf{a}(N), \psi(N)] = \Gamma_j[\mathbf{s}(N), \mathbf{a}(N), \psi(N)] \quad (i = 1, \ldots, \mu) \quad (3.49)$$

In the general case, both sides in (3.49) are functions of all their arguments, in particular, of all $\psi_{ij}(N)$. Substituting (3.48) in (3.49), we come to a set of ordinary differential equations with respect to $a_1, \ldots, a_\mu$, where $1 \leq \mu \leq m$.

To expend the quasistationary approximation upon stability conditions, consider equation (3.31). In the notation of equations (3.47) and (3.48), equation (3.31) takes the form

$$\delta_G(\delta_G W) = \sum_{j=1}^{\mu}\sum_{k=1}^{\mu} \frac{\partial H_j}{\partial a_k} \delta a_j \delta a_k + \sum_{j=1}^{\mu}\sum_{k=1}^{\mu}\sum_{i=1}^{\nu} \frac{\partial H_j}{\partial \psi_{ik}} \delta a_j \delta \psi_{ik} \quad (3.50)$$

It is necessary to transform its right-hand side in a quadratic form with respect to G-variations δa_j. Noting that

$$\omega_{ik}(\mathbf{a} + \Delta \mathbf{a}, N) = \omega_{ik}(\mathbf{a}, N - \Delta N_{ij}) + O(|\Delta \mathbf{a}|^2)$$

and taking into account equation (3.45), we obtain

$$\delta\psi_{jk} = -\frac{\partial\psi_{jk}}{\partial N}\left(\frac{da_k}{dN}\right)^{-1}\delta a_k$$

Therefore, equation (3.50) may be transformed as follows:

$$\delta_G(\delta_G W) = \sum_{j=1}^{\mu}\sum_{k=1}^{\mu}\left[\frac{\partial H_j}{\partial a_k} - \left(\frac{da_k}{dN}\right)^{-1}\sum_{i=1}^{\nu_k}\frac{\partial H_j}{\partial\psi_{ik}}\frac{\partial\psi_{ik}}{\partial N}\right]\delta a_j\delta a_k \qquad (3.51)$$

The set of fatigue cracks is stable with respect to G-coordinates $a_1,\ldots,a_\mu$ if the quadratic form (3.51) is definitively positive in the first ortant of variations, i.e., at all $\delta a_j \geq 0, j = 1,\ldots,\mu$. The right-hand side of (3.51) can be calculated with the use of (3.47), (3.48) and (3.49).

3.9   Single-Parameter Fatigue Cracks

As an example, let us consider a single-parameter crack with a single microdamage measure ($\mu = \nu = 1, 0 \leq \omega \leq 1, h(\omega) \equiv \omega$). Instead of (3.48) we have

$$\psi(N) \approx \omega_{ff}(N) + \lambda_p\left(\frac{da}{dN}\right)^{-1}f[s(N),a]$$

where λ_p is the length of the process zone. For simplification we put $a + \lambda^0 \approx a$. Instead of (3.49) we obtain

$$G(s,a,\psi) = \Gamma(s,a,\psi)$$

In typical cases G does not depend on ψ, while Γ depends on ψ only. For example, let

$$\Gamma = \Gamma_0 F(\psi)$$

with $\Gamma_0 = \text{const}, F(0) = 1, F(1) = 0, F'(\psi) < 0$. The above equations result in the following differential equation with respect to $a(N)$:

$$\frac{da}{dN} \approx \frac{\lambda_p f[s(N),a]}{F^{-1}(G/\Gamma_0) - \omega_{ff}} \qquad (3.52)$$

Equation (3.52) includes, as a special case, practically all empirical and semi-empirical equations of stable growth for single-parameter fatigue cracks. The nominator of its right-hand side takes into account the process of damage accumulation. The denominator is connected to the stability conditions. Actually, the crack growth rate is bounded at $G < \Gamma_0 F[\omega_{ff}(N)]$. Considering the stability conditions in detail, we must formulate this conclusion more precisely. Equation (3.51) for this special case takes the form:

$$\delta_G(\delta_G W) = \left[\frac{dG}{da} - \left(\frac{da}{dN} \right)^{-1} \frac{d\Gamma}{d\psi} \frac{d\psi}{dN} \right] (\delta a)^2$$

Noting that in the quasistationary approximation

$$\frac{d\psi}{dN} \approx \frac{\omega_{ff} - \psi}{\Delta N} \approx \frac{da}{dN} \frac{\omega_{ff} - \psi}{\lambda_p}$$

we obtain the stability condition:

$$\frac{dG}{da} < \frac{\Gamma_0(\omega_{ff} - \psi)}{\lambda_p} \frac{dF}{d\psi} \tag{3.53}$$

It is evident that this condition does not coincide with the condition of positiviness of the denominator in (3.52). For example, when $F(\psi) = 1 - \psi$, equation (3.53) together with condition $G = \Gamma$ gives

$$\frac{dG}{da} < \frac{\Gamma_0}{\lambda_p} \left(1 - \frac{G}{\Gamma_0} - \omega_{ff} \right) \tag{3.54}$$

Meantime the "obvious" condition is

$$G < \Gamma_0(1 - \omega_{ff}) \tag{3.55}$$

The difference between (3.54) and (3.55) is illustrated in Figure 3.15. The critical size a_c corresponds to (3.54) with the equality sign. The crack size that is found from the equality corresponding to (3.55) is larger than a_c. For example, when $dG/da \equiv G/a$ (the linear dependence G on a), the critical magnitude of G according to (3.54) is

$$G = \frac{\Gamma_0(1 - \omega_{ff})}{1 + \lambda_p/a} \tag{3.56}$$

The correction term λ_p/a may be significant for low-cycle fatigue. In high-cycle fatigue λ_p/a is usually small compared with unity. In general, it is more of a theoretical interest because it demonstrates that stability conditions may be violated before the crack growth rate goes to infinity.

In special problems, equation (3.52) may be rather complicated. First, the process zone length depends both on s and a. An additional equation, a finite or differential one, must be added to describe this dependence. Second, the damage accumulation is controlled by the acting stresses near the crack tip and/or on its prolongation. This signifies that the effective concentration factor at the tip enters implicitly in (3.52). An additional equation with respect to the effective stress concentration factor is to be included to get a closed set of equations. In any case, the final equations are much simpler than the initial set of functional equations and inequalities.

Similar simplifications can be performed for bodies with a multiparameter crack or a set of cracks. Under rather evident assumptions, equation (3.52) remains valid in the multiparameter case:

$$\frac{da_j}{dN} = \frac{\lambda_j f_j(s, a)}{F_j^{-1}(G_j/\Gamma_j^0) - \omega_{j,ff}}$$

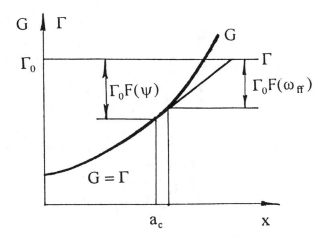

FIGURE 3.15
Relationship between the generalized forces in the vicinity of loss of stability.

Here f_j and G_j, generally, depend on all G-coordinates $a_1, \ldots, a_m$. It means that we arrive at a set of connected differenial equations with respect to $a_1(N), \ldots, a_\mu(N)$.

3.10 Summary

A theory of fatigue crack growth based on the synthesis of fracture mechanics and mechanics of dispersed damage under cyclic and/or sustained loading was developed in this chapter. The notions of fracture mechanics were presented in the multi-parameter version with the use of the principle of virtual work for systems with unilateral constraints. The theory of fatigue crack growth was developed in a general form with the possibility of applications to a broad range of engineering problems. Compared with common models of solid mechanics, this theory contains a number of internal variables characterizing dispersed damage in a cracked body as well as a set of parameters describing the current state at the moving crack tips. Despite its generality, the theory is quite transparent and well organized. Using certain ad hoc assumptions, the set of governing equations of the theory can be reduced to differential equations similar to those widely used in engineering practice. This opens a way of assessing the material parameters of the theory.

Chapter 4

FATIGUE CRACK GROWTH
IN LINEAR ELASTIC BODIES

4.1 Modeling of Material Properties

The aim of this chapter and, partially, of subsequent chapters is to apply the theory developed in Chapter 3 to various materials, loading and environmental conditions. We are going to demonstrate the flexibility of the theory and its availability to fit various, sometimes vague and controversial experimental data.

The next two chapters are dedicated to cracks propagating in materials which follow Hook's law for stress-strain relations and, at the same time, are subjected to microdamage accumulation. Because of microdamage, such materials are not elastic ones in the precise meaning. Actually, dispersed microdamage results in irreversible change of material properties, at least some of them. For the theory of fatigue, the effect of microdamage on the resistance to crack propagation (such as the specific fracture work) is of primary significance. The material fracture toughness with the degrading under cyclic or sustained loading, strictly speaking, cannot be considered as elastic even if we assume, for simplification only, that its compliance does not vary. Taking into account the effect of microdamage on compliance parameters, we come to a material which can be called hypo-elastic. From the general viewpoint, additional variables similar to damage measures are internal parameters. Their introduction allows development of a number of models for continua which exhibit irreversible behavior. Among them, by the way, there are some models describing elasto-plastic and elasto-visco-plastic properties of real materials.

We will treat microdamage in the framework of continuum damage mechanics. Introductory remarks concerning continuum damage mechanics were presented in Sections 1.5 and 2.3. Some applications to modeling of fatigue crack nucleation were discussed in Sections 2.4–2.6. The simplest measure of microdamage is a scalar variable ω. For example, for a planar mode I crack propagating in its initial plane, it is natural to introduce the measure $0 \leq \omega \leq 1$ referring to microcracking in mode I. A similar measure may be used for shear modes. In this case, the measure ω refers to shear mode microcracks. Two or three measures must be introduced for mixed mode cracks. For every damage measure special equations must be based on experimental data. When mode I is considered, the damage accumulation rate near the tip and along its prolongation depends, in the first place, on the range $\Delta\sigma$ of opening stresses acting near the tip. In the case of mode II, the range $\Delta\tau$ of acting shear stresses is the main factor. The rate of the damage measure for microvoid formation depends primarily on the mean tensile stress σ_0 in sustained loading and its range $\Delta\sigma_0$ in cyclic loading.

A typical equation with respect to the scalar damage measure ω for mode I is

$$\frac{\partial\omega}{\partial N} = f(\sigma_{max}, \sigma_{min}) \tag{4.1}$$

where σ_{max} and σ_{min} are extremal normal stresses within a cycle acting in the plane of crack propagation, and $f(\cdot\cdot) \geq 0$. The principal features of damage accumulation can be described by the power-threshold law similar to (2.3),(2.44), etc.

$$\frac{\partial\omega}{\partial N} = \left(\frac{\Delta\sigma - \Delta\sigma_{th}}{\sigma_d}\right)^m \tag{4.2}$$

Here $\Delta\sigma = \sigma_{max} - \sigma_{min}$ is the range of the acting stresses; σ_d is the resistance stress against damage accumulation; $\Delta\sigma_{th}$ is the threshold resistance stress; $m > 0$ is analogous to exponents of fatigue curves. The material parameters $\sigma_d, \Delta\sigma_{th}$, and m, in general, depend on the current damage measure, loading and crack propagation history. For example, if ω is interpreted as the diminishing of the area of the effective cross section due to microcracking, the stress range $\Delta\sigma(1-\omega)^{-1}$ is to be put in (4.2) instead of $\Delta\sigma$. In other words, σ_d and $\Delta\sigma_{th}$ must be replaced by $\sigma_d(1-\omega)$ and $\Delta\sigma_{th}(1-\omega)$. The feedback effect of accumulated damage on the damage growth rate may be described by equation

$$\frac{\partial\omega}{\partial N} = \left(\frac{\Delta\sigma - \Delta\sigma_{th}}{\sigma_d}\right)^m (\omega_* - \omega)^{-n} \tag{4.3}$$

with complementary non-negative material parameters n and ω_*. Usually one may assume $\omega_* = 1$.

Another version of equation (4.2) contains, instead of $\Delta\sigma$, the effective stress range $\Delta\sigma_{eff}$:

$$\frac{\partial\omega}{\partial N} = \left(\frac{\Delta\sigma_{eff} - \Delta\sigma_{th}}{\sigma_d}\right)^m \tag{4.4}$$

Such an approach is appropriate for modeling of the "crack closure effect" and, maybe, of the "shielding effect" [63, 72].

The situation with threshold, closure and related effects in fatigue is not clear uptonow, and experimental data are fragmentary and sometimes contradictory. From the viewpoint of mechanics, we must distinguish the closure effect on microscale, and the closure of macroscopic (long) cracks. The latter effect contributes both to the stress field near the crack tip and to the driving generalized forces. To avoid un-necessary complications, we are going to omit the closure effect almost everywhere in the following numerical analysis. As a rule, in the next sections we follow equations (4.2) and (4.3) assuming that

$$\Delta\sigma_{th} = \Delta\sigma_{th}^0 f(R) \tag{4.5}$$

and $f(0) = 1$. Here $\Delta\sigma_{th}^0$ is the threshold resistance stress at $R = 0$. For example, one may assume $f(R) = (1 - bR)^p$ where $b > 0$ and $p > 0$. Note that R is the ratio of extremal acting stresses near the crack tip that, generally, does not coincide with the common ratio $R = \sigma_\infty^{min}/\sigma_\infty^{max}$ of the extremal applied stresses. In general, all other material parameters entering equations (4.1)-(4.4) depend on R.

The interpretation of values in equations (4.2)-(4.5) as well as in other equations of microdamage accumulation is not clear enough. This concerns, to the same degree, their numerical evaluation. For example, the range $\Delta\sigma_t$ of tip stresses, being evaluated in the framework of elasticity theory, usually exceeds, and significantly, not only the yield limit, but even the ultimate tensile stress. This means that the magnitude of $\Delta\sigma_t$ is, in some aspect, of a representative nature. A similar conclusion immediately follows for related material parameters, $\Delta\sigma_{th}$ and $\Delta\sigma_d$. To fit experimental data, we must take appropiate numerical values for these parameters.

Some preliminary conclusions may be done using the idea of quasistationary approximation that leads to simplified equations of fatigue crack growth (see Sections 3.8 and 3.9). To keep an agreement with experiments, we must take care that a sensible cycle number ΔN should be required on the middle section of the fatigue life to cover a distance λ_g. Here, λ_g is of the order of the material grain size. Consider, for instance, the case when $a/\rho = 10^2$, $m = 4$, $\Delta\sigma_\infty = 100$ MPa. Then $\Delta\sigma_t \approx 2\Delta\sigma_\infty(a/\rho)^{1/2} = 2$ GPa. To produce complete damage in a grain for a single cycle, we must assume $\sigma_d \approx 16$ GPa. Then the estimate for the crack growth rate will be $da/dN \sim 10^{-4}$ m/cycle. This signifies that for metallic alloys we must put $\sigma_d \approx 10$ GPa and even higher. Only in this case we will receive a sensible numerical prediction for crack growth rate (under the condition that the stresses in the process zone are evaluated in the framework of elasticity theory). Numerical values for local strength such as σ_d are of the order of the "theoretical strengh", i.e., of the ultimate stress for a material without flaws. This makes our assumptions of numerical values rather agreeable. In addition, we could mention that the quantities $\sigma_d(\pi\rho_g)^{1/2}$ and $\Delta\sigma_{th}(\pi\rho_g)^{1/2}$ with the dimension of stress intensity factors are (under an appropriate choice of ρ_g) of the same order of magnitude as the related material parameters measured in direct experimentation.

However, if one feels uncomfortable with high magnitudes of σ_d, one could replace equation (4.2) by the following:

$$\frac{\partial\omega}{\partial N} = \frac{1}{N_c}\left(\frac{\Delta\sigma - \Delta\sigma_{th}}{\sigma_c}\right)^m$$

Here N_c is a reference cycle number, and σ_c is the resistance stress corresponding to the complete local damage in N_c cycles of loading. The equivalence with (4.2) follows from the condition $N_c\sigma_c^m = \sigma_d^m$. On the other hand, the acting cyclic stress range of the order $\Delta\sigma_\infty(a/\rho)^{1/2}$ also looks too high. It is a consequence of the assumption that the material remains linear elastic with initial compliance properties despite tip damage. In any case, when we develop a model of fatigue crack growth as an ancestor of linear fracture mechanics, we have to tolerate some physical inconsistencies.

The next item to discuss is the influence of damage on fracture toughness given with the specific fracture work γ (Section 1.6). Assuming that for damaged material this value is connected with that for nondamaged material γ_0 and the damage measure ω, we use the following formula:

$$\gamma = \gamma_0[1 - (\omega/\omega_*)^\alpha]. \qquad (4.6)$$

Here α is a positive exponent, and the critical level ω_* does not, generally, coincide with unity. However, usually we put $\omega_* = 1$. At $\omega = 1$ the magnitude $\gamma = \gamma_0(1 - \omega_*^{-\alpha})$ characterizes the residual fracture work of the completely damaged material. Another form of the relationship between the specific fracture work and the microdamage measure is

$$\gamma = \gamma_0[1 - (\omega/\omega_*)]^\alpha. \qquad (4.7)$$

The choice of a special form of equations for γ is very wide. One of the reasons for choosing them in the presented form is to keep a close relation with the most frequently used empirical equations such as the Paris-Erdogan equation. Turning to equation (3.52) corresponding to the quasistationary approximation, we see that the above assumption follows this idea.

To close the presentation of the model of a material's behavior, we must discuss the influence of microdamage on elastic properties. There is an ample literature dedicated to the influence of microcracks and microvoids on effective elastic moduli. However, the material's compliance enters the theory through the generalized driving forces which, in their turn, depend on the general energy balance in the whole system cracked body-loading. On the other side, taking the effect of microdamage on elastic properties, we unavoidably meet the problems complicated with material's nonhomogeneity and anisotropy which, in addition, vary during the loading and crack growth process. Including these effects requires very laborious and tedious numerical computations. Some attempts to take into account the influence of microdamage on material compliance will be presented in Section 4.9, where the contribution of this influence on generalized forces and crack growth rates is analyzed by numerical simulation. In particular, we will treat the shielding effect as a kind of stress relaxation near the tips. This approach allows overcoming some difficulties in dealing with too high acting stresses.

4.2   Fatigue Crack in Griffith's Problem

The simplest problem of linear fracture mechanics considered primarily by Griffith is the mode I planar crack in an unbounded elastic plate under remotely uniformly distributed tensile stresses σ_∞ (Figure 4.1). This problem is a kind of a test bench in fracture mechanics. We use this problem to develop, as detailed as possible, the essential features of the presented theory of fatigue crack growth. Let the stresses $\sigma_\infty(t)$ vary in time cyclically with the extremal magnitudes $\sigma_\infty^{max}(N)$ and $\sigma_\infty^{min}(N)$ within each cycle. The stress range $\Delta\sigma_\infty(N) = \sigma_\infty^{max}(N) - \sigma_\infty^{min}(N)$, and the

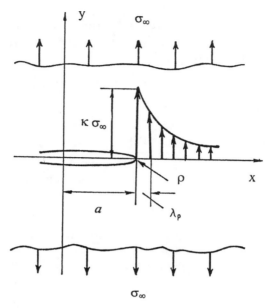

FIGURE 4.1
Fatigue crack in Griffith's problem.

stress ratio $R(N) = \sigma_\infty^{min}(N)/\sigma_\infty^{max}(N)$ enter the analysis. Stress characteristics, as well as half-length of the crack $a(N)$, effective tip radius of curvature $\rho(N)$, and microdamage measure $\psi(N)$ at the tip are considered as continuous functions of N. The microdamage $\omega(x, N)$ along the crack trajectory $\mid x \mid \geq a, y = 0$ is assumed as a scalar function with magnitudes from segment $[0, 1]$ where $\omega = 0$ corresponds to the nondamaged material.

In the framework of the assumptions discussed above, the generalized driving force coincides with the strain energy release rate. It is defined in the plane-stress state as

$$G = \frac{K^2}{E} \tag{4.8}$$

and in the plane-strain state as

$$G = \frac{K^2(1 - \nu^2)}{E} \tag{4.9}$$

Here K is the stress intensity factor K_I for the mode I (index is omitted), E is Young's modulus and ν is Poisson's ratio. Most results discussed later can be referred to as more general conditions of loading and geometry of the body. To include such generalizations, it is sufficient to assume that

$$K = Y\sigma_\infty(\pi a)^{1/2} \tag{4.10}$$

where the correction factor Y is of the order of unity. It depends on geometry and, in some problems, on Poisson's ratio.

The generalized resistance force is determined by equation $\delta A_f = -\Gamma \delta a$. Referring all the forces to a unit of width, we see that $\Gamma = \gamma$, where γ is the specific fracture work. Thus we have

$$\Gamma = \Gamma_0[1 - (\psi/\omega_*)^\alpha] \qquad (4.11)$$

when equation (4.6) is used, and

$$\Gamma = \Gamma_0[1 - (\psi/\omega_*)]^\alpha \qquad (4.12)$$

when we follow equation (4.7). In both cases $\Gamma_0 = \gamma_0$, where γ_0 is the specific fracture work for nondamaged material, ψ is the damage measure at the tip, i.e.,

$$\psi(N) = \omega[N, a(N)] \qquad (4.13)$$

The damage accumulation process is governed with equation (4.1) or its specific case, equation (4.3) with initial conditions stated on $\omega(N, x)$ along all the crack trajectories. Usually one may assume that the initial state of the material is nondamaged. Then $\omega(0, x) = 0$ at all the domain of the body.

To evaluate the damage ahead of the crack tip, the stress distribution at $| x | \geq a, y = 0$ must be known. Let us schematize the crack as a flattened elliptical hole. The distribution around an elliptical hole in an unbounded plane has been studied by many authors. Necessary references and numerical data can be found in manuals on the elasticity theory. The form of the final formula depends on the method of solution. Using the Kolosov-Muskhelishvili approach, we obtain that the stress $\sigma_y(x, y)$ at $| x | \geq a, y = 0$ follows the equation

$$\frac{\sigma(x)}{\sigma_\infty} = \frac{\xi^2 + \epsilon}{\xi^2 - \epsilon} + \frac{(1 - \epsilon)^2[\xi^4 + 3\xi^2 + \epsilon(\xi^2 - 1)]}{2(\xi^2 - \epsilon)^3} \qquad (4.14)$$

Here the notation $\sigma(x) \equiv \sigma_y(x, 0)$ is used as well as the notation

$$\xi = \frac{(x/a) + [(x/a)^2 + (\rho/a) - 1]^{1/2}}{1 + (\rho/a)^{1/2}}, \qquad \epsilon = \frac{a - b}{a + b} \qquad (4.15)$$

Putting $x = a$ in (4.15), we obtain $\xi = 1$. Then equation (4.14) gives the well-known Neuber's formula for the stress concentration factor near an elliptical hole under remote transverse tension:

$$\kappa = 1 + 2\left(\frac{a}{\rho}\right)^{1/2} \qquad (4.16)$$

Formulas (4.14)-(4.16) are illustrated in Figure 4.2 where the normalized stress σ/σ_∞ is plotted against $(x - a)/a$ for various ratios a/ρ beginning from the circular hole until $a/\rho = 50$.

To close the set of equations, we need an equation with respect to the effective radius $\rho(N)$. This radius is in some sense fictitious; it takes into account the complex fractographic picture near the crack tip. The tip sharpens during the crack growth, and begins to blunt during the growth retardation. In addition, blunting can be the result of microdamage accumulation near the tip. The equation with respect to $\rho(N)$ is to remain valid if the tip is fixed. Assuming that the derivative

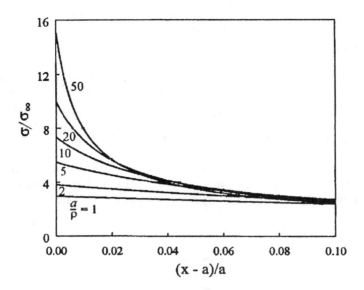

FIGURE 4.2
Stress distribution ahead of the crack tip for various tip curvatures.

$d\rho/dN$ is a linear function of the fatigue growth rate da/dN and of the damage accumulation rate $d\psi/dN$, we obtain the simplest equation

$$\frac{d\rho}{dN} = \frac{\rho_s - \rho}{\lambda_\rho}\frac{da}{dN} + (\rho_b - \rho)\frac{d\psi}{dN}. \qquad (4.17)$$

In this equation ρ_s is the radius of the "sharp" crack, ρ_b is that of the "blunt" one, and λ_ρ is a characteristic scale. In further analysis ρ_s, ρ_b and λ_ρ are supposed to be material constants (at given temperature and other environmental conditions as well as at the fixed stress ratio). Linearity of the right-hand side of (4.17) with respect to derivatives da/dN and $d\psi/dN$ is not necessary. It ought to be noted that the derivative $d\psi/dN$ in (4.17) is taken for the moving crack tip. When more information is available on the conditions near the moving crack tip, equation (4.17) might be transformed and generalized to fit experimental data more adequately.

Consider in detail the process of numerical simulation of fatigue crack growth. The initial state of the system is certainly sub-equilibrium, and at $N = 0$ the inequality $G < \Gamma$ holds. To find the cycle number N_* corresponding to the crack growth initiation, we have to integrate equation (4.3) until the equality $G = \Gamma$ is primarily attained. Using equations (4.6) and (4.11), the cycle number N_* can be found as a root of the equation

$$\psi(N) = \omega_* \left[1 - \frac{G(N)}{\Gamma_0}\right]^{1/\alpha} \qquad (4.18)$$

at $a = a_0 = $ const and $\Gamma_0 = $ const. Due to stress distribution ahead of the crack, the crack begins to propagate at $N > N_*$ remaining in a stable equilibrium state. If there is no decrease of loading level and/or no increase of material resistance along the crack path, the crack propagates continuously. The crack growth follows the pattern schematically shown in Figures 3.12 and 3.13. At a certain cycle number $N > N_{**}$ equation (4.18) has no real roots. The number N_{**} corresponds to the final failure.

The computational program is rather complicated. It includes several iteration procedures; but numerical results are quite stable and do not consume too much processor time. The flow-chart of computations is depicted in Figure 4.3. Assuming the crack size a and the effective curvature radius ρ are known, we compute by equation (4.14) the tensile stress $\sigma(x)$ distributed at $\mid x \mid \geq a, y = 0$. Then, integrating equation (4.3), we evaluate, in the first approximation, $\omega(x)$ and ψ. The next step is to correct the magnitude of ρ and the stress distribution $\sigma(x)$. It is the first iteration loop. Then we evaluate the generalized forces G and Γ. In the equilibrium state they must satisfy equation (4.18). Therefore, we compute ψ_* and

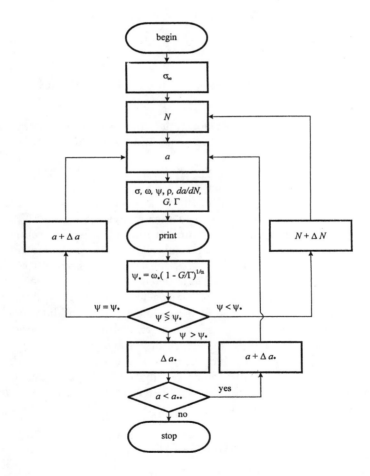

FIGURE 4.3
Flow-chart of computational procedure for fatigue cracks in Griffith's problem.

compare ψ with ψ_*, where ψ_* is defined by the right-hand side of (4.18). At $\psi < \psi_*$ we advance N in the step ΔN and repeat all the sequences of computations. If $\psi = \psi_*$ (with the error stated beforehand), we advance the crack tip in the step Δa. The evaluated magnitude of da/dN is then used to obtain a better approximation for the magnitudes of $\omega(x)$ and ρ. After several iterations, the final magnitudes of all variables of interest are recorded. If $\psi > \psi_*$, it signifies either the final failure or a finite jump-wise advancement of the crack tip. In the latter case the jump size Δa_* is estimated from equation (3.43).

The detailed computational scheme is more complicated than shown in Figure 4.3. In addition to the "external" iteration with respect to ΔN, several "internal" iterations are needed to evaluate $\omega(x, N)$, $\psi(N)$, and $\rho(N)$. After each cycle of "internal" iterations, one must perform the iteration with respect to ΔN. To avoid too fine a mesh, integration schemes are preferable. For example, to evaluate $\omega(x, N)$ in the presence of a moving crack tip, the solution of equation (4.1) is to be presented in the form

$$\omega(x, N + \Delta N) = \omega(x, N) + \int_{N}^{N+\Delta N} f[\sigma_{max}(\xi, \nu), \sigma_{min}(\xi, \nu)]d\nu \qquad (4.19)$$

Here $\xi = x - a(N)$ is the distance from the moving tip to a fixed material point.

The evaluation of $\psi(N) \equiv \omega[a(N), N]$ is particularly sensitive to small perturbations because of rapidly changing stresses in the vicinity of crack tips. When Δa and ΔN are sufficiently small, we may treat da/dN as a constant at $N \in [N, N + \Delta N]$. Then

$$\xi = \Delta a - \left(\frac{da}{dN}\right)(\nu - N)$$

and we can treat the stresses within the segment $[a, a + \Delta a]$ as functions of ξ. Equation (4.19) gives

$$\psi(N + \Delta N) = \omega(a + \Delta a, N) + \left(\frac{da}{dN}\right)^{-1} \int_{0}^{\Delta a} f[\sigma_{max}(\xi), \sigma_{min}(\xi)]d\xi \qquad (4.20)$$

Under certain circumstances the integrand in (4.20), can be calculated analytically. In any case, the numerical integration may be performed for a finer mesh than the mesh with the generally assumed step Δa. The integrals in (4.19) and (4.20), obviously, depend on $a(\nu)$, $\rho(\nu)$, and $\omega(\xi, \nu)$. Thus, to evaluate $\rho(N + \Delta N)$, $\omega(x, N + \Delta N)$ and $\psi(N + \Delta N)$ we also need an iteration procedure.

Some features of crack growth can be observed by direct analytical calculations. For example, on the stage of initial damage accumulation, when $a = a_0 = const$, equation (4.17) describes tip blunting at $\rho < \rho_b$, and tip sharpening at $\rho > \rho_b$:

$$\frac{d\rho}{dN} = (\rho_b - \rho)\frac{d\psi}{dN} \qquad (4.21)$$

On the stage of continuous crack growth, both terms in the right-hand side of (4.17) must be taken into account. Let us compare the contribution of these terms in the kinetics of ρ. Differentiating $\psi(N)$ with respect to N and taking into account equations (4.9) and (4.18), we obtain

$$\frac{d\psi}{dN} = -\frac{\omega_*}{\alpha a_{cr}} \frac{da}{dN} \left(1 - \frac{a}{a_{cr}}\right)^{(1-\alpha)/\alpha} \tag{4.22}$$

where a_{cr} is the critical crack size under monotonous loading. In fatigue $a < a_{cr}$; hence equation (4.22) shows that the tip damage measure is a decreasing function of N at any $da/dN > 0$.

Substitution of (4.22) in (4.17) gives

$$\frac{d\rho}{dN} = \left[\frac{\rho_s - \rho}{\lambda_\rho} - \frac{\rho_b - \rho}{a_{cr}} \frac{\omega_*}{\alpha} \left(1 - \frac{a}{a_{cr}}\right)^{(1-\alpha)/\alpha}\right] \frac{da}{dN} \tag{4.23}$$

We see that $d\rho/dN$ is proportional to the crack growth rate da/dN, though it depends also on the current magnitude of ρ. Usually we have $\omega_{**} \sim \alpha \sim 1$. Then at $a \ll a_{cr}$ the first term in the brackets is of the order $(\rho_s - \rho)/\lambda_\rho$, the second one of the order $(\rho_b - \rho)/a_{cr}$. Hence, during the most of the crack growth stage, the first term is larger than the second one. Then equation (4.23) takes the approximate form

$$\frac{d\rho}{dN} \approx \left(\frac{\rho_s - \rho}{\lambda_\rho}\right) \frac{da}{dN}$$

i.e., it describes tip sharpening at any $\rho > \rho_s$, $da/dN > 0$.

On the final stage of the fatigue life, the correction contributed by the second term in the brackets of (4.23) is of the order λ_ρ/a_{cr} compared with unity and even less. Thus, the final rupture occurs (in the framework of the discussed model) with tip radius close to ρ_s.

The objective of further numerical study is to analyze the effect of loading and material parameters on the process of crack growth and damage accumulation. The basic numerical data (if nothing contrary is noted) are the following: $E = 200$ GPa, $\nu = 0.3$, $\gamma_0 = 20$ kJ/m^2, $\alpha = \omega_* = 1$, $\sigma_d = 10$ GPa, $\Delta\sigma_{th} = 200$ MPa, $m = n = 4$, $\rho_s = 10$ μm, $\rho_b = \lambda_\rho = 100$ μm. Note that the assumed data correspond to the fracture toughness

$$K_{IC} = \left(\frac{\gamma_0 E}{1 - \nu^2}\right)^{1/2} \tag{4.24}$$

that is about 66.3 MPa $\cdot$ m$^{1/2}$. The steady periodic loading is considered with the applied stress range $\Delta\sigma_\infty = 200$ MPa and $R = 0$.

4.3   Influence of Initial Conditions

A large variety of parameters enter equations (4.2)-(4.18). Some of them do not significantly affect the cycle number at the start of crack growth, fatigue crack growth rate, and total fatigue life. The influence of other parameters is more or less essential though predicable, at least qualitatively. Some of the factors need a more detailed analysis, especially when they have influence on crack growth characteristics which enter, at least implicitly, empirical equations. As an example, the power exponent m_p in the Paris-Erdogan equation

$$\frac{da}{dN} = c \left(\Delta K \right)^{m_p} \tag{4.25}$$

ought to be mentioned. Another characteristic is the constant c in the right-hand side of (4.25). Its meaning is rather vague and the variation interval covers several orders of magnitudes. A more convenient form of (4.25) is

$$\frac{da}{dN} = \lambda \left(\frac{\Delta K}{K_d} \right)^{m_p} \tag{4.26}$$

where K_d is of the dimension of the stress intensity factor. It may be identified with the standard fracture toughness characteristic K_{IC} when λ, being of the dimension of length, is properly chosen.

The numerical results using the presented model were firstly published in the paper [30]. The following numerical studies are of a more detailed character. We begin this study from the group of parameters that may be considered as those of initial conditions [30]. Among them are the initial crack size, the initial effective radius at the crack tip and the initial level of microdamage along the crack trajectory, in particular, near the tip.

The influence of the initial crack size a_0 is illustrated in Figures 4.4 and 4.5. The listed set in Section 4.3 of the numerical data is used in computations, and only the initial size a_0 is subjected to variation: $a_0 = 0.1, 1, 10$ mm (lines 1, 2, 3, respectively). The initial tip radius is taken as $\rho_0 = 50$ μm. The time history $a(N)$ is presented in Figure 4.4a. The crack behavior is predicable: the longer the initial crack, the shorter the durations of the initiation stage and of the total fatigue life.

Most important for engineering analysis is the crack growth rate diagram. The computed diagrams are presented in Figure 4.4b where the tip crack growth rates da/dN for three various initial crack sizes a_0 are plotted against the range ΔK of the stress intensity factor. The curves expose a sigmoidal character well-known to experimenters. However, the initial sections of the curves do not refer to the material resistance threshold. Actually, the threshold ΔK_{th}, expressed in terms of stress intensity factors, has the order of magnitude of

$$\Delta K_{th} = (\Delta \sigma_{th}/2)(\pi \lambda_{th})^{1/2} \tag{4.27}$$

The right-hand size of (4.27) contains a characteristic length parameter λ_{th}. Its choice is rather arbitrary. It may be equal to the "sharp" radius ρ_s, the "blunt" radius ρ_b, or the length λ of the process zone. Substituting in (4.27) $\Delta \sigma_{th} = 200$ MPa, $\lambda_{th} = 100$ μm, we obtain $\Delta K_{th} \approx 2$ MPa·m$^{1/2}$ while the stress intensity factor range ΔK at $a_0 = 1$ mm is $\Delta K = \Delta \sigma_\infty (\pi a_0)^{1/2} \approx 11$ MPa·m$^{1/2}$. The origins of the curves in Figure 4.4b correspond approximately to initial magnitudes of ΔK. All the curves go lower in their initial section than might be expected by interpolation of the central section slope into the domain of smaller ΔK. The initial behavior of crack growth rates may be explained as a result of the beginning crack

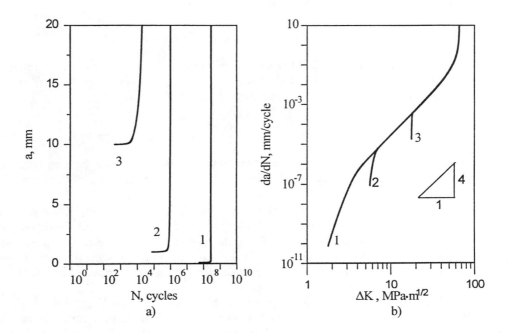

FIGURE 4.4
Crack size versus cycle number and crack growth rate diagrams for initial crack sizes $a_0 = 0.1$, 1, and 10 mm (lines 1, 2, 3 respectively).

developing under a comparatively low stress concentration. Later on, the stress concentration has a tendency to rapidly increase.

With increasing of ΔK all the curves tend to merge. It is quite natural because the memory of the system with respect to loading and crack growth history is limited, especially when damage in remote domains of the body is low. The slope of the middle section of the diagram (which is sometimes referred to as the Paris stage) is close to $m = 4$. Therefore, the exponents m_p in (4.25) and (4.26) may be attributed to the exponent in equations of damage accumulation. The final rupture occurs near the critical magnitude of $K_{max} = K_C \approx 66.3$ MPa $\cdot$ m$^{1/2}$. Note that in this example $R = 0$ and, therefore, $\Delta K \equiv K_{max}$.

More "intimate" features of fatigue crack growth are illustrated in Figure 4.5 where the effective tip radius ρ and the tip damage measure ψ are plotted versus the cycle number for $a_0 = 0.1$, 1, and 10 mm. When the crack tip is fixed, the cracks are blunting from the initial tip radius $\rho_0 = 50$ μm though the "blunt" tip radius $\rho_b = 100$ μm is not attained. With the crack growth, sharpening of the tip begins. The final crack propagation occurs at the "sharp" radius close to $\rho_s = 10$ mm (Figure 4.5a).

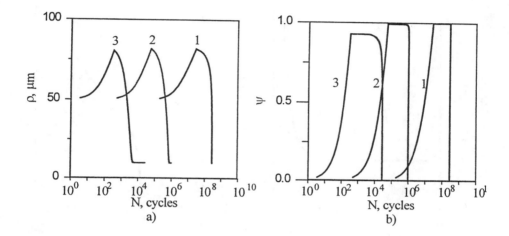

FIGURE 4.5
Effective tip radius and tip damage measure versus cycle number for various
initial crack sizes (notation as in Figure 4.4).

The time history $\psi(N)$ of the tip damage is shown in Figure 4.5b. The ascending branches of the curves correspond to the stage when damage is accumulating at the fixed tip. Crack propagation begins at the damage close to the critical level $\psi_* = 1$. The tip damage decreases rapidly when we are approaching the final rupture.

It is also of interest to study the influence of the initial tip radius ρ_0 on further crack propagation. Some numerical results are presented in Figures 4.6 and 4.7. These results are obtained for $a_0 = 1$ mm. The initial radius is assumed $\rho_0 = 50, 100, 200$ μm. The crack growth rates (Figure 4.6b) are rather close for all the considered initial conditions, and the slope in the middle stage is close to $m = 4$. However, the distribution of the total life between the initiation stage and crack growth stage differs significantly. This is easy to see considering the evolution of the effective tip radius and damage measure during the fatigue life (Figure 4.7). Note that, in contrast to Figures 4.4 and 4.5 where cycle numbers are plotted in a logarithmic scale, a uniform scale for N is used in Figures 4.6 and 4.7. It is why the initiation stage is difficult to observe in Figure 4.7. Looking at Figures 4.4b, and 4.6b, we observe that all the three stages of fatigue crack growth are described

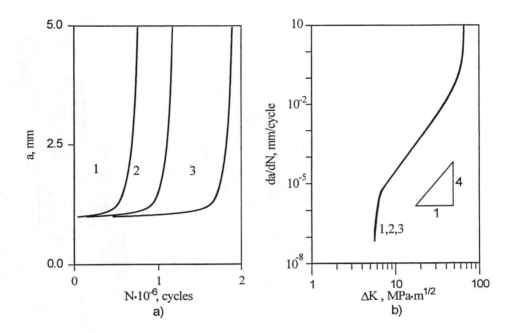

FIGURE 4.6
Crack size versus the cycle number and crack growth diagrams for initial tip
radii $\rho_0 = 50$, **100, and 200** μ**m (lines 1, 2, 3 respectively).**

with the discussed model: the initial stage crack growth rates, the middle stage corresponding to the Paris approximation, and the final stage near the quasibrittle fracture. All these features can be observed in the following analysis, too.

4.4   Influence of Loading Conditions

Experimental results of fatigue crack growth as well as the numerical data for the engineering design are usually presented in the form of crack growth rate diagrams plotting the rate da/dN against the range ΔK of the stress intensity factor. However, ΔK is not the only loading parameter affecting fatigue crack growth. Apart from temperature, loading frequency and cycle shape, the maximum applied stress σ_∞^{max} is to be taken into account along with the range $\Delta\sigma_\infty$. In particular, the final sections of growth rate diagrams are controlled (through stability conditions) by $K_{max} = \sigma_\infty^{max}(\pi a)^{1/2}$. The applied stress ratio $R = \sigma_\infty^{min}/\sigma_\infty^{max}$ or, more precisely,

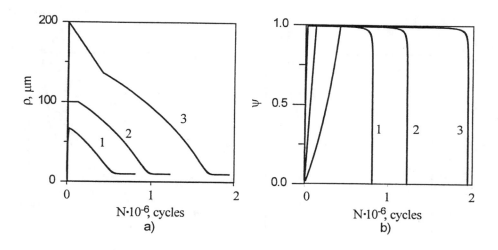

FIGURE 4.7
Effective tip radius and tip damage measure versus cycle number for various initial tip radii (notation as in Figure 4.6).

the acting stress ratio affects the crack growth rate mainly through their influence on material parameters, in particular, on the threshold resistance stress $\Delta\sigma_{th}$. Last but not least, the closure depends on the relationship between $K_{min} = \sigma_\infty^{min}(\pi a)^{1/2}$ and a certain material parameter referred to as crack closure (or crack opening).

The aim of this section is to compare crack growth rate diagrams plotted with the use of different control parameters. We vary the ratio $R = \sigma_\infty^{min}/\sigma_\infty^{max} = K_{min}/K_{max}$ neglecting, however, the threshold and closure effects which will be discussed later, in Sections 4.6 and 4.7. Equation (4.5) is used here for the threshold stress with the function $f(R) = 1 - 0.5R$.

Usually, numerical results are presented in the form of the relationship between da/dN and ΔK treating $R = \sigma_\infty^{min}/\sigma_\infty^{max} = K_{min}/K_{max}$ as a parameter. In Figure 4.8a such diagrams are presented for the basic data listed in Section 4.3. The applied stress range is taken as $\Delta\sigma_\infty = 200$ MPa and the stress ratio $R = 0$, 0.25, 0.5 and 0.75. This corresponds to $\sigma_\infty^{max} = \Delta\sigma_\infty(1 - R)^{-1}$ equal to 200, 300, 400, and 600 MPa. The initial crack length is the same, $a_0 = 1$ mm, and all curves begin from the same point. The influence of the threshold is practically absent. The reason is the high stress concentration at the tip. Therefore, the acting stresses are much higher than the assumed threshold stress. The lines corresponding to different R begin to split until the final failure that occurs near $K_{max} = \Delta K(1 - R)^{-1}$.

A similar diagram is plotted in Figure 4.8b , where the maximal stress intensity factor K_{max} is considered as an independent variable and R as a parameter. Here an opposite picture is observed: the lines are split at an earlier stage and tend to merge during the further process of crack propagation.

The third type of plotting is to present da/dN as a function of the range

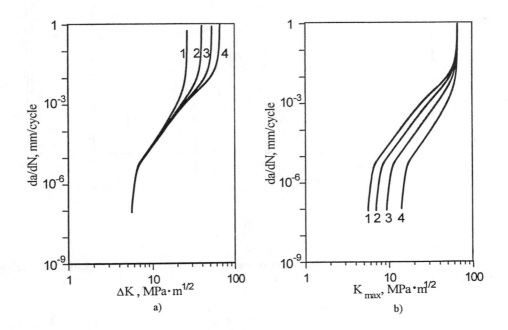

FIGURE 4.8
Influence of the applied stress ratio $R = \sigma_\infty^{min}/\sigma_\infty^{max}$ on the crack growth rate for $R = 0$, 0.25, 0.5 and 0.75 (lines 1, 2, 3, and 4, respectively).

$\Delta G = G_{max} - G_{min}$ of the driving force G. Using (4.9) we obtain

$$\Delta G = \frac{1 - \nu^2}{E} \Delta K (K_{max} + K_{min})$$

In some sense, ΔG takes into account both the range ΔK and the maximal stress intensity factor. The divergence of curves is significant. Therefore, all hopes to use ΔG as the most appropriate loading parameter seem to be lost. As a result, we must conclude that the standard approach presenting crack growth rate in the function of stress intensity range is the most adequate among the three ones presented in Figures 4.8-4.9.

All the above computations were performed for the fixed magnitude $\Delta\sigma_\infty$ and various but fixed R. In practice, loading processes usually are of much complex character including nonsteady and random components. The subject of fatigue life analysis under nonsteady loading is out of the framework of the present study. However, it is of interest to observe how nonstationarities of loading conditions affect fatigue crack growth.

Two nonsteady loading regimes are presented in Figure 4.10. In the first case

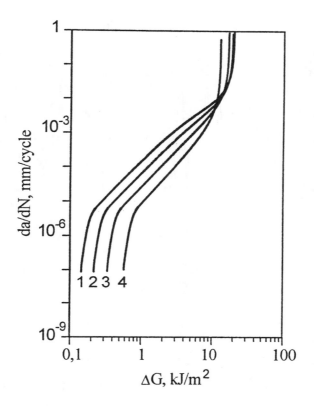

FIGURE 4.9
Fatigue crack growth rate in the function of the range of generalized force (notation as in Figure 4.8).

$R = 0$ under varying $\Delta\sigma_\infty$. The regime depicted in Figure 4.10a consists of the load decreasing stage AB from $\Delta\sigma_\infty = 300$ MPa to $\Delta\sigma_\infty = 100$ MPa, and backward stage BC until the final failure. The first stage covers 10^6 cycles, and the second one is much longer to cover most of the crack growth rate diagram. The loading regime shown in Figure 4.10b consists of the stress growing stage AB. After attaining the level $\Delta\sigma_\infty = 300$ MPa, the stress range is kept constant until the final failure.

Computational results are presented in Figure 4.11. In Figure 4.11a we observe a temporary growth of da/dN close to the initiation stage (point A). Then the point corresponding to the current da/dN returns to the "regular" curve and moves backward with the decreasing stress level. At point B the return movement begins, and this movement practically everywhere follows the "regular" crack growth rate curve.

A similar diagram in Figure 4.11b exhibits a significant deviation from the "regular" curve on the initial stage when the stress level is increasing. Then we observe merging of both curves.

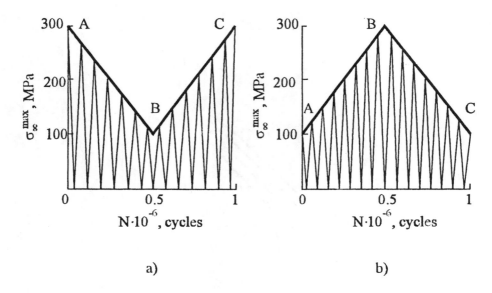

a)   b)

FIGURE 4.10
Schematical presentation of two nonsteady loading regimes.

4.5   Influence of Damage Accumulation Process

Since most of the fatigue life runs under almost constant microdamage measure at the tip, it is of interest to study the influence of the parameters entering the damage accumulation equations. In particular, concerning equation (4.3), we have to analyze the effects caused by the change of σ_d, $\Delta\sigma_{th}$, m and n. The influence of σ_d and $\Delta\sigma_{th}$ is rather evident. It can be followed, for example, with the use of the quasistationary approximation (see Section 4.8).

The effect caused by the change of the power exponent m is of special interest. A close correlation has already been mentioned between m and the exponents entering the Paris-Erdogan, Forman, etc. equations. Therefore, the connection between m in equation (4.3) and the exponent m_p in equations (4.25) and (4.26) is to be studied.

Another item of interest is the influence of the exponent n. It is a question of applied interest because the equation of damage accumulation becomes much simpler if its right-hand side does not depend on ω. In terms of equation (4.3) it is the case $n = 0$. On the other side, the case $n = m$ corresponds to the classical Rabotnov-Kachanov model of continuum damage mechanics.

The influence of exponent m is illustrated in Figures 4.12 and 4.13. All these results are obtained for the case $m = n$. Figure 4.12a shows the crack growth at $a_0 = 1$ mm, $\rho_0 = 50$ μm, and various m. The magnitude of m affects very strongly the cycle number N_* at the growth start as well as on the critical cycle number N_{**} at the final rupture (Figure 4.12a). The total fatigue life N_{**} and the duration of the crack growth stage $N_{**} - N_*$ depend on m significantly. Roughly

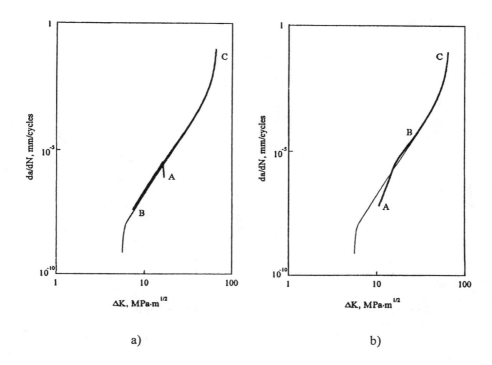

a)   b)

FIGURE 4.11
Crack growth rate diagram for regimes sketched in Figure 4.10,a and b.

speaking, the increase of m twice results into the increase of N_* and N_{**} in two orders of magnitude. However, this effect strongly depends on the applied stress level. In the considered numerical example, we have $\Delta\sigma_\infty/\sigma_d \sim 10^{-2}$. With the stress concentration factor of the order of 10 this results in $\Delta\sigma/\sigma_d \sim 10^{-1}$. This is in agreement with numerical results, especially if we refer to those obtained in the quasistationary approximation (Section 4.8). The crack growth rate diagrams are presented in Figure 4.12b. The rate da/dN, as could be expected, depends on m very strongly. All the features discussed in connection with the previous numerical studies are present here also. The slope of the middle part of the curves may be identified with the exponent m_p in equations (4.25) and (4.26).

Additional information is presented in Figure 4.13. The crack tip radius ρ is plotted in Figure 4.13a versus the cycle number N. The initial sections of the curves, as in the earlier examples, correspond to the initiation stage when the crack tips are fixed. Three sections are seen in these curves, although their comparative duration significantly varies with m. When m is small, e.g., $m = 2$, the middle part of curves $\rho(N)$ is short compared with the case of larger m. On the central sections of the curves, the radius ρ is rather large, although it is smaller than the "blunt" radius $\rho_b = 100$ μm. At smaller m, a significant section of the fatigue life

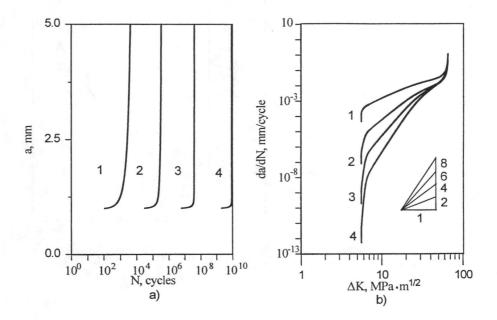

FIGURE 4.12
Crack size histories and crack growth rate diagrams for power exponents $m = 2, 4, 6, 8$ (lines 1, 2, 3, 4, respectively).

takes place with ρ that is close to the "sharp" radius $\rho_s = 10\mu$m. At larger m such a stage of crack growth is practically absent. The damage measure $\psi(N)$ is plotted in Figure 4.13b. Crack propagation occurs at the damage level very close to the critical one, $\psi = 1$. However, the feature of final sections of curves varies with increase of m. The descending sections are rather long for smaller m. But the curves fall almost abruptly at $m = 8$. A similar tendency can be observed in the curves depicting time histories of the effective tip radius ρ.

The influence of n is illustrated in Figures 4.14 and 4.15. Figure 4.14a shows the crack growth at $a_0 = 1$ mm, $m = 4$, $\rho_0 = 50$ μm for three various magnitudes of n. Namely, lines $1, 2, 3$ are plotted for $n = 0$, 2, 4, respectively. As it is easy to expect, the increase of n shortens the fatigue life, however, within the same order of magnitude. Correspondingly, the crack growth rate diagrams (Figure 4.14b) expose a comparatively low scatter, especially in the $\log - \log$ scale. The slope of the curves also differs slightly being close to $m = 4$ within the middle sections of the diagrams. The diagrams for ψ and ρ (Figure 4.15) change only quantitatively.

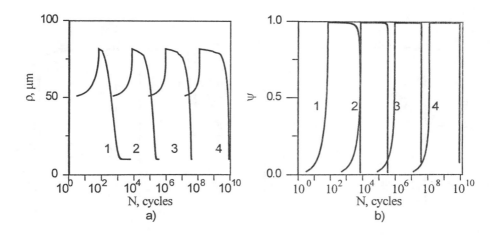

FIGURE 4.13
Effective tip radius and tip damage measure versus cycle number for various power exponents (notation as in Figure 4.12).

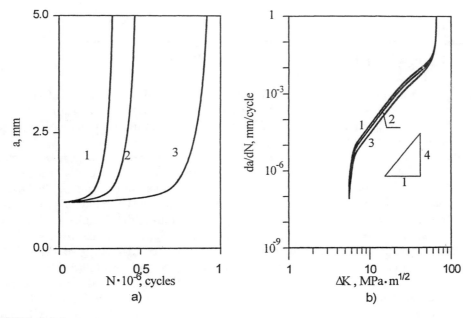

FIGURE 4.14
Influence of power exponents n on crack growth at $m = 4$, $n = 0, 2, 4$ (lines 1, 2, 3, respectively).

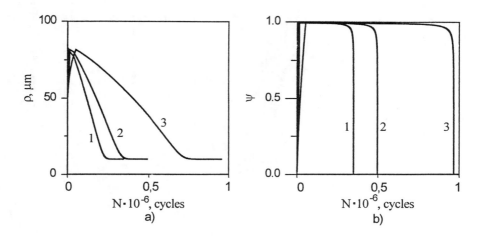

FIGURE 4.15
Influence of the power exponent n on the effective tip radius and tip damage measure for $m = 4$, $n = 0, 2, 4$ (notation as in Figure 4.14).

4.6 Threshold and Related Effects

The problem of fatigue thresholds is usually formulated in terms of common fatigue curves as well as in terms of fatigue crack growth rate diagrams. The initial branches of diagrams corresponding to crack growth initiation and early growth are of primary interest. When crack dimensions are small, being, e.g., of the order of grain size, crack growth must be considered in the framework of micromechanics. Then statistical scatter of grain properties and randomness of distribution of microcracks, micropores and microinclusions are to be taken into account. This approach has been discussed in Sections 2.6 and 2.7. Hereafter the behavior of macroscopic cracks is considered. The discussed models are purely deterministic. However, the initial crack size, or the applied stress level (or both), are assumed small. As a result, crack growth rates are very small, say, of the order 10^{-12} m/cycle and even less. Under certain circumstances, such cracks may be interpreted as nonpropagating ones. In other cases, the initial sections of crack growth rate diagrams must be studied. Various threshold effects are expected here that are influenced by stress ratio and material properties, in particular, by the material's sensitivity to damage accumulation under cyclic loading. Some features of threshold effects have been already discussed in the above sections. In particular, the threshold resistance stress enters equations (4.2)-(4.3). However, the problem of threshold effects is rather vague, and experimental data are fragmentary and sometimes controversial.

We must distinguish several types of threshold effects. First, it is a natural resistance of any material against microdamage accumulation. To begin this process, the tip stress range $\Delta\sigma$ has to overcome a certain threshold $\Delta\sigma_{th}$. This kind of threshold effect is an intrinsic material property.

Another effect is referred to as the interaction between the macroscopic crack and the array of microscopic cracks situated in the process zone. The effective stress intensity factor assessed in the presence of such an array occurs less than that for nondamaged material [54]. In such a connection, it is frequently said on the shielding effect, though this term is used also in another context [63]. To understand the nature of shielding due to microdamage, it is sufficient to note that microdamage produces an increase of material compliance. As a result, we have the diminishment of stresses in the process zone, the retardation of the process of damage accumulation and, finally, the retardation of crack growth. Another source of shielding is the dilatance deformation that is also produced by localized damage. To describe this set of phenomena, rather sophisticated models are necessary. Some of them will be discussed later, in Sections 4.10-4.11 and 6.10.

The third threshold effect is related to the crack closure studied by many experimenters. The closure effect, however, is of macromechanical nature and will be discussed separately. By the threshold effect of the third type, we mean the increase of material resistance to damage due to closure of microcracks and microvoids ahead of the crack tip.

There is a distinct border between the listed threshold effects. The intrinsic threshold effect can be introduced with the threshold resistance stress range $\Delta\sigma_{th}$ as has been done in equations (4.2), (4.3), etc. To fit experimental data, a suitable function $f(R)$ is to be chosen in equation (4.5). In fact, this effect has been taken into account in all the numerical examples presented above. But to study this effect in detail (as well as other threshold phenomena) we must consider comparatively short cracks and comparatively low load levels.

A few additional numerical examples are given in Figures 4.16 and 4.17. They are obtained for the following data: $E = 200$ GPa, $\nu = 0.3$, $\gamma_0 = 20$ kJ/m^2, $\sigma_d = 10$ GPa, $m = 4$, $n = 0$, $\alpha = 1$, $\rho_s = 10$ μm, $\rho_b = \lambda_\rho = 100$ μm. These data are similar to those used, in particular, in Figures 4.4-4.7. Compared with the above computations, we assume the threshold stress $\Delta\sigma_{th} = \Delta\sigma_{th}^0(1 - R)$, i.e., $f(R) = 1 - R$ with a higher resistance level, $\Delta\sigma_{th}^0 = 400$ MPa.

Figure 4.16 is drawn for an initially short crack ($a_0 = 0.5$ mm) at $\rho_0 = 50$ μm and three load levels, $\Delta\sigma_\infty = 75$, 100, and 125 MPa (lines 1, 2, and 3). The stress ratio is $R = 0.2$ in all the cases. The rate diagrams are similar to those given above; however, at $\Delta\sigma_\infty = 75$ MPa the crack propagation threshold counted in terms of stress intensity factor range is equal to several units of MPa$\cdot$m$^{1/2}$. At a lower load, say, at $\Delta\sigma_\infty = 60$ MPa no crack propagation takes place. As to the fatigue life, it varies in the vicinity of threshold in two orders of magnitude. At $\Delta\sigma_\infty = 75$ MPa we have $N_* = 1.9 \cdot 10^7$ and $N_{**} = 2.6 \cdot 10^8$, while at $\Delta\sigma_\infty = 125$ MPa we obtain $N_* = 2.0 \cdot 10^5$ and $N_{**} = 2.1 \cdot 10^6$, respectively.

Another way to study the vicinity of a threshold is to consider short cracks. Figure 4.17 is drawn for $\Delta\sigma_\infty = 150$ MPa, $R = 0.2$. The initial crack size is $a_0 = 0.05$, 0.1 and 0.5 mm (lines 1, 2, and 3). The cycle number until crack initiation varies from $N_* = 2.5 \cdot 10^8$ till $N_* = 6.1 \cdot 10^4$, and the total life from

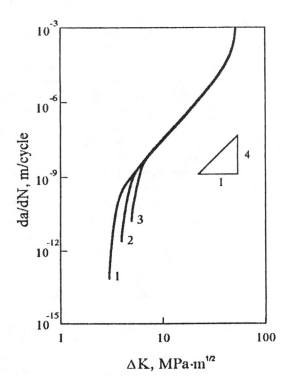

FIGURE 4.16
Crack growth rate diagrams at $a_0 = 0.5$ mm and $\Delta\sigma_\infty = 75, 100,$ and 125 MPa (lines 1, 2, and 3, respectively).

$N_{**} = 2.0 \cdot 10^9$ until $N_{**} = 6.48 \cdot 10^5$ when we go from $a_0 = 0.05$ mm to $a_0 = 0.5$ mm.

Note that the theoretical threshold stress intensity factor, following equation (4.27), is

$$\Delta K_{th} = (\Delta\sigma_{th}/2)(\pi\rho)^{1/2}$$

where ρ takes the magnitudes between ρ_s and ρ_b. For the assumed numerical data we have $\Delta K_{th} \approx 0.89$ MPa $\cdot$ m$^{1/2}$ in the first case and $\Delta K_{th} \approx 2.85$ MPa $\cdot$ m$^{1/2}$ in the second case. Looking on Figure 4.17, we observe that the "tail" of line 1 is near the threshold referred to as the "blunt" tip radius ρ_b. It is understandable since the crack growth begins at the effective tip radius close to ρ_b. If the saturation level of damage is already attained at $G < \Gamma$, no crack growth takes place.

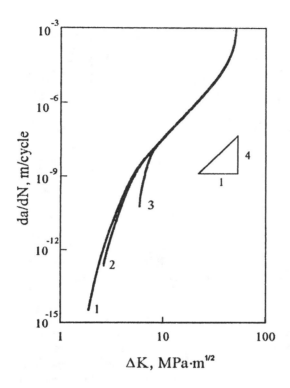

FIGURE 4.17
Crack growth rate diagrams at $\Delta\sigma_\infty = 150$ **MPa,** $a_0 = 0.05, 0.1,$ **and** 0.5
mm (lines 1, 2, and 3, respectively).

There is some evidence [49, 111] that the compressive stage of cycle also contributes to microdamage, and, therefore, affects the crack propagation. For example, evaluating the threshold magnitude ΔK_{th} by a standard test procedure, we sometimes observe the return to crack growth replacing the initial loading regime with the symmetrical cyclic loading.

Two typical loading regimes used for the fatigue threshold assessment are schematically shown in Figure 4.18. The test may be performed at constant minimal stress σ_∞^{min} (Figure 4.18a) until the rate da/dN drops to a very low magnitude, say, $da/dN = 10^{-12}$ m/cycle. Then the crack is conditionally interpreted as arrested. The corresponding cycle number is denoted N_a. However, the crack growth can be recovered if, keeping the magnitude of σ_∞^{max} attained at the "arrest", we change the stress ratio from $R \geq 0$ to $R = -1$. In case $R = 0$ this signifies doubling

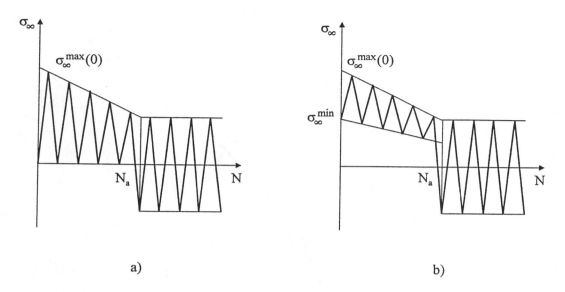

a)   b)

FIGURE 4.18
Schemes of testing to estimate the threshold of fatigue crack growth: (a) σ_∞^{min} = const; (b) at R = const; crack arrest is referred to as the rate $da/dN = 10^{-12}$ m/cycle.

of $\Delta\sigma_\infty$ which may (and may not) initiate a new stage of crack growth. Another regime is shown in Figure 4.18b. There the stress ratio is kept constant; meantime the maximal applied stress σ_∞^{max} is decreasing. If after the "arrest" at $N = N_a$ we change the stress ratio from $R \geq 0$ to $R = -1$, the crack may begin to grow again. Evidently, such situations can occur only in cases when the threshold stresses are not too sensitive to R, and the compression part of the cycles is not negligible. For example, if the function $f(R)$ takes doubled and higher magnitudes at the switch from $R = 0$ to $R = -1$, the double increase of the stress range does not result in the recovery of crack growth.

Figures 4.19 and 4.20 present the results for the case when the first stage of loading occurs at $R = 0$. The following numerical data are used: $E = 200$ GPa, $\nu = 0.3$, $\gamma_0 = 20$ kJ/m^2, $\sigma_d = 10$ GPa, $\Delta\sigma_{th}^0 = 200$ MPa, $m = 4$, $\rho_s = 10\mu$m, $\rho_b = \lambda_\rho = 100$ μm. The function of the stress ratio R in (4.5) is taken as $f(R) = 1 - 0.5R$ which provides the effect of the crack growth recovery. The parameters of the loading regime are $\sigma_\infty^{max}(N) = \sigma_\infty^{max}(0)[1 - (N/N_0)]$ where $\sigma_\infty^{max}(0) = 300$ MPa, $N_0 = 1.5 \cdot 10^4$, $\sigma_\infty^{min} = 0$. The initial conditions are $a_0 = 1$ mm, $\rho_0 = 50$ μm. The crack growth history is shown in Figure 4.19a, and the growth rate history in Figure 4.19b. The crack begins to grow at $N_* = 3.2 \cdot 10^4$ cycles and its growth rate drops until 10^{-12} m/cycle at $N_a = 1.38 \cdot 10^4$ cycles. The threshold stress might be estimated as $\sigma_\infty^{max}(N_a) = 24.0$ MPa. After the conditional arrest, we observe a recovery of crack growth with a rather low rate which, when we are approaching the final failure, rapidly increases.

The crack growth rate diagram is presented in Figure 4.20a. Two curves are

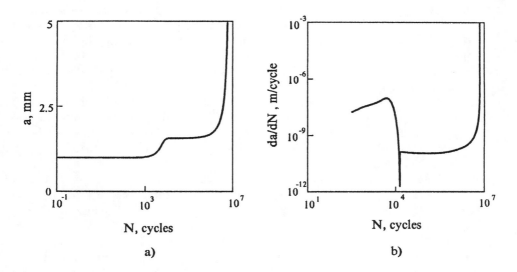

FIGURE 4.19
Crack size (a) and its rate (b) for the loading regime schematically presented in Figure 4.18a.

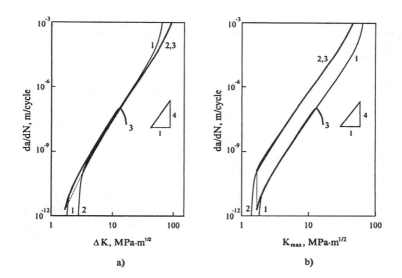

FIGURE 4.20
Crack growth rate diagrams for the case presented in Figure 4.18a (solid lines); lines 1 and 2 are drawn for $R = 0$ and $R = -1$, respectively.

drawn there corresponding to the regular regimes at $\Delta\sigma_\infty^{min} = 0$, i.e., $R = 0$, and $R = -1$ (lines 1 and 2, respectively). The solid curve 3 is drawn for the nonsteady regime. The curve begins close to the numeral 3. Its descending branch corresponds to the first stage of loading. The return point of this curve may be interpreted as the growth arrest (the growth rate is dropped here lower than 10^{-12} m/cycle). Then we come to the ascending stage with the rate staying very close to the regular curve for $R = -1$.

To make the picture clearer, a diagram of crack growth is represented in Figure 4.20b with use, as a control parameter, of the maximal stress intensity factor K_{max}. In these coordinates, the basic curves 1 and 2 become distinguishable upon all stages of crack propagation, and all stages of evolution of da/dN during the loading history are easy to observe.

The same numerical data are used for Figures 4.21 and 4.22 that are drawn for the case when on the first stage $R = 0.5$. The maximal stresses are diminishing as $\sigma_\infty^{max}(N) = \sigma_\infty^{max}(0)[1 - (N/N_0)]$ where $\sigma_\infty^{max}(0) = 300$ MPa, $N_0 = 3 \cdot 10^5$. The crack growth initiates at $N_* = 5.7 \cdot 10^3$ cycles. The maximal applied stress at the crack arrest is $\sigma_\infty^{max}(N_a) \approx 38.6$ MPa. The functions $a(N)$ and $da(N)/dN$ are depicted in Figure 4.21. They look similar to those in Figure 4.19. The crack growth rate diagrams are plotted in Figure 4.22 where ΔK and K_{max} are used as control parameters. Curve 1 is drawn for the "regular" regime with the fixed $R = 0.5$, curve 2 for $R = -1$. The first stage of nonsteady regime depicted by the solid curve 3 includes a short branch of initial growth of da/dN. Then the growth rate begins to drop until the apparent arrest. After coming to the symmetrical cycle loading, we observe an increase of da/dN, and we practically join curve 2.

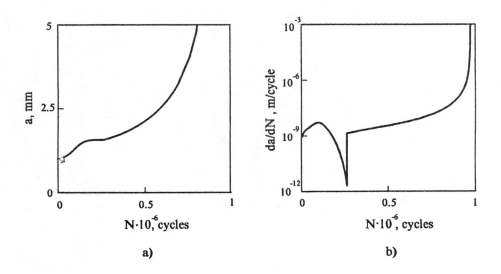

FIGURE 4.21
Crack size (a) and its rate (b) for the loading regime schematically presented in Figure 4.18b.

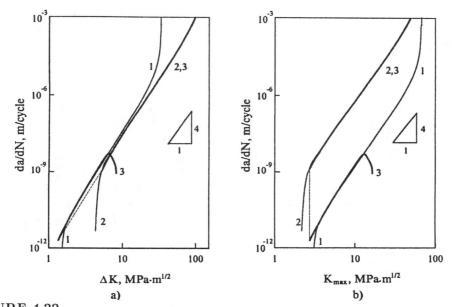

FIGURE 4.22
Crack growth rate diagram for the case presented in Figure 4.18b (solid lines);
lines 1 and 2 are drawn for σ_∞^{min} = const and $R = -1$, respectively.

The above numerical study demonstrates that no universal, standard fatigue crack growth rate diagram exists for a given material under a given temperature, etc. Though the representative points on diagrams are wandering in the vicinity of "regular" curves, the deviation from the latter happens be considerable.

Note that these phenomena are not referred to as crack arrest due to overloading which is a result of plastic straining, tip blunting, etc. These phenomena will be discussed in Chapter 6, in the context of interaction between high-cycle and low-cycle fatigue.

The simplest way to introduce the shielding effect in the model of fatigue crack growth is a phenomenological approach. Due to damage, the relaxation of acting stresses occurs along the crack path. The most intensive relaxation takes place near the tips where the damage level is close to saturation. As a result, we observe displacing of stress maxima ahead of the tips with the diminishing of maximal stresses compared with the tip stresses evaluated in the framework of the classical model of elasticity theory.

Let us assume the effective stress range at $\mid x \mid \geq a$ given as

$$\Delta\sigma_{eff} = \Delta\sigma g(\omega) \qquad (4.28)$$

where the range $\Delta\sigma$ corresponds to the initial nondamaged state of the material; $g(\omega)$ is a function of the local damage measure. At $\omega = 0$ we have $\Delta\sigma_{eff} = \Delta\sigma$. When the material is completely damaged, $\Delta\sigma_{eff}$ tends to zero. For example, we may assume that

$$g(\omega) = (1 - c\omega)^q \qquad (4.29)$$

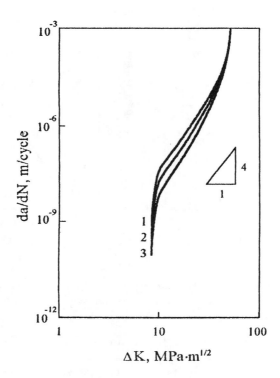

FIGURE 4.23
Crack growth rate diagrams: in absence of shielding effect (1); in cases of moderate (2); and strong (3) shielding effect.

with two material parameters, $c > 0$ and $q > 0$.

Some numerical results are shown in Figures 4.23 and 4.24. They were obtained for $\Delta\sigma_\infty = 150$ MPa, $R = 0.2$; other data are similar to the basic ones used in Section 4.2. The crack growth rate diagram in Figure 4.23 is obtained with the use of equations (4.4) and (4.29) for $c = 0.5$ and three values of the exponent in (4.29): $q = 0$ (no shielding), $q = 0.5$ (moderate shielding), and $q = 1$ (intensive shielding). Lines 1, 2, and 3 are drawn for the above three levels of shielding. It is seen in Figure 4.23 that the influence of shielding is significant. This influence covers practically all of the field of the diagram. As a result, the total fatigue life varies in a wide interval. When no shielding is taken into account, the cycle number at crack growth initiation is $N_* = 6.9 \cdot 10^3$, and the total fatigue life $N_{**} = 9.1 \cdot 10^4$. At $q = 0.5$ we obtain $N_* = 1.8 \cdot 10^4$ and $N_{**} = 2.1 \cdot 10^5$. At $q = 1$ we obtain $N_* = 6.1 \cdot 10^4$ and $N_{**} = 6.4 \cdot 10^5$, respectively, i.e., the critical cycle numbers increase in an order of magnitude.

The source of such a strong shielding effect becomes understandable when we look on the effective stress distribution ahead of the crack at various q. The effective

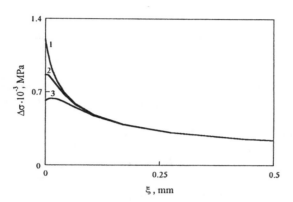

FIGURE 4.24
**Stress distribution ahead of the crack tip at crack growth initiation;
notation as in Figure 4.23.**

stress range $\Delta\sigma_{eff}$ computed according to (4.28) and (4.29) is drawn in Figure 4.24
as a function of the distance $\xi = x - a_0$. Lines 1, 2, and 3 correspond for $q = 0$,
0.5 and 1. They are drawn for the moment of crack growth start, i.e., at $N = N_*$.
When no shielding effect is present (line 1), we observe the usual stress distribution.
At $q = 0.5$ (line 2) we observe an essential stress relaxation, and the maximum
stress range is situated a little ahead of the tip. At $q = 1$ practically no stress
concentration is observed.

It ought to be emphasized that Figure 4.24 is drawn for the initial state in the
process of crack propagation. When a crack begins to grow, the damage level in
the process zone diminishes, and then the contribution of shielding becomes not so
impressive. It seems that the numerical value $c = 0.5$ in equation (4.29) is taken
too high. In any case, the above numerical results demonstrate that the interaction
between the processes of micro- and macrolevel might be included in the model in
a rather simple way.

As to the microscale closure effect, it might be inserted into the model described, for example, by equation (4.5). Actually, the influence of microcracking on
the effective stress $\Delta\sigma_{eff}$ depends on the stress ratio R. Then instead of (4.28) we
obtain

$$\Delta\sigma_{eff} = \Delta\sigma f(R)g(\omega) \tag{4.30}$$

Here $f(R)$ is a function of R with $f'(R) \geq 0$.

Note that contrary to the intrinsic threshold effect which is included in the
threshold stress $\Delta\sigma_{th}$, the microscale closure effect enters the equation of damage
accumulation with acting stresses. In practice, it is difficult to separate the listed

threshold effects. If shielding is negligible, both effective stresses and threshold stresses depend only on R. Not to go into additional complications, we, as a rule, include all the effects into the resistance threshold stress.

4.7   Crack Closure Effect

Along with the threshold effects, the closure effect is important in case of low minimal stresses within a cycle, especially when they are compressive and when sufficiently long cracks are concerned. Sometimes one considers these effects together. However, there is a significant difference between the two effects. When we say threshold effects, we mean phenomena naturally belonging to the properties of the material in its current state. Actually, the resistance to damage accumulation is an intrinsic material property. To the contrary, the crack closure is mainly of an external nature. The influence of crack closure on crack growth enters the fatigue problems with global deformations of a specimen or structural component. Thus, this effect depends on applied loads, body and crack geometry [44]. In this aspect, crack closure in their essential parts is a problem of elasticity theory (and of other domains of continuum mechanics). From the viewpoint of fracture mechanics, we may treat the problem of crack closure as an auxiliary one.

Compared to equation (4.30) for the microscale closure effect, the macroscale closure effect must be described in terms of acting stresses σ_∞. To keep connection with the common presentation of experimental results, let us assume the following presentation for the effective stress range:

$$\Delta\sigma_\infty^{eff} = \begin{cases} \sigma_\infty^{max} - \sigma_{cl}, & \sigma_\infty^{min} < \sigma_{cl} \\[2mm] \sigma_\infty^{max} - \sigma_\infty^{min}, & \sigma_\infty^{min} \geq \sigma_{cl} \end{cases} \tag{4.31}$$

Here σ_{cl} is the closure stress which may be positive, negative, or equal to zero.

The most important variables affecting σ_{cl} are the crack length a and the stress ratio R. A simplest dependence may be assumed as follows:

$$\sigma_{cl} = \sigma_{cl}^0 (1 - b_1 R)^{p_1} + \sigma_{cl}^\infty [1 - \exp(-a/a_\infty)] \tag{4.32}$$

Here σ_{cl}^0 and σ_{cl}^∞ are two characteristic closure stresses, a_∞ is a characteristic crack length, b_1 and p_1 are non-negative material related parameters. When $a \ll a_\infty$ equation (4.32) gives

$$\sigma_{cl} \approx \sigma_{cl}^0 (1 - b_1 R)^{p_1} \tag{4.33}$$

which means that σ_{cl}^0 is equal to the closure stress for very short cracks at $R = 0$. Equation (4.33) is similar to equation (4.5) for the threshold resistance stress $\Delta\sigma_{th}$; however the latter is measured in acting stresses while σ_{cl} is referred to as applied stresses. When $a \gg a_\infty$, i.e., a crack is long enough, we obtain

$$\sigma_{cl} = \sigma_{cl}^0 (1 - b_1 R)^{p_1} + \sigma_{cl}^\infty \tag{4.34}$$

The first term in the right-hand side of (4.34) appears to be small compared with the second term. It means that σ_{cl}^{∞} is near to the closure stress for long cracks.

It is possible to include the influence of effective tip curvature on the threshold stress. For example, at $\sigma_{cl}^{\infty} > 0$ equation (4.33) may be taken in the form

$$\sigma_{cl} = \sigma_{cl}^0 (1 - b_1 R)^{p_1} \exp(-\rho/\rho_{\infty})$$

with the characteristic tip radius ρ_{∞}. Then σ_{cl}^0 is to be interpreted as the closure stress for short and sharp cracks. At $\rho \gg \rho_{\infty}$ no closure occurs for short cracks at all.

Obviously, this approach is equivalent to the standard empirical description of the closure effect when the effective stress intensity factor range $\Delta K_{eff} = K_{max} - K_{cl}$ is used. Here K_{cl} is supposed to be a material related parameter. Following (4.32), one may consider this parameter depending on the crack length:

$$K_{cl} = K_{cl}^0 (1 - b_1 R)^{p_1} + K_{cl}^{\infty} [1 - \exp(-a/a_{\infty})] \tag{4.35}$$

with characteristic stress intensity factors K_{cl}^0 and K_{∞}^{cl}.

To illustrate the applicability of the crack closure concept in the framework of the theory developed in this book, let us present a few numerical examples. As above, we follow the principal governing relationship $G \lesseqgtr \Gamma$ together with formulas (4.9) and (4.11) for the generalized forces G and Γ, and equation (4.17) for the effective crack radius ρ. The only difference is that the equation of damage accumulation is taken as (4.4) with $\Delta\sigma_{eff}$ corresponding to the effective applied stress range (4.31).

The same numerical data are used in Figures 4.25-4.27. Equation (4.31) is taken for the crack closure stress at $\sigma_{cl}^0 = -100$ MPa, $\sigma_{cl}^{\infty} = 200$ MPa, $a_{\infty} = 10$ mm, $b_1 = 1$, $p_1 = 1$. The threshold effect is also included in the analysis with $\Delta\sigma_{th} = \Delta\sigma_{th}^0 (1 - 0.5R)$ at $\Delta\sigma_{th}^0 = 200$ MPa. The initial conditions are $a_0 = 0.1$ mm, $\rho_0 = 50$ μm at zero damage along all the crack trajectory.

Figure 4.25 is obtained for $R = 0$ and various applied stress levels: $\Delta\sigma_{\infty} = 100, 150, 200, 250,$ and 300 MPa (lines 1, 2, 3, 4, and 5, respectively). When the stress level is low, the closure may arrest crack growth when the crack length becomes sufficiently long. It is illustrated by line 1. This phenomenon can be observed both looking on the crack growth history (Figure 4.25a) and the growth rate diagram (Figure 4.25b). The influence of crack closure at higher stresses is not so drastic. It is invisible when we consider crack growth history but is distinctly seen on the growth rate diagram. In the vicinity of $a = a_{\infty}$ a temporary retardation of crack growth is observed as well as other uncommon patterns. These features are mostly controlled by the second term of the right-hand side of (4.32), being typical only for long cracks.

The threshold effect is also taken into account in Figure 4.25. It is natural that this effect is essential on the earlier stage of crack growth. The total fatigue life depends on both effects, and the threshold effect is mostly responsible for its variation. Actually, the duration N_* of the initiation stage varies from $0.28 \cdot 10^7$ until $0.24 \cdot 10^5$ when the applied stress range changes from 150 MPa to 500 MPa. Thus, diminishing of this stage is measured in two orders of magnitude. The total life N_{**} varies from $0.75 \cdot 10^7$ till $0.13 \cdot 10^6$, i.e., in one order of magnitude. When the initial crack length is large enough, the rates da/dN corresponding to equal

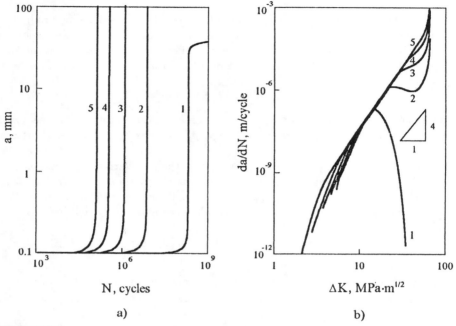

a)   b)

FIGURE 4.25
Influence of crack closure on crack size history (a) and growth rate diagram (b)
at $\Delta\sigma_\infty = 100, 150, 200, 250,$ and 300 MPa (lines 1, 2, 3, 4, and 5, respectively).

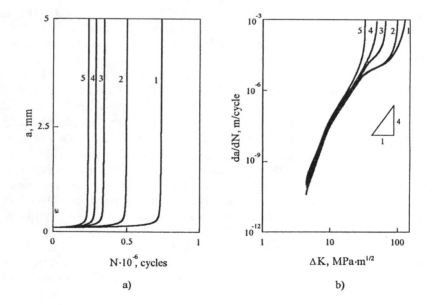

a)   b)

FIGURE 4.26
The same as in Figure 4.25 at $\Delta\sigma_\infty = 250$ MPa and $R = -1, -0.5, 0, 0.25,$ and
0.5 (lines 1, 2, 3, 4, and 5, respectively).

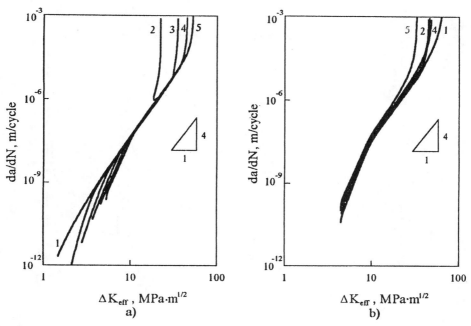

FIGURE 4.27
Influence of crack closure in terms of effective stress intensity factor for the same data as in Figure 4.25a and in Figure 4.26b.

ΔK are close (Figure 4.25b). Note that the slope of the middle sections of these diagrams remains close to $m = 4$.

The influence of stress ratio is illustrated in Figure 4.26 which is plotted for $\Delta \sigma_\infty = 200$ MPa and $R = -1$, -0.5, 0, 0.25, and 0.5 (lines 1, 2, 3, 4, and 5, respectively). The significant separation of lines in Figure 4.26,b near the final failure is a result of the growing influence of the maximal stress $\sigma_\infty^{max} = \Delta \sigma_\infty (1 - R)^{-1}$ as well as of the closure effect. The threshold effect in this case plays a comparatively modest role, and the variation of fatigue life is small compared with Figure 4.25. In particular, at $R = -1$ we have $N_* = 0.13 \cdot 10^6$, $N_{**} = 0.74 \cdot 10^6$ while at $R = 0.5$ we have $N_* = 0.41 \cdot 10^5$, $N_{**} = 0.24 \cdot 10^6$. Note that compared with Figure 4.25a where cycles are plotted in log-scale, the uniform scale is used in Figure 4.26a.

In fatigue tests, to diminish the divergence of results due to the closure effect, the stress intensity factor range ΔK is usually replaced by its effective value

$$\Delta K_{eff} = \begin{cases} K_{max} - K_{cl}, & K_{min} < K_{cl} \\ K_{max} - K_{min}, & K_{min} \geq K_{cl} \end{cases} \tag{4.36}$$

Here K_{cl} is the stress intensity factor (4.35) corresponding to the closure stress (4.32):

$$K_{cl} = \sigma_{cl} (\pi a)^{1/2} \tag{4.37}$$

Plotting the crack growth range da/dN versus the effective stress intensity fac-

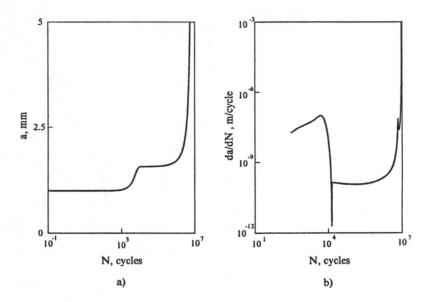

FIGURE 4.28
Influence of crack closure on crack size (a) and growth rate (b) histories for the loading regime schematically presented in Figure 4.18a.

tor range ΔK_{eff}, we may expect a more compact situation of diagrams, especially in their middle sections. Figure 4.27a is plotted for various $\Delta \sigma_\infty$, and Figure 4.27b for various R. However, the divergence remains significant both on the earlier and later stages of fatigue growth. The divergence in the middle part of the diagrams is moderate.

The macroscopic closure effect may become drastic when essentially unsteady loading is concerned. Hereafter this phenomenon is illustrated in Figures 4.28–4.34 obtained by the numerical simulation of the experimental procedures used in the study of threshold phenomena. These regimes have been already discussed in Section 4.7 in the context of the application of the theory to unsteady loading processes. However, only intrinsic threshold resistance to damage accumulation has been taken into account there. Other effects, in particular the macroscale crack closure effect, may be of significance for long cracks.

Let the loading process be followed by the pattern schematically illustrated in Figure 4.18a. The process consists of two stages. The first stage is loading at the constant minimum applied stress $\sigma_\infty^{min} = 0$ MPa and the maximum applied stress σ_∞^{max} varying as

$$\sigma_\infty^{max} = \sigma_\infty^0 (1 - N/N_0) \qquad (4.38)$$

Here $\sigma_\infty^0 = 300$ MPa, and $N_0 = 1.5 \cdot 10^4$. The crack arrest is referred to as the rate $da/dN = 10^{-12}$ m/cycle. The corresponding cycle number is N_a. After that the loading is switched to a symmetrical one ($R = -1$) with the range equal to $2\sigma_\infty^a$

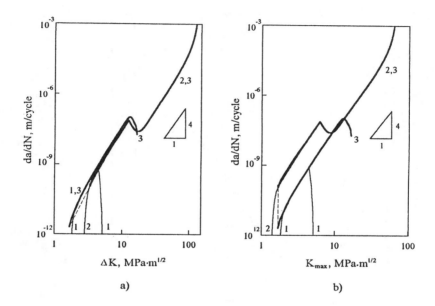

a)   b)

FIGURE 4.29
Crack growth diagrams corresponding to Figure 4.28 (solid lines); lines 1 and 2 are drawn for $R = 0$ and $R = -1$, respectively.

where σ_∞^a is taken according to (4.38) at $N = N_a$.

The same numerical data are used in the presented computations as before. The only specification is that a larger initial crack size is taken, $a_0 = 1$ mm. Crack growth history (Figure 4.28a) contains the initiation stage till $N_* = 0.32 \cdot 10^3$, the growth stage until $N_a = 1.38 \cdot 10^4$, the temporary crack arrest stage, and the slow recovery of crack growth until the final failure at $N_{**} = 0.94 \cdot 10^7$. All these stages are observable in Figure 4.28b. There is a rather unexpected peculiarity close to the end, namely, a temporary drop of the crack growth rate. It is a result of complex interaction between several mechanisms affecting damage accumulation in the presence of the crack closure effect.

The crack growth rate is plotted in Figures 4.29a and 4.29b against the range ΔK and maximum K_{max} of the stress intensity factor. The solid line is drawn for the unsteady loading regime. The travel of the state point begins at point 3. Two light lines, 1 and 2, correspond to the steady regimes, at $R = 0$ and $R = -1$, respectively (line 1 has a returning branch). As, for example, in Figure 4.20, the point of the current relationship between da/dN and ΔK (or K_{max}) begins to move along line 3 in the vicinity of line 1. Then the point goes down until the

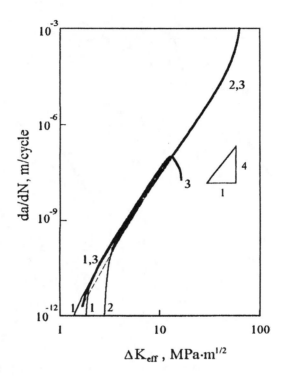

FIGURE 4.30
Crack growth rate diagrams in terms of effective stress intensity factor for the case presented in Figures 4.28 and 4.29.

state assumed as the crack arrest. After turning to another regime, the point approaches line 2 and follows this line until the final failure. Note that the segment of nonmonotonous varying of da/dN in paths 2 and 3 corresponds to the "thorn" in Figure 4.28b.

Plotting da/dN against the effective stress intensity range ΔK_{eff} given in (4.36), we obtain a more compact, agreeable diagram, without any peculiarities that remain hidden in the closure stress intensity factor (4.37). This diagram is presented in Figure 4.30.

Similar results were obtained for the case when loading follows the scheme shown in Figure 4.18b. Figures 4.31 and 4.32 are drawn for the case when equation (4.38) is valid for the first stage of loading at R kept constant. The following numerical data are assumed for the first stage: $\sigma_\infty^0 = 300$ MPa, $N_0 = 3.0 \cdot 10^5$,

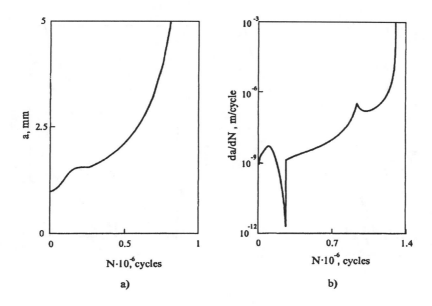

FIGURE 4.31
Influence of crack closure on crack size (a) and growth rate (b) histories for the
loading regime schematically presented in Figure 4,18b.

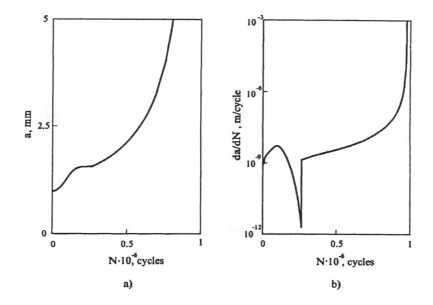

FIGURE 4.32
Evolution of the crack size and growth rate in the absence of macroscale closure
effect for the same data as on Figure 4.31.

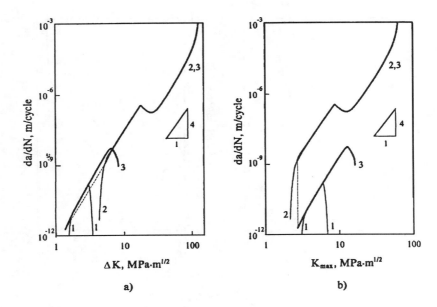

FIGURE 4.33
Influence of crack closure on growth rate diagrams for the loading regime schematically presented in Figure 4.18b.

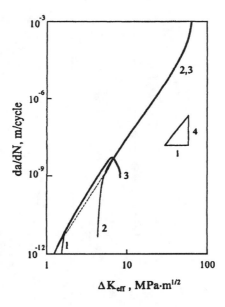

FIGURE 4.34
Crack growth rate diagram in terms of effective stress intensity factor for the same data as in Figure 4.33.

$R = 0.5$. The crack size and growth rate histories shown in Figure 4.31 are similar to that in Figure 4.28. Due to the use of uniform scale for N (instead of log scale in Figure 4.28), the "thorn" corresponding to the switch on the crack closure mechanism is more distinctive here.

For comparison, the crack growth in absence of crack closure is shown in Figure 4.32. The crack history is very similar. The initiation stage takes the same part of the fatigue life, $N_* = 0.57 \cdot 10^4$, and the total life decreases from $N_{**} = 1.30 \cdot 10^6$ when crack closure is taken into account till $N_{**} = 0.97 \cdot 10^6$ when this effect is neglected. Crack growth history changes more significantly; in particular, the "thorn" vanishes (Figure 4.32b). Correspondingly, the peculiarities of the crack growth rate diagram vanish if the macroscale closure effect is not included.

Figures 4.33 and 4.34 present crack growth rate diagrams in function of various load parameters, ΔK, K_{max}, and ΔK_{eff} in the presence of both effects. The same notation is used here as in Figures 4.29 and 4.30.

Concluding this section, we note that among a number of sources which are usually attributed to threshold effects, the intrinsic threshold effect is perhaps the most important when comparatively short cracks are considered. In the absence of reliable experimental data it is expedient to model other threshold effects by introduction of the appropriate threshold resistance stress range in equations of microdamage accumulation. Crack closure is important for longer cracks, especially when compression components are present in the loading process. In principle, all these phenomena may be covered by the presented theory.

4.8 Differential Equations of Fatigue Crack Growth

The approximate approach in Section 3.7 called the quasistationary approximation is applied in this section to fatigue crack growth in Griffith's problem. The objective is, starting from general constitutive equations, to obtain simple equations as close as possible to equations of fatigue crack growth used in the treatment of experimental data and engineering design. A number of simplifying assumptions have been used to come to such equations. It is the reason why these equations cannot be considered as an approximate solution of the complete problem. They are rather of demonstrative character. However, they might be considered as rather consistent equations for fitting experimental data. One of the features of these equations is that they cover all the stages of fatigue crack propagation and exhibit the explicit connection between micro- and macromechanical parameters.

The discussed problem is a single-parametrical one. Its quasistationary version completely enters the patterns discussed in Section 3.8 in a more general context. Assume for simplification that the process zone size λ_p and the effective curvature radius ρ are material constants and the tensile stress $\sigma(x) \equiv \sigma_y(x, 0)$ is uniformly distributed in the process zone $a \leq |x| \leq a + \lambda_p, y = 0$. We use Neuber's equation for the tip stress σ_t:

$$\sigma_t = \sigma_\infty \left[1 + 2 \left(\frac{a}{\rho} \right)^{1/2} \right] \tag{4.39}$$

Outside this zone we assume $\sigma = \sigma_\infty$, i.e., we refer all stress concentrations to the process zone. Let the damage measure in this zone be governed by equation (4.2). Then

$$\frac{d\psi}{dN} = \left(\frac{\Delta\sigma_t - \Delta\sigma_{th}}{\sigma_d} \right)^m \tag{4.40}$$

and the tip damage is

$$\psi(N) \approx \omega_{ff}(N) + \lambda_p \left(\frac{da}{dN} \right)^{-1} \left(\frac{\Delta\sigma_t - \Delta\sigma_{th}}{\sigma_d} \right)^m \tag{4.41}$$

Here $\omega_{ff}(N)$ is the far field damage measure; the estimate for the cycle number required for the crack tip crossing the process zone is:

$$\Delta N \approx \lambda_p \left(\frac{da}{dN} \right)^{-1} \tag{4.42}$$

Substituting (4.42) in the equilibrium condition of the system with respect to G-coordinate a, we obtain the differential equation

$$\frac{da}{dN} = \lambda_p \left(\frac{\Delta\sigma_t - \Delta\sigma_{th}}{\sigma_d} \right)^m \left[\left(1 - \frac{G_{max}}{\Gamma_0} \right)^{1/\alpha} - \omega_{ff}(N) \right]^{-1} \tag{4.43}$$

Here equation (4.11) is used for the generalized resistance force, and G_{max} is the maximal generalized driving force within a cycle. Similarly, using equation (4.12) for the resistance force we obtain

$$\frac{da}{dN} = \lambda_p \left(\frac{\Delta\sigma_t - \Delta\sigma_{th}}{\sigma_d} \right)^m \left[1 - \left(\frac{G_{max}}{\Gamma_0} \right)^{1/\alpha} - \omega_{ff}(N) \right]^{-1} \tag{4.44}$$

It is evident that the difference between (4.43) and (4.44) becomes significant only when we approach the final stage of crack growth, i.e., at $G_{max} \sim \Gamma_0$ or $\omega_{ff}(N) \sim 1$.

It is of interest to present the right-hand sides of equations (4.43) and (4.44) completely in terms of stress intensity factors. For this purpose we use formula (4.39). When $a \gg \rho$ we may put $\sigma_t \approx 2\sigma_\infty (a/\rho)^{1/2}$. This allows introduction of the range ΔK of the stress intensity factor in equations (4.41), (4.43), and (4.44). Actually, the approximate equality holds:

$$\Delta\sigma_t \approx 2\Delta K (\pi\rho)^{-1/2} \tag{4.45}$$

Therefore equation (4.43) takes the form

$$\frac{da}{dN} = \lambda_p \left(\frac{\Delta K - \Delta K_{th}}{K_d} \right)^m \left[\left(1 - \frac{K_{max}^2}{K_C^2} \right)^{1/\alpha} - \omega_{ff}(N) \right]^{-1} \tag{4.46}$$

Here the notation is introduced for material parameters which characterize the resistance to microdamage accumulation in terms of stress intensity factors:

$$K_d = (\sigma_d/2)(\pi\rho)^{1/2}, \qquad \Delta K_{th} = (\Delta\sigma_{th}/2)(\pi\rho)^{1/2} \qquad (4.47)$$

The relationship between (4.45) and (4.47) is evident. In addition, the generalized forces G and Γ_0 are introduced in (4.46) in terms of stress intensity factors, too. Similarly, instead of (4.44) we obtain

$$\frac{da}{dN} = \lambda_p \left(\frac{\Delta K - \Delta K_{th}}{K_d}\right)^m \left[1 - \left(\frac{K_{max}^2}{K_C^2}\right)^{1/\alpha} - \omega_{ff}(N)\right]^{-1} \qquad (4.48)$$

It is here proper to remember that equations (4.46) and (4.48) are valid at $\Delta K > \Delta K_{th}$ (in the opposite case we must put $da/dN \equiv 0$). Equation (4.46) becomes invalid at $K_{max} \geq K_C(1 - \omega_{ff}^{\alpha/2})$, and equation (4.48) at $K_{max} \geq K_C(1 - \omega_{ff})^{\alpha/2}$. The far field damage term ω_{ff} takes into account, at least partially, the history-dependent component of fatigue.

Turning to a more general equation of microdamage accumulation, equation (4.3) at $n > 0$, we have to follow the scheme presented with equations (3.46)-(3.48). As

$$h(\omega) = \frac{(1 - \omega)^{n+1}}{n + 1} + \text{const}$$

we obtain that equation (3.47) takes the form

$$(1 - \omega_{ff})^{n+1} = (1 - \psi)^{n+1} + \lambda_p(n + 1)\left(\frac{da}{dN}\right)^{-1}\left(\frac{\Delta\sigma_t - \Delta\sigma_{th}}{\sigma_d}\right)^m \qquad (4.49)$$

Assume equation (4.12) for the resistance force at $\omega_* = 1$. Then equation (4.48) is to be generalized as follows:

$$\frac{da}{dN} = \lambda_p(n + 1)\left(\frac{\Delta K - \Delta K_{th}}{K_d}\right)^m\left[(1 - \omega_{ff})^{n+1} - \left(\frac{K_{max}^2}{K_c^2}\right)^{(n+1)/\alpha}\right]^{-1} \qquad (4.50)$$

The characteristic length increases, according to (4.50) to $\lambda_p(n + 1)$; meantime the characteristic stress intensity factors follow, as earlier, equations (4.47).

We followed the scheme sketched above in Sections 3.8 and 3.9. However, we might change the calculation a little arriving, however, to the same equations. Actually, considering a crack advancement during a short interval ΔN, we will neglect the partial time derivative in the formula for the material derivative of the damage measure. Then

$$\frac{d\omega}{dN} = \frac{\partial\omega}{\partial N} + \frac{\partial\omega}{\partial x}\frac{da}{dN} \approx \frac{\partial\omega}{\partial x}\frac{da}{dN}$$

Equation (4.2) is to be replaced by the following:

$$(1 - \omega)^n \frac{d\omega}{dx} = \left(\frac{da}{dN}\right)^{-1} \left(\frac{\Delta\sigma_t - \Delta\sigma_{th}}{\sigma_d}\right)^m$$

Integrating this equation in the interval $x \in [a, a + \lambda_p]$ where λ_p is the length of the process zone, we arrive at equation (4.49) and then at equation (4.50).

It ought to be stressed that the choice of the analytical form for equations (4.2), (4.3), (4.4), etc., is rather arbitrary. We might, for illustration, use the following equations for damage accumulation and its influence on the resistance to crack propagation:

$$\frac{d\psi}{dN} = \cosh\left(\frac{\Delta K - \Delta K_{th}}{K_d}\right), \qquad \Gamma = \Gamma_0(1 - \tanh\psi)$$

Then the equation of fatigue crack growth takes the form

$$\frac{da}{dN} = \lambda_p \cosh\left(\frac{\Delta K - \Delta K_{th}}{K_d}\right)\left[\tanh^{-1}\left(1 - \frac{K_{max}^2}{K_C^2}\right) - \omega_{ff}(N)\right]^{-1}$$

and so on.

4.9 Comparison with Semi-Empirical Equations

Equations (4.46), (4.48) and (4.50) are of a similar structure as most empirical and semi-empirical equations; remaining, however, theoretical ones. To obtain the Paris-Erdogan equation (4.25) from equations (4.46) and (4.48), it is necessary to put $K^2 \ll K_C^2$, $\Delta K \gg \Delta K_{th}$, $\omega_{ff}(N) \ll 1$. It means that we must not approach too closely to the quasi-brittle fracture border, as well as to neglect the influence of damage threshold and far field damage. A little more general equation

$$\frac{da}{dN} = \lambda_p \left(\frac{\Delta K - \Delta K_{th}}{K_d}\right)^m \tag{4.51}$$

will be quite agreeable for practical design. Opposite to equation (4.25), equation (4.51) does not contain parameters of indistinct physical meaning and, moreover, of a peculiar dimension such as the constant c in the right-hand side of (4.25). Another simplified version of equation (4.46) has the form

$$\frac{da}{dN} = \lambda_c \left(\frac{\Delta K - \Delta K_{th}}{K_C}\right)^m \left(1 - \frac{K_{max}^2}{K_C^2}\right)^{-1} \tag{4.52}$$

where a new length parameter is introduced, $\lambda_c = \lambda_p(K_C/K_d)^m$, and $\alpha = 1$.

The contribution of each mechanism that controls fatigue crack growth can be followed in (4.46) and (4.48). The terms in the first parentheses of its right-hand side originate from equation (4.2). They correspond to the contribution of microdamage accumulated in the tip zone. If we agree with this interpretation, we may identify the Paris-Erdogan power exponent m_p with the power exponent m in the equation of microdamage accumulation. This conclusion, although following from a very simplified version of the theory, is in agreement with numerical treatment of its unabridged version. Actually, looking on the crack growth rate diagrams predicted by the suggested theory, we see that the slope of the middle, most important section of the crack growth rate curves is close to the power exponent m. This conclusion may be done considering Figures 4.4, 4.6, et al. The terms in the second parentheses of (4.46) and (4.48) correspond to the contribution of the general balance of forces and energy in the system cracked body-loading. When $K_{max} \ll K_C$, $\omega_{ff}(N) \ll 1$, the influence of this factor may be ignored. When K_{max} is approaching K_C, the influence of this term becomes more and more significant. At $\omega_{ff}(N) \ll 1$, $\lambda_p \ll a$ instability takes place at $K \approx K_C$. In the general case the critical fracture toughness parameter in fatigue is to be determined as it is shown in Sections 3.8 and 3.9. Let, for example, $\alpha = 1$. Then we may follow equation (3.56) that results in

$$K_{CF} \approx K_C \left(\frac{1 - \omega_{ff}}{1 + \lambda_p/a} \right)^{1/2}$$

It is of interest to find the relationship between the parameters entering equations (4.46), (4.48) and (4.50) and the parameters of the equation (4.17). It has been stated above that the exponent m_p is close to the exponent m in equations (4.2)-(4.4). It is a strong argument in favor of the power law (or, in a more general case, of the threshold-power law) used to describe the process of damage accumulation. As to the constant c (Figure 4.35), it may be identified with $\lambda_p K_d^{-m}$ or $\lambda_p(n+1)K_d^{-m}$. Substituting K_d from (4.47), we obtain for the latter case:

$$c = \lambda_p(n+1) \left[\frac{2}{\sigma_d(\pi\rho)^{1/2}} \right]^m \tag{4.53}$$

It is evident that the magnitude c strongly depends on the exponent m, especially when n grows with m, for example when $n = m$. Typical numerical values for c are of the order of magnitude $10^{-10} - 10^{-14}$ when ΔK is measured in MPa $\cdot$ m$^{1/2}$. As one can see from (4.53), this parameter is very sensitive to the numerical data concerning the resistance stress σ_d and the effective tip radius ρ. However, equation (4.53) predicts rather sensible magnitudes for c when appropriate numerical data are inserted in its right-hand side. Let, for example, $m = 4$, $n = 0$, $\lambda_p = 10^{-3}$m, $\rho = 10^{-4}$m, $\sigma_d = 10$ GPa. Then $c \sim 10^{-11}$. Moreover, at $n = 0$ equation (4.53) exhibits a linear relationship between $\log c$ and m. In case, when, following the equation by Petch (1953), we assume $\sigma_d\rho^{1/2} \approx$ const, we see a reason to expect a linear regression relationship of $\log c$ on m for a wide variety of metallic alloys. Such a possibility already has been mentioned by some authors. As to the direct comparison between equation (4.53) with the results of numerical solution of the unabridged problem, the agreement may be satisfactory only when we consider the middle section of the crack growth rate diagrams (Figure 4.35).

The weak point of the simplified approach developed in this section is the limitations put on the length parameters λ_p and ρ. In fact, it is difficult to agree that they are material constants. In the framework of the thin plastic zone model both the length of the tip zone λ and crack tip opening displacement δ are proportional to the crack size a. In addition, they depend on the applied stress σ_∞.

If we follow the linear elastic model, we may estimate the length of the process

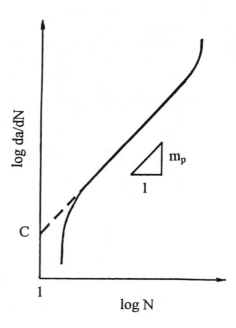

FIGURE 4.35
Estimation of parameters in the Paris-Erdogan equation.

zone from the condition of static equivalency. Actually, dividing the field ahead of the crack into the near and far fields, we substitute the real stress distribution with a schematic one such as shown in Figure 4.36a. The static equivalency of the opening tensile stress $\sigma(x)$ on the segment $(a, a + \lambda_p + \lambda_1)$ results in the equation

$$(\kappa \lambda_p + \lambda_1)\sigma_\infty = \int\limits_a^{a+\lambda_p+\lambda_1} \sigma(x)dx \qquad (4.54)$$

where κ is the tip stress concentration factor. The tip zone length λ_p, evidently, depends on the assumed length λ_1.

Some numerical results obtained with the use of (4.14) and (4.54) are shown in Figure 4.36b. The influence of λ_1 is insignificant even at moderate λ_1/a. As could be expected from elementary considerations, the tip zone length λ_p is approximately proportional to $(a\rho)^{1/2}$. It means that, assuming $\rho = \text{const}$, we have to agree that λ_p grows as $a^{1/2}$. Then the right-hand side in (4.46), (4.48) and (4.50) will explicitly depend on a. However, ρ also varies in the process of crack growth, and usually decreases with increasing a. This means a partial compensation of the error made by the assumption that ρ and λ are constant.

Consider in detail the time history of the characteristic length $(a\rho)^{1/2}$. At the beginning of crack growth, the effective tip radius is close to ρ_b. When, for example, $a_0 = 1$ mm, $\rho_b = 100$ μm, we obtain that $(a_0\rho_b)^{1/2} \approx 0.32$ mm. At the finish of crack growth, the effective tip radius approaches ρ_s. However, the crack size may

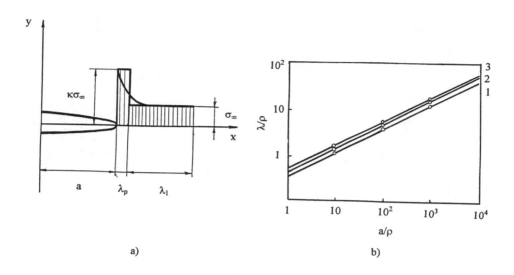

a)   b)

FIGURE 4.36
Estimation of the process zone length: (a) schematization of stress distribution;
(b) process zone length as a function of other geometrical parameters at $\lambda_1/a =$
1, 5, 10 (lines 1, 2, 3, respectively)

increase too rapidly, and this results in the growth of $(a\rho_s)^{1/2}$. Such behavior is illustrated in Figure 4.37 drawn for the same numerical data as for Figures 4.4 and 4.6. There the magnitudes of $(a\rho)^{1/2}$ are plotted versus the cycle number throughout the entire fatigue life for $a_0 = 1$ mm, $R = 0.2$, $\Delta\sigma_\infty = 100$, 150, 200, and 250 MPa (lines 1, 2, 3, and 4 respectively). Before the growth initiation, we observe increasing $(a\rho)^{1/2}$ which is caused by crack tip blunting. Then a slow drop of $(a\rho)^{1/2}$ occurs until the final stage when $(a\rho)^{1/2}$ begins to grow rapidly. Summarizing all this, we may conclude that the characteristic length is of the order $(a_0\rho_b)^{1/2}$ during the major part of crack propagation. However, this length depends on a_0, which results in the dependence of c in the Paris equation (4.25) on the initial crack size. This effect may be not observable on the background of scattering of experimental data, moreover when a growth rate diagram is plotted in a $\log - \log$ scale.

A more general question arises about the nature of parameters $K_d, \Delta K_{th}$ and $\sigma_d, \Delta\sigma_{th}$ characterizing the resistance to microdamage accumulation. If we assume that σ_d and $\Delta\sigma_{th}$ are material constants, then K_d and ΔK_{th} will vary during the crack growth, and vice versa. Since the effective tip curvature, due to blunting and sharpening, is not constant, and a special equation is to be added to describe the variability of this curvature, the situation becomes more complicated. Simplified approaches based on the quasistationary approximation remain, nevertheless, to be of interest since they show the correspondence between the general theory and popular semi-empirical equations. Another advantage is the possibility of building a link between purely theoretical variables (some of which are of a microscale nature)

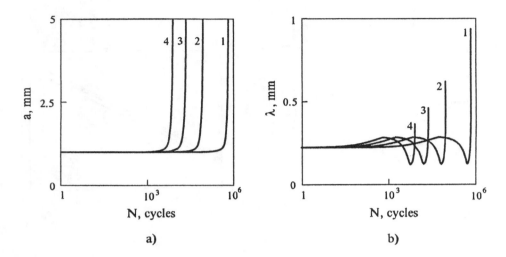

FIGURE 4.37
Time history of crack length (a) and process zone size (b) at $\Delta\sigma_\infty = 100, 150,$
200, and 250 MPa.

with parameters measured directly in macroscopic experimentation. In principle,
using up-to-date identification techniques, one could assess all the parameters of
the model from macroscopic experimental data. However, it is expected that identification equations will be badly conditioned, especially when parameters describing
damage accumulation are concerned.

4.10   Interaction between Microdamage and Material Properties

In the previous sections fatigue crack propagation has been considered in the framework of analytical and numerical-analytical approaches. Such approaches required
a number of simplifications. For example, the effect of microdamage on the compliance of materials has been neglected. At first glance, this assumption seems
acceptable. It is used implicitly in linear fracture mechanics and in some models of
nonlinear fracture mechanics, say in the Dugdale model. Nevertheless, change in
deformative properties can significantly affect the stress distribution in the vicinity
of the crack as well as, in a lesser degree, in the far field. The change of stress distribution near the crack tips results in the change of microdamage level in the near
field and, consequently, in the change of the material resistance to crack growth.
In turn, the crack growth rates vary, and that reciprocally affects the stress-strain
field near the moving tip, etc. In addition, microdamage accumulation is accom-

panied by the generation of deformations that are not explicitly connected with microscopic stresses averaged upon the macroscopic scale. Later on, these deformations are called swelling strains. Analogous deformations such as dilatancy play an important part in the micromechanics of plasticity. The influence of dilatant strains on fatigue crack growth may also be considerable; these strains may affect, at least indirectly, via the stress-strain field.

Interaction of the listed factors was discussed earlier (Section 3.1) in a qualitative way. The main direct and feedback connections, in particular, the influence of microdamage and crack growth on the generalized forces, have been included in the former analysis. Two new components will appear now in that scheme: the change of material compliance due to microdamage accumulation, and swelling. Introduction of these components makes the statement and solution of the problem essentially more difficult. Deformative properties of a body become nonhomogeneous and, generally, anisotropic. In addition, these properties vary in time. No analytical solutions are available in such complicated situations. The only way to study the interaction of all involved factors is computational experimentation. It could facilitate the understanding of the phenomena if we perform a thorough parametric study. Such experiments, of course, need the introduction of additional assumptions concerning the properties of materials subjected to microdamage.

Introduce a model of a material that behaves under single monotonous loading in an "almost" linear elastic way. The word "almost" means that the strain tensor ϵ and the stress tensor σ are connected as

$$d\epsilon = d(\kappa \cdot \cdot \sigma) + d\eta \tag{4.55}$$

Here κ is the compliance tensor of fourth rank, and the sign $\cdot\cdot$ denotes the tensorial multiplication with the double convolution. Compared with Hook's law $\epsilon = \kappa \cdot \cdot \sigma$, equation (4.55) takes into account that the compliance tensor is varying during the loading and deformation process. The additional term $d\eta$ enters the right-hand side of (4.55) where η is the dilatance part of the strain tensor. The quasiplastic strain $d\epsilon_p$ enters in (4.55) with the differential

$$d\epsilon_p = d\kappa \cdot \cdot \sigma + d\eta \tag{4.56}$$

Under a single monotonous loading the influence of the quasiplastic strain is negligible, but it becomes significant under cyclic loading. When unloading takes place, we use instead of (4.55)

$$d\epsilon = \kappa \cdot \cdot d\sigma \tag{4.57}$$

where the compliance tensor κ corresponds to the state attained at the end of the last loading. When the process $\sigma(t)$ is non-proportional, a special criterion is to be introduced to distinguish up- and down-loading stages. In general, equations (4.55) and (4.57) may be considered as those of an incremental theory of plasticity with internal parameters characterizing microdamage.

The next step is to find a way to take into account the properties given by equations (4.55) and (4.57) when generalized forces are concerned. For the one-parameter case the generalized driving force is given in (3.21) as

$$G = \int\limits_{S'_j} p'_\alpha \frac{\partial u_\alpha}{\partial a} dS + \int\limits_{V'_j} x_\alpha \frac{\partial u_\alpha}{\partial a} dV - \frac{\partial}{\partial a} \int\limits_{V'_j} w \, dS. \qquad (4.58)$$

The first two terms in the right-hand side are calculated as usual. The only difference is that the displacement fields are to be found taking into account the the constitutive equations given in (4.55) and (4.57). The last term in (4.58) contains the density w of the internal work performed from the time moment when a body has been in the initial (unstressed) state, up to the considered time moment. The stability condition of the system cracked body-loading contains the differences between the driving and resistance generalized forces. Checking stability is to be performed in moments when these differences are maximal. As a rule, one may assume that these moments coincide with those when the corresponding driving forces attain their maximum. In the case of cyclic loading, checking must be performed in the moments when unloading begins. The density of internal work will be computed for the corresponding states of the body.

In the beginning, consider a one-dimensional relationship between the stress σ and the strain ϵ (Figure 4.38). Assume that the microdamage accumulation occurs only during the half-cycle of loading above a certain threshold. The latter, generally, depends on the microdamage level, and on other parameters, in particular, the cycle stress ratio R. At that stage the compliance tensor κ changes, and the dilatance measure η increases. Let us schematize a segment of the stress-strain diagram within each loading cycle by a straight line with the slope $1/\kappa$, and the horizontal segment that corresponds to the strain increment because of the compliance change and the dilatance deformation during the considered half-cycle.

The increment of the density of internal work within N-th cycle at the end of the loading stage is equal to the shaded area in Figure 4.38. Thus, we have

$$\Delta w(N) = \sigma_{N-1}\kappa_{N-1}(\sigma_N - \sigma_{N-1}) + \left(\tfrac{1}{2}\kappa_{N-1}(\sigma_N - \sigma_{N-1})^2 \text{sign}(\sigma_N - \sigma_{N-1})\right) +$$

$$+\sigma_N \left[(\kappa_N - \kappa_{N-1})\sigma_N + (\eta_N - \eta_{N-1})\right]$$

$$(4.59)$$

Here κ_{N-1} is the compliance at the beginning of N-th cycle; κ_N is the compliance at the end of this cycle; σ_N is the maximal stress of the cycle; η_N is the measure of swelling achieved at this cycle. The multiplier that depends on the sign of $\sigma_N - \sigma_{N-1}$ takes into account the segments of loading during which the maximum stresses are decreasing (see the triangle BDF in Figure 4.38).

Neglecting the terms which have the order of squares of stress increments, rewrite (4.59) in the form

$$\Delta w(N) = \sigma_{N-1}\kappa_{N-1}(\sigma_N - \sigma_{N-1})+$$

$$(4.60)$$

$$+\sigma_N \left[(\kappa_N - \kappa_{N-1})\sigma_N + (\eta_N - \eta_{N-1})\right]$$

The work density of internal forces at the N-th cycle of loading is to be found by summation of the increments $\Delta w(N)$ upon N cycles:

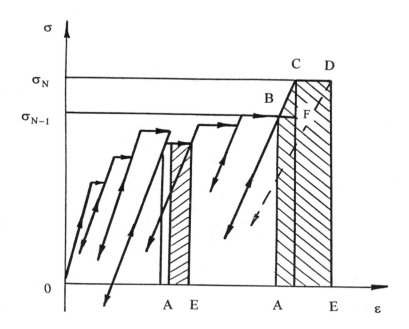

FIGURE 4.38
Stress-strain relationship taking into account microdamage accumulation.

$$w(N) = \sum_{N_1=1}^{N} \Delta w(N_1) \tag{4.61}$$

Now we approach a more general case of cyclic loading. Let the stresses in each point of a body vary in such a way that, at the times chosen for checking stability, their magnitudes are given with the tensor $\sigma(N)$. When this tensor varies sufficiently slowly with the cycle number N, we may treat the loading process as a differentiable function of N. Introduce a similar assumption concerning the dilatance strain tensor $\eta(N)$ and the compliance tensor $\kappa(N)$. Then, instead of (4.60) and (4.61), we obtain:

$$w = \int_0^N \sigma \cdot \left[\frac{\partial}{\partial N_1}(\kappa \cdot \cdot \sigma) + \frac{\partial \eta}{\partial N_1} \right] dN_1 \tag{4.62}$$

If the loading consists of a set of differentiable segments, the work density w can be found with summing of integrals similar to those given in (4.62). When the contribution of a single cycle is to be considered as significant, e.g., in the study of fatigue under overloading, additional terms are to be calculated with the use of equation (4.60).

To estimate the indirect (due to the change of compliance and dilatant deformation) effect of microdamage on fatigue crack growth, consider an opening mode crack under applied cyclic stress $\sigma_\infty(t)$. Let the material in the initial state be

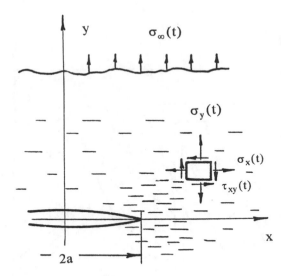

FIGURE 4.39
Microcracking ahead of the crack tip.

linear elastic and isotropic with Young's modulus E, shear modulus G and Poisson's ratio ν. During the loading process these properties are subjected to change. The character of this change depends on the choice of microdamage measures and the assumed mechanism of microdamage accumulation. For example, when a set of spherical microvoids occur in the material under loading, and these microvoids are homogeneously distributed in sufficiently small but macroscopic domains, the damaged material remains macroscopically isotropic (although it generally becomes nongomogenous). In the general case, the damaged material acquires anisotropy.

For the opening mode crack, assume that microdamage is represented by an array of a large number of microcracks coplanar to the main, macroscopic crack (Figure 4.39). Introduce the scalar microdamage measure ω equal to the ratio of the total area of microcracks in a unit volume to a certain critical area corresponding to multiple cracking, and as a result, to the drop of fracture toughness to zero. The postulate for microdamage accumulation following the power-threshold law is given in equation (4.2).

The damaged material is macroscopically orthotropic. For the plane-stress state the constitutive equations within the loading stage take the form

$$\epsilon_x = \frac{\sigma_x}{E_x} - \nu_{xy}\frac{\sigma_y}{E_y} + \eta_x$$

$$\epsilon_y = \frac{\sigma_y}{E_y} - \nu_{yx}\frac{\sigma_x}{E_x} + \eta_y \qquad (4.63)$$

$$\gamma_{xy} = \frac{\tau_{xy}}{G_{xy}} + \eta_{xy}$$

where E_x and E_y are Young's moduli, G_{xy} is the shear modulus, ν_{xy} and ν_{yx} are Poisson's ratios. Assume that microdamage affects the magnitude of modulus E_y, but not that of moduli E_x and G_{xy}. Among Poisson's ratios, only ν_{xy} is assumed dependent on ω to preserve the symmetry condition for the compliances as $\nu_{xy}/E_y = \nu_{yx}/E_x$. In further calculations we set

$$E_x = E, \quad E_y = (1 - \omega^\beta)E, \quad G_{xy} = G$$

$$\nu_{xy} = (1 - \omega^\beta)\nu, \quad \nu_{yx} = \nu \tag{4.64}$$

$$\eta_x = \eta_{xy} = 0, \quad \eta_y = \eta_0 \omega^{\beta_1}$$

with $\beta > 0, \beta_1 > 0, \eta_0 = $ const. Last, assume that the specific fracture work depends on microdamage in such a way that equation (4.11) remains valid for the generalized resistance force.

Equations (4.55), (4.63) and (4.64) are among the simplest ones. One could replace them by others, describing the mechanism of microdamage and the influence of microcracking on the material compliance in a more sophisticated way. For example, we can assume that microcracks are formed under the effect of maximum tensile stresses and are oriented accordingly. Then instead of (4.63) and (4.64) we come to more complicated constitutive equations for a medium with nonhomogeneous local orthotropy. Another approach is to replace equation (4.64) with more accurate equations for effective elastic parameters of the damaged material. Such equations can be found in the publications on micromechanics where a wide scope of models has been proposed. Among them are models of regular arrays of microcracks, stochastic models, models based on the self-consistent field approach, etc. However, the simplified model discussed here seems to be sufficient for the evaluation of the main effects of microdamage on fatigue crack growth: first, the fracture toughness degradation (that was included in all the preceding analyses), and, second, the compliance degradation and the resulting change of stress-strain fields near the crack tips.

4.11 Discussion of Numerical Results

Computations were performed in the paper by Bolotin and Kovekh [28] using the finite element method in space and the finite difference method in time. The finite element mesh was generated automatically with refinement in the vicinity of tips to provide necessary accuracy. The chosen cycle number steps provided a sufficiently small change of the damage measure and compliance properties within each block of loading.

The computational algorithm consisted of the following moduli: formation of the finite element mesh; computation of stress, strain and displacement fields for a given distribution of deformative properties; computation of the microdamage fields; re-evaluation of the stiffness matrix taking into account microdamage; iterative procedure in the frame of the former moduli; computation of the driving and resistance generalized forces; control of stability and the evaluation of the cycle number corresponding to the advancement of the crack tip in the next node of the mesh.

Special comments are needed, probably, concerning the evaluation of the driving force G. A finite difference equivalent of equation (4.58) for two-dimensional problems in the absence of body forces is

$$G\Delta a = \int_{\partial\Omega} p_\alpha^I(u^{II} - u^I)dS - \int_\Omega (w^{II} - w^I)d\Omega \qquad (4.65)$$

The crack length increment Δa is used here as a finite-difference equivalent of the variation δa. Indexes I and II relate to two identical and indentically loaded bodies with the crack sizes a and $a + \Delta a$, respectively. Equation (4.65) contains explicitly the isochronic variation since all the quantities entering the right-hand side correspond to the same cycle number. To realize equation (4.65), the parallel solution of two problems must be performed with a sensible use of intermediate numerical data related to one of the problems.

The crack advancement in (4.65) was taken equal to the minimum longitudinal step $\Delta a = 10^{-2}a_0$ where a_0 is the initial half-length of the crack. The main time step was taken as $\Delta N = 250$. The computation at the fixed crack tip has been performed until the equality $G = \Gamma$ was reached, and then both cracks were moved one step ahead. At the initial stage of microdamage accumulation, one increment of the crack length approximately corresponded to $5 \cdot 10^3$ cycles, i.e., to 20 time steps. The time mesh was refined automatically by approaching the relationship $G = \Gamma_0$. The computational error estimated by test problems appears to be between 2.3 and 3.3%.

Some results of the computational experiment are presented in Figures 4.40-4.44. The crack was modeled as a narrow slit with the initial clearance equal to a double minimum tranverse step of the mesh. It means that the effective tip radius is assumed constant and of the same length order. The distribution of the tensile stress σ_y on the prolongation of the crack at $N < N_*$, i.e., before the growth initiation, is shown in Figure 4.40. The uniform distribution is assumed within the first mesh step. Then the interpolation upon the nodes is performed. Figures 4.40a and 4.40b are plotted for the cases $\beta = 1$ and $\beta = 2$, respectively. The first case corresponds to a stronger influence of microdamage on the compliance. Numbers $1, 2, 3, 4$ and 5 at curves in Figure 4.40a correspond to cycle numbers $N = 0, 2 \cdot 10^3, 4 \cdot 10^3, 6 \cdot 10^3$ and $8 \cdot 10^3$. At the last cycle number the equality $G = \Gamma$ is primarily attained and the crack begins to grow. Numbers 1, 2, 3, and 4 in Figure 4.40b relate to $N = 0, 2 \cdot 10^3, 4 \cdot 10^3$ and $5.75 \cdot 10^3$. The diminishing of stresses near the tip is not so considerable in the second case.

The influence of microdamage on the generalized forces G and Γ is illustrated in Figure 4.41. The generalized forces are related to the resistance force Γ_0 for the nondamaged material. Numbers near the lines correspond to various magnitudes of the power exponent β in (4.64). In particular, Young's modulus across the cracking planes is $E_y = E(1 - \omega^\beta)$ where ω is the damage measure. Lines 1, 2 and 3 correspond to $\beta = 0.5$, 1 and 2, respectively. Line 4 is drawn at $\beta \to \infty$, i.e., under the assumption that microdamage does not affect deformative properties. In the first case the effect of microdamage on the resistance force is most considerable. However, the change of the driving force due to microdamage in the considered

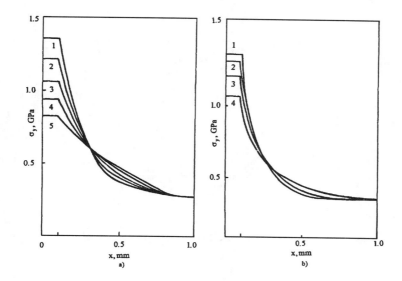

FIGURE 4.40
Stress distribution ahead of the tip before the start of crack growth: (a) $\alpha = \beta = 1$; (b) $\alpha = 1$, $\beta = 2$.

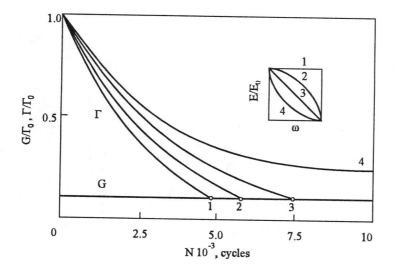

FIGURE 4.41
Influence of microdamage on generalized forces and duration of the initiation stage; lines 1, 2, 3, 4 correspond to $\beta = 0.5$, 1, 2, and $\beta \to \infty$, respectively.

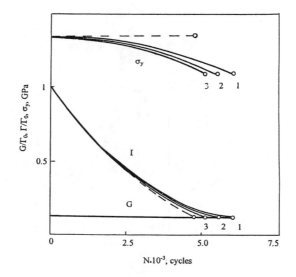

FIGURE 4.42
Influence of dilatance on generalized forces, stresses in the process zone, and duration of the initiation stage; lines 1, 2, and 3 are drawn for $\eta_0 = 0$, 10^{-3}, and $2 \cdot 10^{-3}$, respectively.

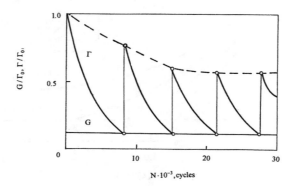

FIGURE 4.43
Beginning of crack propagation in terms of generalized forces.

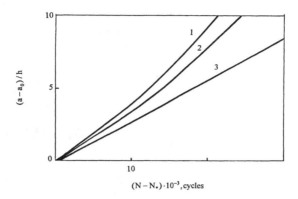

FIGURE 4.44
Crack growth at $\beta = 1$, 2, and $\beta \to \infty$ (lines 1, 2, 3, respectively).

numerical example does not exceed several percent and is not shown in Figure 4.41.

The influence of microdamage on the duration of the initiation stage, i.e. on the cycle number N_* at the first arrival to the equality $G = \Gamma$, is also seen in Figure 4.41. This influence is significant. When the sensitivity of the material to microdamage increases (in this example, when the power exponent β diminishes), the cycle number N_* at the start of crack growth increases in 1.5 times and even more. There are reasons to suppose that the real sensitivity of metals and alloys is such that $\beta \geq 1$ (see insertion into Figure 4.41).

Figure 4.41 corresponds to the fixed parameters characterizing the dilatancy phenomenon: $\eta_0 = 10^{-3}$, $\beta_1 = 1$. The influence of dilatance on the generalized forces and the duration of the initiation stage are shown in the Figure 4.42. Computations are performed for $\beta = 2$, $\beta_1 = 1$ at the various values η_0. Line 1 corresponds to the case when $\eta_0 = 0$, i.e., dilatance strains are absent. Lines 2 and 3 are drawn for $\eta_0 = 10^{-3}$ and $2 \cdot 10^{-3}$, respectively, and the dashed line for $\beta \to \infty$. In the latter case the influence of microdamage on deformative properties is not taken into account. As is shown in Figure 4.42, the effect of dilatance, generally, is small compared with the direct influence of microdamage. The reason is explained in the upper part of Figure 4.42 that shows the maximal stress σ_y at the crack tip in the function of N and η_0.

The results of computation including the initial stage of crack propagation are presented in Figure 4.43. The evolution of generalized forces is shown on several first steps of the finite-difference approximation. The first step corresponds to the initiation stage. After attaining the equality $G = \Gamma$ the tip advances in one step $\Delta a = 0.01 a_0$. The generalized forces take magnitudes corresponding to the new position of the tip. The microdamage measure ω is assumed equal to its average value within the next finite element. At the next step damage accumulation proceeds, and the resistance force again diminishes until the equality $G = \Gamma$ is attained. Then the next advancement of the tip takes place, etc.

Using the results of computation, the crack growth histories at various β are plotted (Figure 4.44). Numbers $1, 2$, and 3 correspond to power exponents $\beta = 1$, $\beta = 2$ and $\beta \to \infty$, i.e., to the consequential decrease of the influence of microdamage on material deformative properties. For all numerical examples it was assumed $\eta_0 = 10^{-3}$, $\beta_1 = 1$.

4.12 Assessment of Parameters for Microscale Models

There are a number of material parameters entering the models of fatigue crack initiation and propagation. Some of them, such as E, ν, and γ_0, are of a macroscale nature; they are either known beforehand or may be easily estimated by standard tests. Other parameters are of a microscale nature, and most of them cannot be assessed directly. Even when we neglect threshold and closure effects, the number of microscale parameters is large. Among them are parameters σ_d, $\Delta\sigma_{th}$, and m in equation (4.2) of microdamage accumulation, exponent α in equation (4.11) for the generalized resistance force, and parameters ρ_s, ρ_b, and λ_ρ in the equation that describes the evolution of the state at the crack tip.

Some of microscale parameters are connected, in the framework of the proposed theory, with material parameters which in turn may be assessed directly. For example, exponent m is close to the Paris exponent m_p in equation (4.24). Resistance stresses σ_d and $\Delta\sigma_{th}$ can refer, as in equations (4.46), to the corresponding stress intensity factors K_d and ΔK_{th} if the information about the effective tip radius is available.

In general, the assessment of parameters for microscale models from experiments performed on the macrolevel is a typical problem of parameter identification. The idea of a solution is illustrated in Figure 4.45. There are some typical numerical values on the crack growth rate diagram which can be used for the assessment of microscale parameters. Among them are the threshold 1 and critical 2 magnitudes of ΔK, the slope 3 and the position 4 of the middle section of the diagram. In addition, there are some data of the transition sections such as the radius 5 of the arc in Figure 4.43 as well as coordinates 6 and 7 of the curvature center. In case of necessity, similar data are available for another transition section. But even without the latter data, we have seven numerical values to obtain seven identification equations, with respect to σ_d, $\Delta\sigma_{th}$, m, ρ_s, ρ_b, λ_ρ, and α. Unfortunately, as it follows from numerical experimentations, this set of equations is badly conditioned.

Additional information can be obtained from standard fatigue curves. However, most of the available data on fatigue curves are incomplete and have no exact interpretation from the viewpoint of solid mechanics. Fatigue tests are subjected to codes and specifications; but very frequently it is not clear what the initial conditions are, and what is understood when one talks about limit states. Regarding initial conditions, we mean the surface roughness which varies in a wide scale or,

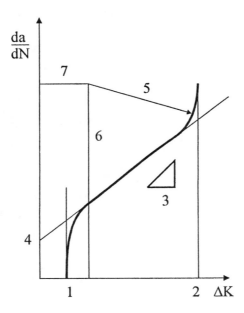

FIGURE 4.45
Typical set of data for assessment of model parameters.

in the case of the initial notch, we also mean the shape of the notch, and specimen geometry. As to the limit states, the initiation of a visible crack is interpreted sometimes as such a state. In other cases this state is referred to a macroscopic crack of a given size or to the final rupture of a specimen.

The relationship between crack growth histories at different load levels (under other equal circumstances) and the fatigue curves is illustrated in Figure 4.46. The initial crack size a_0 is assumed constant in Figure 4.46a. In the case of "smooth" specimen, a_0 is just a characteristic of surface roughness. Other notations are: a_* is the size of visible cracks, which may be only a little larger than the initial size a_0; and a_{**} is the crack size corresponding to a critical state. Both sizes are assumed in Figure 4.46a as constant.

Two fatigue curves are depicted in Figure 4.46b: that corresponding to the visible size a_* and that corresponding to the critical size a_{**}. The first curve can be referred to as the cycle number N_* at crack growth start, the second curve to as the critical cycle number N_{**}. There are several numerical data in Figure 4.46b which are available for parameter assessment. Together with the data taken from growth rate diagrams, they seem to be sufficient to stabilize the identification procedure.

For coarse estimates, the equations of quasistationary approximation are useful. Let, for example, $G_{max} \ll \Gamma_0$, $\omega_{ff} \ll 1$, $\lambda_p = $ const. Equation (4.42) for the tip damage takes the form

$$\frac{d\psi}{dN} = \left(\frac{\Delta\sigma_t - \Delta\sigma_{th}}{\sigma_d} \right)^m \qquad (4.66)$$

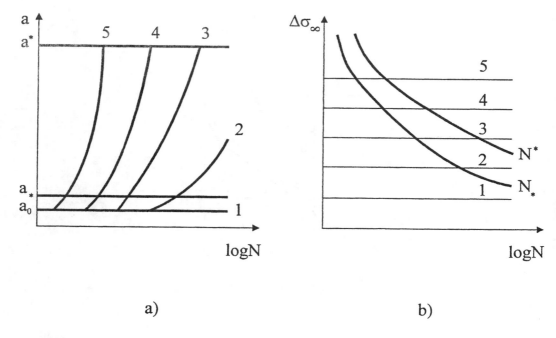

a)   b)

FIGURE 4.46
Relationship between crack growth histories (a) and standard fatigue curves (b); lines 1 ... 5 correspond to five increasing load levels.

and equation (4.42) for the crack growth rate

$$\frac{da}{dN} = \lambda_p \left(\frac{\Delta\sigma_t - \Delta\sigma_{th}}{\sigma_d} \right)^m \tag{4.67}$$

Here $\Delta\sigma_t$ is the range of the stress at the tip estimated as

$$\Delta\sigma_t \approx 2Y\Delta\sigma_\infty \left(\frac{a}{\rho} \right)^{1/2} \tag{4.68}$$

where $\Delta\sigma_\infty$ is the range of the applied or reference stress, Y is the correction factor, ρ is the effective tip radius. For the sake of simplicity, we assume that $\Delta\sigma_\infty$, Y, and ρ are constant, though they, as a rule, vary during the crack growth process. For edge cracks in mode I we have $Y \approx 1.12$. When a crack propagates from a notch, σ_t must be evaluated to account for stress concentration at the tip.

An estimate for N_* follows from equations (4.66) and (4.68)

$$N_* \sim \left[\frac{\sigma_d}{2Y\Delta\sigma_\infty(a/\rho)^{1/2} - \Delta\sigma_{th}} \right]^m \tag{4.69}$$

where $a_0 \le a \le a_*$, $\rho_s \le \rho \le \rho_b$. Probably, it is more appropriate to put $a = a_*$, $\rho = \rho_b$. Then the fatigue limit stress (in the conventional sense) is estimated as

$$\Delta\sigma_{TH} \sim \frac{\Delta\sigma_{th}}{2Y} \left(\frac{\rho_b}{a_*}\right)^{1/2} \tag{4.70}$$

Equations (4.69) and (4.70) show that the fatigue limit stress is proportional to threshold resistance stress $\Delta\sigma_{th}$ in the equation of microdamage accumulation. Moreover, at $\rho_b \approx a_*$ which is typical for comparatively coarse surface roughness, and $Y \approx 1$, this equation yields to $\Delta\sigma_{TH} \sim \Delta\sigma_{th}$. Note that $\Delta\sigma_{TH}$ refers to applied stresses while $\Delta\sigma_{th}$ is a microscale parameter. Therefore, this conclusion is not very evident though it looks trivial.

Another conclusion concerning equation (4.69) is that the exponent m which is also a microscale parameter coincides (in the given approximation) with the exponent of fatigue curve relating the cycle number N_* to the range $\Delta\sigma_\infty$ of the reference stress:

$$N_* \sim \left(\frac{\sigma_D}{\Delta\sigma_\infty - \Delta\sigma_{TH}}\right)^m$$

Here σ_D is connected with σ_d with an equation similar to (4.70), i.e.,

$$\sigma_D \sim \frac{\sigma_d}{2Y} \left(\frac{\rho_b}{a_*}\right)^{1/2} \tag{4.71}$$

Turning to equation (4.67), note that in the case when λ_p, Y, ρ, not saying σ_d, $\Delta\sigma_{th}$, and m are constant, its solution is elementary. To obtain more transparent results, assume that $\Delta\sigma_t \gg \Delta\sigma_{th}$, $\rho = \rho_b$. Then we come to equation

$$\frac{da}{dN} = \lambda_p \left(\frac{2Y\Delta\sigma_\infty}{\sigma_d}\right)^m \left(\frac{a}{\rho_b}\right)^{m/2}$$

Its solution for boundary conditions $a(N_*) = a_*$, $a(N_{**}) = a_{**}$ at $m \neq 2$ is

$$\frac{(a_{**}/\rho_b)^{1-(m/2)} - (a_*/\rho_b)^{1-(m/2)}}{1 - (m/2)} = \frac{\lambda_p}{\rho_b} \left(\frac{2Y\Delta\sigma_\infty}{\sigma_d}\right)^m (N_{**} - N_*) \tag{4.72}$$

At $m = 2$ we have, respectively,

$$\log\left(\frac{a_{**}}{a_*}\right) = \frac{\lambda_p}{\rho_b} \left(\frac{2Y\Delta\sigma_\infty}{\sigma_d}\right)^2 (N_{**} - N_*) \tag{4.73}$$

Let us consider equation (4.72) at $a_{**} \gg a_*$, $m > 2$. Under this condition we obtain that

$$N_{**} \approx N_* + \frac{a_*/\lambda_p}{(m/2) - 1} \left(\frac{\sigma_D}{\Delta\sigma_\infty}\right)^m \tag{4.74}$$

where σ_D is a characteristic stress similar to given in (4.71).

Equation (4.74) has the same structure as common empirical equations such as given in (1.2). It is essential that the exponent m from the equation of microdamage accumulation enters equation (4.74) also.

It seems paradoxical that the limit size a_{**} when it is much larger than a_* is not present in (4.74). There is not such a case at $m = 2$. In particular, equation (4.73) which is valid at $m = 2$ gives

$$N_{**} \approx N_* + \frac{a_*}{\lambda_p} \left(\frac{\sigma_D}{\Delta\sigma_\infty} \right)^2 \log \left(\frac{a_{**}}{a_*} \right)$$

For most materials, however, $m > 2$.

One could use some of the above equations for the assessment of microscale parameters. Together with equations for crack growth rate they provide a more stable procedure of assessment. In particular, equations (4.70) and (4.71) are useful since they connect microscale parameters with parameters estimated in direct macroscopical tests.

4.13 Summary

The theory of fatigue crack growth presented in the previous chapter was applied here to the simplest case of one-parameter mode I planar crack in an unbounded body whose material is linear elastic though subjected to damage accumulation. This model was used as a benchmark model to illustrate versatility of the proposed theory and its compatibility with experimental results and semi-empirical equations. In particular, the influence of loading conditions (nonsteady regimes included), initial crack parameters, and material parameters was studied. Special attention was paid to threshold, shielding, and closure effects, and the models were discussed to take these effects into account. Using the quasistationary approximation technique and some additional assumptions, the set of constitutive equations was reduced to differential equations similar to the Paris-Erdogan equation. An interpretation of this equation was given in terms of solid mechanics. The interaction of damage, compliance, and fracture toughness fields was studied, and the contribution of each factor was evaluated.

Chapter 5

FATIGUE CRACK GROWTH IN LINEAR ELASTIC BODIES (CONT'D)

5.1   Stress Distribution Near the Crack Tips

This chapter, as the former one, deals with fatigue cracks in bodies whose material is linear elastic. Material is subjected to microdamage that changes its mechanical properties, at least those which are responsible for the resistance to crack propagation. Compared with the previous chapter completely dedicated to fatigue cracks in Griffith's problem, a number of more complicated problems are considered here. Among them are the problems of crack propagation associated with kinking and branching, and crack growth through materials with randomly distributed properties. On the other side, we omit here some effects studied in detail in Sections 4.6 and 4.7 such as the crack closure effect. It is done here for the sake of brevity. Any problem discussed in this chapter may be easily supplemented by taking into account tip shielding, crack closure, etc. Concerning the intrinsic threshold effect due to material resistance to damage accumulation, it is included in all the problems discussed hereafter.

The general approach presented in Chapter 3 is applicable to a wide range of problems. However, the approach to most of them is associated with considerable computational difficulties. The main difficulties originate not from fracture and fatigue mechanics but from the related topics of solid mechanics. Actually, to evaluate the damage fields along crack trajectories, one needs to know stress-strain fields in cracked bodies. The problem is complicated because of the necessity of treating cracks as very narrow slits or hollows with finite (usually, varying in time) tip curvatures. The number of comprehensive solutions of such problems is rather

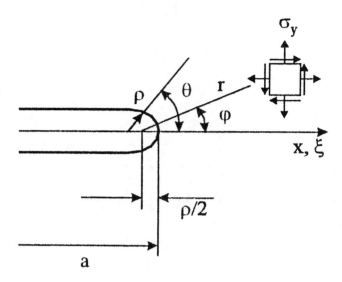

FIGURE 5.1
Evaluation of the relationship between stress concentration and stress intensity factors.

small, and the final results are mostly presented in terms of stress concentration factors [115]. This means that only stresses at the tips are evaluated without references to the stress distribution at the tip prolongation. Even concerning stress concentration factors, the amount of available information is rather modest compared with the up-to-date data on stress intensity factors [106, 135, 141]. Hence a natural idea arises to use the information on stress intensity factors to evaluate stress distribution near the cracks with finite tip curvatures. This approach was used by a number of authors.

The simplest case is delivered by Neuber's equation (4.16) for the stress concentration factor κ at the larger semi-axis of an elliptical hole in a plane. The direct comparison of equations for $\kappa = \sigma_{max}/\sigma_\infty$ and K_I results in

$$\kappa \approx \frac{2K_I}{\sigma_\infty (\pi\rho)^{1/2}} = 2 \left(\frac{a}{\rho}\right)^{1/2} \tag{5.1}$$

where the relationship $(\rho/a)^{1/2} \ll 1$ is assumed.

Another approach was suggested by Rice (1968) who considered the stress energy release rate in a plane stress state around a mode I crack with the tip radius $\rho(\theta)$ (Figure 5.1). Denoting by $\sigma_\theta(\theta)$ the circumferential normal stress close to the tip, Rice assumed that

$$G = \frac{K_I^2}{E} = \frac{1}{2E} \int\limits_{-\pi/2}^{\pi/2} \sigma_\theta^2(\theta)\rho(\theta) \cos\theta d\theta \tag{5.2}$$

Equation (5.2) also connects stress concentration and stress intensity factors, and the numerical coefficient in this relationship is of the order given in (5.1).

Similar relationships can be obtained with the use of classical solutions of elasticity theory. For example, equations for the shear stress concentration factors $\kappa = \tau_{\max}/\tau_\infty$ in mode II cracks

$$\kappa \approx \left(\frac{a}{3\rho}\right)^{1/2}$$

and for mode III cracks

$$\kappa \approx \left(\frac{a}{\rho}\right)^{1/2}$$

could be found in the literature. In all these cases the stress concentration factors are of the order $(a/\rho)^{1/2}$.

As has been mentioned by many authors, the magnitudes of the correction factor Z in the generalized Neuber's formula

$$\kappa = 1 + 2Z(a/\rho)^{1/2} \tag{5.3}$$

and the correction factor Y for stress intensity factor

$$K_I = Ys(\pi a)^{1/2} \tag{5.4}$$

are close for a wide variety of notch and crack problems. Here s is a nominal or reference stress, e.g., the remotely applied or net cross section stress, and the stress concentration factor κ in (5.3) refer to the same stress s. The approximate equality $Z \approx Y$ holds with an agreeable accuracy for sharp notches and, consequently, for cracks, though for shallow notches the discrepancy attains 10% and more. A systematical study of the relationship between formulas (5.3) and (5.4) was performed by Shin, Man and Wang [133] who analyzed more than 20 types of notches in wide intervals of a/ρ. The first term in (5.3) takes into account the regular stress component and usually is negligible compared with the second term. This fact has been already used in Chapter 4 when the connection was discussed between the suggested theory of fatigue crack propagation and the common semi-empirical equations.

Another item of interest is the stress distribution ahead of crack tips and, in general, along the future crack trajectories. Sometimes the ready solutions of elasticity theory can be used. Among them are the solutions for an elliptical hole in an unbounded plate and for a rotational ellipsoid in an unbounded three-dimensional body.

The analogy with stress distribution in the vicinity of mathematical cuts known in linear fracture mechanics as the Inglis–Williams equations has been used by Creager and Paris (1967). The stress σ_y ahead of the notch in the uniform remote tension is given as

$$\sigma_y = \sigma_\infty + \frac{K_I}{(2\pi r)^{1/2}}\left[\left(\frac{\rho}{2r}\right)\cos\frac{3\varphi}{2} + \cos\frac{\varphi}{2}\left(1 + \sin\frac{\varphi}{2}\sin\frac{3\varphi}{2}\right)\right]$$

The first term represents the remote stress, and the second is proportional to K_I. Note that the origin of the polar coordinates r, φ is taken at the point $x = a - \rho/2$ which does not coincide with the center of the notch contour (Figure 5.1). Evidently the above analogy may be extended upon other stress components as well as upon a wider class of crack and loading cases.

A number of authors suggested much simpler empirical (at $x \geq a$, $y = 0$) equations connecting the stress σ_y ahead of the crack with $\sigma_{\max} = \kappa s$ where s is the reference stress in (5.4). One of these equations has the form

$$\sigma_y = \frac{\sigma_{\max}}{(1 + 4\xi/\rho)^{1/2}} \tag{5.5}$$

where $\xi = x - a$, i.e., ξ is the coordinate measured from the crack tip. Equation (5.5) is sometimes attributed to Neuber. It presents good results for the plane problems both near the tip of the notch and further, up to $\xi = 10\rho$. This distance is usually sufficient to estimate the damage field, especially when the far field component may be considered as negligible. The numerical factor at ξ/ρ has been argued by some authors, and, alternatively, rather cumbersome equations have been proposed. Their survey and the comparison with available analytical and finite element method solutions can be found in the paper [133]. Quite another approach is based on the coarse approximation similar to that in Figure 4.36. The prolongation of the crack is presented as a process zone with stresses equal to their magnitude at the crack tip, while the far field is presented with nominal stresses. At least, such an approximation allows reducing the problem to simple differential equations (see Section 4.8).

5.2   Some Generalizations of Griffith's Problem in Fatigue

The mode I crack in the unbounded linear elastic body has been studied in detail in Chapter 4 with special attention to the influence of initial conditions, loading regimes, and material properties as well as to more "intimate" mechanisms such as the evolution of damage and stress concentration at the moving crack tip. Some generalizations of this problem are considered in this section. As a silent agreement, it is usually assumed that a universal relationship exists for a given material and loading conditions between the stress intensity factor range ΔK_I and the crack growth rate da/dN in the mode I. The aim of this section is to check this assumption in the framework of the suggested theory.

Three problems of practical interest are shown in Figure 5.2. It is a plate of finite width w with a central crack under remote uniform tension (a); a similar and similarly loaded plate with an edge crack (b); and a similar plate under bending (c). In all these cases the stress distribution is presented by equations (5.4) and (5.5) where s is the reference stress. Therefore, stress along the crack prolongation is

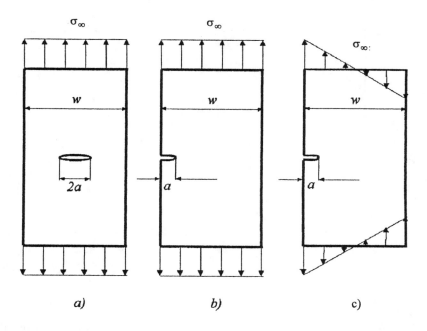

FIGURE 5.2
Fatigue cracks in plate: (a) central crack under tension; (b) edge crack under tension; (c) edge crack under bending.

$$\Delta\sigma = \frac{Y(a/w)(\pi a)^{1/2}\Delta s}{(1 + 4\xi/\rho)^{1/2}}, \qquad (5.6)$$

where the subscript in $\Delta\sigma_y$ is omitted. When $\Delta\sigma$ being determined by (5.6) becomes less than the nominal stress, the latter is used instead of (5.6). This is a sound schematization because the far field damage is comparatively small at all the stages of crack propagation.

Some numerical results are shown in Figures 5.3–5.5. Figure 5.3 is drawn for the case pictured in Figure 5.2.a. The model given in equations (4.3), (4.11) and (4.17) is used for the following numerical data: $E = 200$ GPa, $\nu = 0.3$, $\gamma_0 = 20$ kJ/m^2, $\sigma_d = 20$ GPa, $\Delta\sigma_{th} = 250$ MPa, $m = 4, \alpha = 1, n = 0, \rho_s = 10$ μm, $\rho_b = \lambda_p = 100$ μm, $\Delta\sigma_\infty = 200$ MPa, $R = 0$. Evidently, these data are the same as in Figures 4.4-4.7. The only difference is that the width of the plate is finite. The initial magnitude of ratio $\zeta = 2a/w$ is subjected to variation as $\zeta_0 = 2a_0/w = 0.01$, 0.02, and 0.04. For the plate width $w = 100$ mm, it means that the initial crack size a_0 is assumed equal to 0.5, 1 and 2 mm. The nominal stress is just the net cross-section stress, i.e., $s = \sigma_\infty w/(w - 2a)$. The correction factor in (5.6) is taken in polynomial form [106]

$$Y = 1 - 0.128\zeta - 0.288\zeta^2 + 1.523\zeta^3$$

that is valid until $2a/w = 0.7$ with the error less than 0.6%.

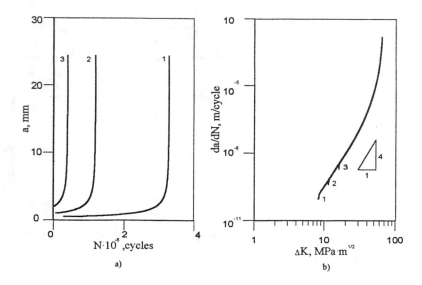

FIGURE 5.3
Crack size versus cycle number (a) and growth rate diagram (b) for plates with
central cracks under tension; numbers 1, 2, 3 correspond to $2a_0/w = 0.01, 0.02$,
and 0.04.

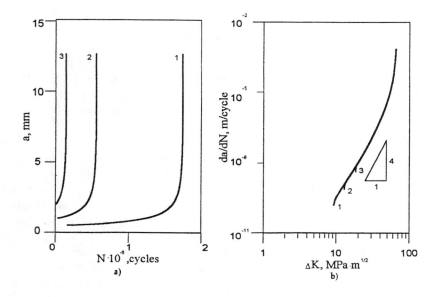

FIGURE 5.4
The same as in Figure 5.3 for plates with edge cracks under remote tension;
numbers 1, 2, 3 correspond to $a_0/w = 0.01, 0.02$, and 0.04.

The crack growth histories are shown in Figure 5.3a. Lines 1, 2, 3 are drawn for $2a_0/w = 0.01$, 0.02, and 0.04, respectively. It is evident that the total fatigue life varies with $2a_0/w$ several times. The growth rate diagrams (Figure 5.3b) do not show such a strong effect, especially being drawn in log − log scale. The only difference is observed at the near-threshold stage. Taking into account a high scatter of experimental data, we might agree that a common crack growth rate diagram is valid for a wide interval of $2a/w$.

Similar numerical results are obtained for the edge crack (Figure 5.4). Assume, as a reference stress, the material net cross-section stress evaluated by the elementary strength analysis:

$$\frac{s}{\sigma_\infty} = \frac{w}{w-a} + \frac{wa}{12}\frac{w+a-2x}{(w-a)^2}$$

The first term in the right-hand side represents tension, the second one bending due to eccentricity. The correction factor is

$$Y = 1.12 - 0.231\zeta + 10.55\zeta^2 - 21.72\zeta^3 + 30.39\zeta^4$$

where $\zeta = a/w$. This formula gives the error not exceeding 0.5% at $a/w \leq 0.6$. Numbers 1, 2, 3 correspond to the initial crack length ratios $a_0/w = 0.005$, 0.01, and 0.02 which means 0.5, 1, and 2 mm for a plate width $w = 100$ mm. The picture is very similar to that in the case of the central crack.

The case of bending (Figure 5.2c) is illustrated in Figure 5.5. The same equation (5.6) is used as above with reference stress $s_\infty = 6M/w^2$ (M is the bending moment per unit of the plate thickness). The nominal stress distribution is

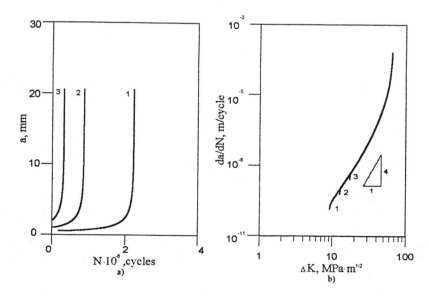

FIGURE 5.5
The same as in Figure 5.4 for plates under remote bending.

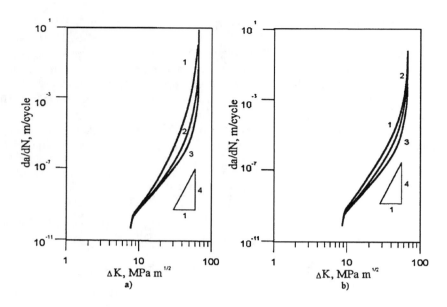

FIGURE 5.6
**Growth rate diagrams for plates shown in Figure 5.2,a and 5.2,b at $a_0 =$
1 mm; lines 1, 2, and 3 correspond to $w =$10, 100 mm, and $w \to \infty$.**

$$\frac{s}{\sigma_\infty} = \frac{w}{w-a} + \frac{w^2(w+a-2x)}{(w-a)^3}$$

and the correction factor

$$Y = 1.222 - 1.140\zeta + 7.33\zeta^2 - 13.08\zeta^3 + 14.0\zeta^4$$

with the error less than 0.2% at $a/w \le 0.6$.

The same conclusions could be made in this case, too. Moreover, the crack growth rate diagrams for all three cases are very close, especially in log − log scale. The discrepancy vanishes when we compare plates of various widths but at fixed initial crack size. Initial parts of the crack growth diagrams are very close. Then the lines separate and merge only at the final stage of crack propagation. It is illustrated in Figure 5.6 obtained for cases (a) and (b) in Figure 5.2. The initial crack size $a_0 = 1$ mm and the crack width varies as $w = 10, 100$ mm and $w \to \infty$ (lines 1, 2 and 3, respectively). The divergence of curves is rather significant; however, it is not so drastic taking into consideration the natural scatter of experimental rate da/dN.

The results of the numerical solutions may look too optimistic. Actually, they are obtained for the fixed applied stress range and fixed stress ratio; in addition, some effects accompanying crack growth such as the closure effect are not taken into account. There is no doubt that, including these effects, we will observe a significant scatter of crack growth rates. In any case, we may treat the crack growth diagram for a crack in an unbounded body as a "principal" diagram representing all the main features of the mode I crack both qualitatively and quantitatively.

5.3   Circular Planar Crack in Tension

As another simple example, let us consider a planar circular crack in an unbounded body. Present this crack as a strongly flattened axially symmetrical ellipsoid with the principal plane coinciding with x, y plane (Figure 5.7). The radius a of the crack is equal to the length of the larger semi-axis, and the crack opening displacement in the center to the double length of the smaller semi-axis $c \ll a$. The effective curvature radius at the crack front $\rho = (ac)^{1/2}$ is small compared with a. We consider this radius as a characteristic of stress concentration near the damaged crack front.

Let the body be subjected to remotely applied cyclic stresses $\sigma_\infty(t)$ directed along the z-axis. The crack grows in the x, y plane remaining circular. At $\rho \to 0$ this crack transforms into the corresponding model of linear fracture mechanics, i.e., in a circular mathematical cut with radius a. The stress intensity factors for this crack and loading conditions are

$$K_I = 2\sigma_\infty(a/\pi)^{1/2}, K_{II} = K_{III} = 0$$

Stress distribution in the vicinity of an elliptical hollow is studied in elasticity theory in all details. The final equations depend on the used analytical technique and are very cumbersome even for axisymmetrical ellipsoids. The relatively simple equation for the stress σ_z at $z = 0$ is given in [110].

$$\frac{\sigma_z}{\sigma_\infty} = 1 + (12B + \alpha C)E - \frac{12B + \alpha C}{\sinh u} + \frac{A + 4B}{\sinh^3 u}$$

with notation

$$A = D[-6gt^2 + (4\alpha - 2)g + 4t^4 - 4\alpha t^2]$$

$$B = D[(2 - \alpha)g + 2t^4 + (\alpha - 3)t^2], \quad C = D(6g - 12t^2)$$

$$D = \frac{t^4 \sinh u_0}{6[(\alpha - 4)g^2 + 8gt^2 - 6gt^4 + 4t^6 - 4t^4]}$$

$$E = \cotan^{-1}(\sinh u), \quad g = \cosh^4 u_0 \sinh u_0 [\cotan^{-1}(\sinh u_0)] - \cosh^2 u_0 \sinh^2 u_0$$

Here $\alpha = 2(1 - \nu), u_0 = \cosh^{-1}[a/(a^2 - c^2)^{1/2}], u = \cosh^{-1}[r/(a^2 - c^2)^{1/2}], t = \cosh u_0; r$ is the planar radius, $r \geq a$. Distribution of stress $\sigma_z(r)$ is shown in Figure 5.8 for various a/ρ.

As in the case of Griffith's crack, we neglect the influence of microdamage on the driving force G. Then, using Irwin's equation, we obtain

$$G = \frac{8\sigma_\infty^2 a^2(1 - \nu^2)}{E} \tag{5.7}$$

To evaluate the resistance force Γ, we consider the work spent on the crack propagation as $\delta W_f = -\gamma(a + \delta a)^2 + \pi\gamma a^2 = -2\pi\gamma a + O(\delta a^2)$. Then

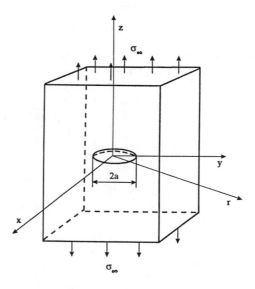

FIGURE 5.7
Circular planar crack as rotational ellipsoid.

05-08

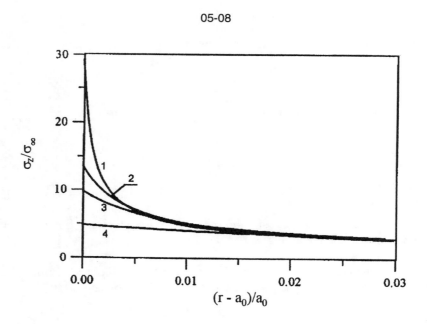

FIGURE 5.8
Opening stress distribution ahead of circular crack at $a/\rho = 10$, 50, 100, and 500 (lines 1, 2, 3, and 4).

$$\Gamma = 2\pi\gamma a \qquad (5.8)$$

Note that the dimension of the generalized forces in (5.7) and (5.8) differs from, for example, that given in (4.11). Both forces refer in (5.7) and (5.8) to the unit of area. Therefore they are measured in J/m^2 (instead of J/m for thorough cracks in plates whose thickness is assumed equal to unity).

Coming to fatigue, we assume equation (4.2) for microdamage accumulation and equation (4.11) for the influence of microdamage on the resistance force. Similarly, we assume equation (4.17) to describe the change of the effective tip radius during crack propagation.

Some computed results are presented in Figure 5.9. The following numerical data are used: $E = 200$ GPa, $\nu = 0.3$, $\gamma_0 = 25$ kJ/m^2, $\alpha = 1$, $\sigma_d = 10$ GPa, $\Delta\sigma_{th} = 0$, $\rho_s = 10$ μm, $\rho_b = \lambda_\rho = 100$ μm, $\Delta\sigma_\infty = 100$ MPa, $R = 0.25, m = 4$, $\rho_0 = 50$ μm and various initial crack sizes a_0. Lines 1, 2, 3, 4 are drawn in Figure 5.9 for $a_0 = 0.5$, 1, 2.5, and 5 mm, respectively. Time histories of crack growth are shown in Figure 5.9a and growth rate diagrams in Figure 5.9b. Compared with Figures 5.3- 5.6, a log scale is used for the cycle number in Figure 5.9a. As to the crack growth rate diagrams, they differ only in the vicinity of thresholds. All curves contain three typical stages: the section of retarded growth in the beginning, the middle section close to the linear (in the $\log - \log$ scale) law of growth, and the final section before the total failure. It is essential that, similar to cracks in Griffith's problem (Section 4.4), the middle and final sections do not depend on initial conditions. The slope coefficient of the middle section is a little higher than the exponent $m = 4$ entering equation (4.2). Additional information is given in Figure 5.10 where the time histories of the effective tip radius ρ and the tip damage measure ψ are shown. The numeration of curves is the same as in Figure 5.9. The moment of the crack growth start is distinctly identified in Figure 5.10 by the essential change of the behavior of ρ and ψ as functions of N.

Crack growth rate in the middle part of fatigue life is controlled, in the first line, by damage accumulation. Therefore, the rate da/dN strongly depends on the power exponent m in equation (4.2). It is illustrated in Figure 5.11. The same numerical data are used here as in Figures 5.9 and 5.10. The only difference is that the specific fracture work is $\gamma_0 = 8$ kJ/m^2 and the exponent m varies from 2 to 8. Evaluating the slope in the interval $8 \leq \Delta K_I \leq 30$ MPa·m$^{1/2}$, we find that the equivalent of the exponent m_p in equation (4.25) is equal to 3.13, 5.39, 7.59, and 9.28 when $m = 2$, 4, 6, and 8, respectively. It means that the actual power exponent is larger than m although the general tendency $m_p \sim m$ is also observed here. The length parameter in equation (4.26) varies, correspondingly, as follows: $\lambda = 17.1$, 33.6, 58.1, and 89.4 μm. Note that different m refer, as a matter of fact, to materials with different properties.

5.4 Elliptical Planar Cracks. Driving Forces

The next step to generalize the problem discussed in the former section is to consider an elliptical planar crack under cyclic tension across its plane. To simplify this

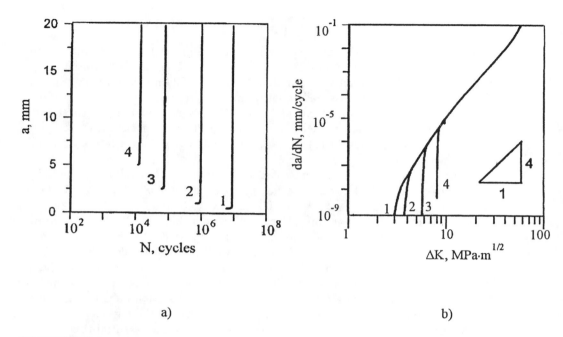

a)   b)

FIGURE 5.9
Circular crack growth histories (a) and growth rate diagrams (b) at $a_0 = 0.5$,
1, 2.5, and 5 mm (lines 1, 2, 3, 4, and 5, respectively).

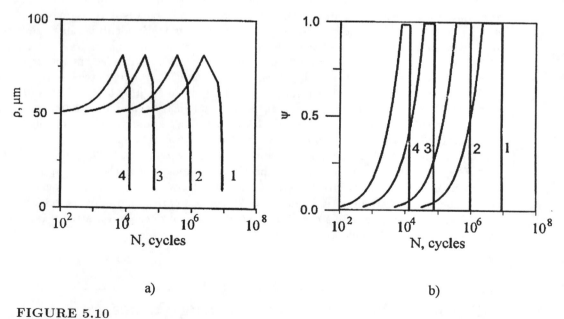

a)   b)

FIGURE 5.10
Time histories of effective tip radius (a) and tip damage measure (b) for circular
cracks (notation as in Figure 5.9).

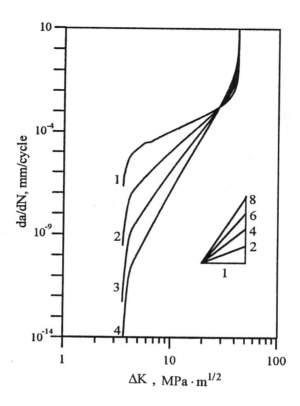

FIGURE 5.11
Effect of damage accumulation exponent ($m = 2, 4, 6, 8$) on growth rate diagrams for circular cracks.

problem, we assume that the crack remains elliptical during the following growth. As a result, we come to the problem with two G-coordinates, namely semi-axes of an ellipse. It is a rather good approximation to real internal cracks. If a crack is far from elliptical, we can apply the two-side estimates using circumscribed and inscribed ellipses [61]. Most crack-like defects in pipes and vessels are considered as half-elliptical ones. An angle crack, such as one located near a corner, is usually approximated as a quarter of an ellipse. It signifies that elliptical (two-parameter) approximation could find a wide application in engineering design.

Consider in detail an internal crack in an unbounded isotropic elastic body under the remotely applied normal stress σ_∞. Denote the semi-axes of the ellipse a and b. The strain energy of a body with such a crack is given in [92] as

$$U = \text{const} - \frac{2\pi\sigma_\infty^2 ab^2(1-\nu)}{3\mu\mathbf{E}(k)} \tag{5.9}$$

Here μ is the shear modulus, $k^2 = 1 - (b/a)^2, b \leq a$. The standard notation of complete elliptical integrals of the first and second kinds is used in (5.9) and later:

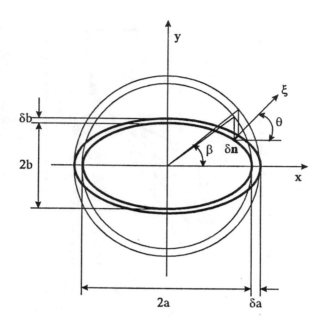

FIGURE 5.12
Elliptical crack and variation of its dimensions.

$$\mathbf{K}(k) = \int\limits_0^{\pi/2} (1 - k^2 \sin^2 \beta)^{-1/2} d\beta, \ \ \mathbf{E}(k) = \int\limits_0^{\pi/2} (1 - k^2 \sin^2 \beta)^{1/2} d\beta \quad (5.10)$$

The semi-axes a and b enter the problem as generalized coordinates, and their variations δa and δb as G-variations (see Figure 5.12 where a quarter of an ellipse is depicted). The summed work of external and internal forces performed on G-variations is presented in the form $\delta_G W_e + \delta_G W_i = G_a \delta a + G_b \delta b$ where G_a and G_b are generalized driving forces. Evidently $\delta_G W_e + \delta_G W_i = -\delta_G U$. Therefore

$$G_a = -\frac{\partial U}{\partial a}, \qquad G_b = -\frac{\partial U}{\partial b} \qquad (5.11)$$

Variables defined in (5.11) may be interpreted as energy release rates referred to as semi-axes. Substituting (5.9) in (5.11), we come to the following equations:

$$G_a = \frac{2\pi \sigma_\infty^2 b^2 (1 - \nu)}{3\mu \mathbf{E}(k)} \left[1 + k'^2 \frac{\mathbf{D}(k)}{\mathbf{E}(k)} \right]$$

$$G_b = \frac{2\pi \sigma_\infty^2 ab (1 - \nu)}{3\mu \mathbf{E}(k)} \left[2 - k'^2 \frac{\mathbf{D}(k)}{\mathbf{E}(k)} \right] \qquad (5.12)$$

As an addition to (5.10), the notation is introduced here

$$\mathbf{D}(k) = \int\limits_{0}^{\pi/2} (1 - k^2 \sin^2 \beta)^{-1/2} \sin^2 \beta d\beta, \qquad k'^2 = 1 - k^2$$

Equations (5.12) are illustrated in Figure 5.13. There the generalized forces refer to the half of the corresponding force for a circular crack given in (5.7):

$$G_0 = \frac{2\sigma_\infty^2 a^2 (1 - \nu)}{\mu}$$

Curves a and b correspond to the generalized forces G_a and G_b. At $a = b$, as could be expected, the curves intersect.

It is to be noted that the generalized forces (5.12) are of the "global" nature compared, say, with the forces calculated with use of stress intensity factors at $x = a$, $y = 0$ and $x = 0$, $y = b$. The stress intensity factor around the contour of an elliptical crack is given in equation (2.56)

$$K(\beta) = \frac{\sigma_\infty (\pi b)^{1/2}}{\mathbf{E}(k)} \left(\sin^2 \beta + \frac{b^2}{a^2} \cos^2 \beta\right)^{1/4} \tag{5.13}$$

where β is the elliptical angle (Figure 5.12). The subscript indicating the fracture mode is omitted. Insert $K(\beta)$ at $\beta = 0$ and $\beta = \pi/2$ in Irwin's equation

$$G(\beta) = \frac{K^2(\beta)(1 - \nu)}{2\mu} \tag{5.14}$$

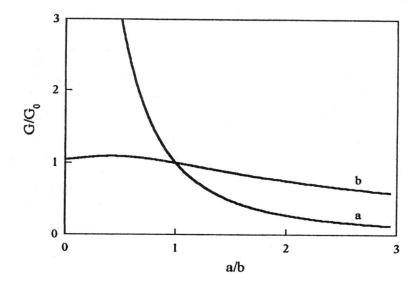

FIGURE 5.13
Normalized generalized driving forces for elliptical crack refer to semi-axes a and b.

Here the shear modulus $\mu = E/2(1 + \nu)$ is used instead of Young's modulus E not to mix the notation for E and $\mathbf{E}(k)$. Combining (5.13) and (5.14), we obtain

$$G(0) = \frac{\pi\sigma_\infty^2 b^2 (1 - \nu)}{2\mu\mathbf{E}^2(k)a}, G(\pi/2) = \frac{\pi\sigma_\infty^2 b(1 - \nu)}{2\mu\mathbf{E}^2(k)} \tag{5.15}$$

Equations (5.12) and (5.15) look different. The numerical difference is not significant, as has been shown in [23].

5.5 Elliptical Planar Cracks. Resistance Forces

Before considering fatigue, let us discuss the quasi-brittle fracture of a body with an internal elliptical crack. For simplification, let us assume the specific fracture work γ being constant along all the contour of the crack. The virtual fracture work $\delta W_f = -\gamma[\pi(a + \delta a)(b + \delta b)] + \gamma\pi ab = -\pi\gamma(b\delta a + a\delta b) + O(\delta a^2)$. Therefore, the generalized resistance forces are

$$\Gamma_a = \pi\gamma b, \qquad \Gamma_b = \pi\gamma a \tag{5.16}$$

Compared with (5.8), these forces are referred to semi-axes. This explains that by putting in (5.16) $a = b$ we obtain half of the force given in (5.8).

The system cracked body-loading will be in a sub-equilibrium state if $G_a < \Gamma_a$, $G_b < \Gamma_b$. Crack growth start with respect to coordinate a begins when the equality $G_a = \Gamma_a$ will be primarily attained under the condition that $G_b < \Gamma_b$. Oppositely, the start with respect to b begins at $G_b = \Gamma_b, G_a < \Gamma_a$. Checking the signs of derivatives $\partial(G_a - \Gamma_a)/\partial a$ and $\partial(G_b - \Gamma_b)/\partial b$, one can see whether these equilibrium states, as well as the double equilibrium state at $G_a = \Gamma_a$ and $G_b = \Gamma_b$, are stable or unstable. At $b < a$ the critical condition is first attained at $G_b = \Gamma_b$, i.e., the crack begins to grow with respect to the minor semi-axis. This fact is known in fracture mechanics. It follows, in particular, from the comparison of stress intensity factors at the ends of semi-axes, and is in agreement with the comparison theorems of linear fracture mechanics [61].

In the study of fatigue crack growth, we again return to the problem of stress distribution ahead of the crack. The approximate approach discussed in Section 5.1 is useful again. Let us assume that equation (5.5) is valid, as a sound approximation, for the tensile stresses $\sigma_z(\xi, \beta)$ in the plane $z = 0$ surrounding an elliptical crack. We interpret ξ as the distance measured from the crack front in the direction of the external normal to this front (see Figure 5.12). Combining equations (5.4), (5.5) and (5.13), we obtain

$$\Delta\sigma_z(\xi, \beta) = \frac{2\Delta\sigma_\infty}{\mathbf{E}(k)} \left(\frac{b}{\rho}\right)^{1/2} \frac{[\sin^2\beta + (b^2/a^2)\cos^2\beta]^{1/4}}{[1 + (4\xi/\rho)]^{1/2}}$$

This equation will be used later to evaluate microdamage accumulated around the crack under cyclic tension.

The next step is to calculate the virtual fracture work

$$\delta W_f = - \int\limits_0^{2\pi} \gamma(\beta) \delta n(\beta) ds \tag{5.17}$$

Here $\gamma(\beta)$ is the specific fracture work at the crack tip taking into account micro-damage, $\delta n(\beta)$ is the virtual increment in the normal direction to the crack contour due to crack extension, and $ds = (a^2 \sin^2 \beta + b^2 \cos^2 \beta)^{1/2} d\beta$ is the elementary arc length. Noting that

$$\delta x = -a \delta \beta \sin \beta + \delta a \cos \beta = \delta n \cos \theta$$

$$\delta y = b \delta \beta \cos \beta + \delta b \sin \beta = \delta n \sin \theta$$

where θ is the angle between the normal n and x-axis (Figure 5.12), we obtain

$$\delta n = \frac{b \delta a \cos^2 \beta + a \delta b \sin^2 \beta}{(a^2 \sin^2 \beta + b^2 \cos^2 \beta)^{1/2}}$$

$$\delta \beta = \frac{(a \delta a - b \delta b) \sin \beta \cos \beta}{a^2 \sin^2 \beta + b^2 \cos^2 \beta}$$

Hence, equation (5.17) takes the form

$$\delta W_f = - \int\limits_0^{2\pi} \gamma(\beta)(b \delta a \cos^2 \beta + a \delta b \sin^2 \beta) d\beta$$

The generalized resistance forces are determined as follows:

$$\Gamma_a = \int\limits_0^{2\pi} \gamma(\beta) b \cos^2 \beta d\beta, \qquad \Gamma_b = \int\limits_0^{2\pi} \gamma(\beta) a \sin^2 \beta d\beta \tag{5.18}$$

When $\gamma = \text{const}$ along all the crack contour, we come to equation (5.16). Generally, the specific fracture work is nonuniformly distributed around the contour. When $b < a$, the stress intensity factor, according to (5.13), takes larger magnitudes at points situated closer to the y-direction. This results in higher stresses $\sigma_z(\xi, \beta)$ in that domain and, therefore, in higher rates of microdamage accumulation. Similar to quasi-brittle fracture, fatigue cracks of elliptical shape tend to propagate in the direction of the minor semi-axis.

5.6   Elliptical Crack Growth in Fatigue

Hereafter equation (4.2) is used for the process of damage accumulation, equation (4.17) for the evolution of the effective tip radius around the crack contour, and equation (4.6) for the specific fracture work as a function of accumulated damage. The numerical procedure is organized as follows. The initial crack dimensions a_0

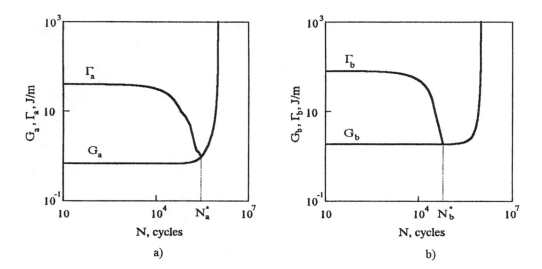

FIGURE 5.14
Relationship between generalized forces as referred to semi-axes a and b during crack initiation and growth.

and b_0 correspond to a subequilibrium state of the system, i.e., $\delta W < 0$ at $N = 0$. Then $G_a < \Gamma_a$, $G_b < \Gamma_b$, where G_a and G_b are given in (5.12), Γ_a and Γ_b in (5.18). The first attainment of the equality $G_a = \Gamma_a$ or $G_b = \Gamma_b$ signifies the start of crack growth with respect to the corresponding semi-axis. Choosing small steps of semi-axes Δa and Δb, we replace a with $a + \Delta a$ (in case $G_a = \Gamma_a$) or b with $b + \Delta b$ (in case $G_b = \Gamma_b$). The computations are continued until the next attainment of equality of generalized forces. The sequence of cycle numbers at each step is recorded. The steps Δa and Δb are to be sufficiently small to allow smoothing with respect to all variables of interest, $a(N)$ and $b(N)$ in particular. The procedure stops when, beginning from a certain cycle number N_*, the increase of a or b results in the relationship $G_a > \Gamma_a$ or $G_b > \Gamma_b$. This means the attainment of an unstable state of the system, i.e., quasi-brittle fracture with noncontrolled crack propagation. Instability occurs also under the conditions $G_a = \Gamma_a$, $\partial G_a/\partial a > \partial \Gamma_a/\partial a$ or $G_b = \Gamma_b$, $\partial G_b/\partial a > \partial \Gamma_b/\partial a$.

The following numerical data are taken: $E = 200$ GPa, $\nu = 0.3$, $\gamma_0 = 25$ kJ/m^2, $\rho_s = 10$ μm, $\rho_b = \lambda_\rho = 100$ μm, $\sigma_d = 15$ GPa, $\Delta\sigma_{th} = 250$ MPa, $\omega_* = 1, m = 4$, $n = 0$, $\alpha = 1$. The loading regime is given with the remote tensile stress range $\Delta\sigma_\infty = 300$ MPa at $R = 0.2$.

The relationship between generalized forces in the process of damage accumulation and crack propagation is illustrated in Figure 5.14 for $a_0 = 1$ mm, $b_0 = 1$ mm, $\rho = 50$ μm. The crack growth initiates in the direction of the minor semi-axis at $N_{b*} = 0.28 \cdot 10^6$. The growth in the direction of the major semi-axis begins much later, at $N_{a*} = 0.65 \cdot 10^6$. A little "unsteady" behavior of $\Gamma_a(N)$ can be observed after the crack begins to grow with respect to other generalized coordinates. It is

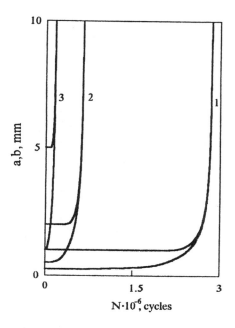

FIGURE 5.15
Time histories of semi-axes of elliptical cracks: (1) $a_0 = 1$ mm, $b_0 = 0.25$ mm; (2) $a_0 = 2$ mm, $b_0 = 0.5$ mm; (3) $a_0 = 5$ mm, $b_0 = 1$ mm

a result of complex interaction between the variables entering governing equations. When the ratios of a and b, da/dN and db/dN vary, the distribution of effective curvatures along the crack perimeter varies also, and that affects the microdamage accumulation rate. As a result, the relationship between generalized forces changes. In turn, this affects growth rates, etc. These complex phenomena can be observed directly when we follow the history of internal variables ρ and ψ in functions both of the cycle number and the elliptical angle.

The time histories of semi-axes $a(N)$ and $b(N)$ for various initial crack sizes are presented in Figure 5.15. Curves 1, 2, 3 are drawn for $a_0 = 1$ mm and $b_0 = 0.25$ mm, $a_0 = 2$ mm and $b_0 = 0.5$ mm, $a_0 = 5$ mm and $b_0 = 1$ mm, respectively. During growth, all cracks tend to the circular shape, i.e., to $a(N) \approx b(N)$. This fact, being similar to that in the quasi-brittle fraction under monotonous loading, is also illustrated in Figure 5.16. In the latter case lines 1, 2, 3 correspond to initial ratios $a_0/b_0 = 1$, 2, and 5, respectively.

In Figure 5.17 the crack growth rate diagrams similar to those in Figures 5.3–5.6, et al. are given. There the rate da/dN is plotted versus the range ΔK_a of $K_a = K(0)$ and the rate db/dN versus the range ΔK_b of $K_b = K(\pi/2)$. Here $K(\beta)$ is defined in (5.13).

It is of interest to compare the diagrams for da/dN and db/dN plotted together. Such a comparision is presented in Figure 5.18 for two cases: $a_0 = 1$ mm, $b_0 = 0.25$ mm, and $a_0 = 2$ mm and $b_0 = 0.5$ mm. The curves exhibit a significant divergence at the initial stage only. Later on they tend to merging. Equations

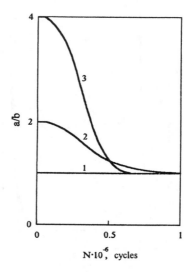

FIGURE 5.16
Time histories of semi-axis ratios: (1) $a_0/b_0 = 1$; (2) $a_0/b_0 = 2$; (3) $a_0/b_0 = 5$.

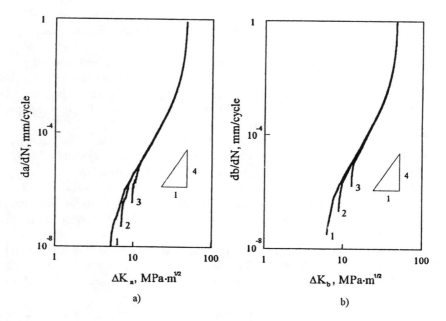

FIGURE 5.17
Growth rate diagrams for semi-axes (notations as in Figure 5.15).

$$\frac{da}{dN} = c_a \Delta K_a^{m_p}, \frac{db}{dN} = c_b \Delta K_b^{m_p}$$

are frequently used in the engineering analysis of elliptical and semi-elliptical cracks. The above numerical results may be considered as indirect evidence of consistency of such an approach.

It is of interest to follow the time history of variables that enter the theory as essential components but cannot be measured directly. Among them there are the effective tip radius ρ and the tip damage measure ψ. Figure 5.19 presents the diagrams for ρ and ψ in the function of N at $a_0 = 1$ mm, $b_0 = 0.5$ mm, $\rho_0 = 50$ μm. Lines 1 are plotted at $\beta = 0$, i.e., for the end of the major semi-axis, lines 2 for $\beta = \pi/4$, and lines 3 for $\beta = \pi/2$. During the initiation stage, both ρ and ψ are monotonously growing functions of N. The crack growth start occurs along the minor semi-axis, and then the effective tip radius begins to decrease tending to the "sharp" radius ρ_s. The start of growth at the major semi-axis (line 1) begins essentially later. The character of function $\psi(N)$ differs in this case from that at $\beta = \pi/2$. First, the damage level is close to $\psi = 1$ on a significant portion of fatigue life. Second, one can observe nonmonotonous change of the damage measure. This observation is rather unexpected. This phenomenon may be explained by the complex interaction of several factors which are partially competitive. Actually, the crack growth is accompanied with the redistribution of stress around the crack and the crack tip sharpening. This affects the process of damage accumulation which,

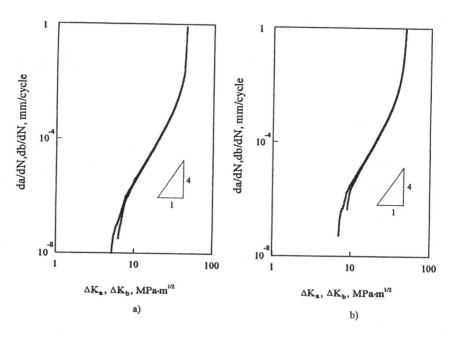

FIGURE 5.18
Combined growth rate diagrams: (a) at $a_0 = 1$ mm, $b_0 = 0.25$ mm; (b) at $a_0 = 2$ mm, $b_0 = 0.5$ mm.

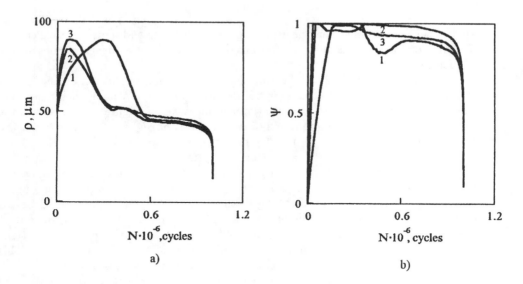

FIGURE 5.19
Effective tip radius (a) and tip damage measure (b) at $\beta = 0, \pi/4$ and $\pi/2$ (lines 1, 2, and 3, respectively).

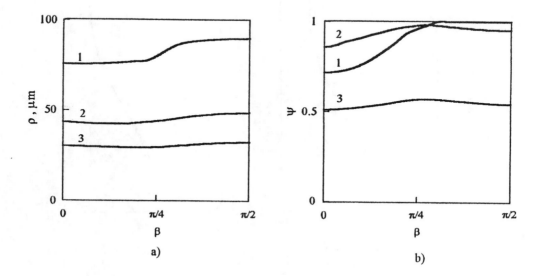

FIGURE 5.20
Angular distribution of (a) effective tip radius and tip damage measure (b) at $N \cdot 10^{-6} = 0.1, 0.45,$ and 0.95 (lines 1, 2, and 3, respectively).

in its turn, controls the process of crack growth. For completeness, line 3 shows the processes at an intermediate point of the crack front.

The studied above phenomena are present in other fatigue problems with two and more G-coordinates. However, their influence on the process of crack growth, moreover measured and discussed in terms of crack growth rate, is insignificant as one may observe in Figures 5.15-5.18. To complete the picture, the angular distribution is shown in Figure 5.20 of the effective tip radius ρ and the tip damage Ψ at three consecutive magnitudes of the cycle number. This distribution corresponds to the initial sizes $a_0 = 1$ mm, $b_0 = 0.5$ mm; other numerical data are the same as in Figures 5.15-5.19.

5.7 Planar Cracks of Arbitrary Shape

Consider a generalization of the problems treated in Sections 5.3-5.6. It is a planar crack of an arbitrary smooth shape in an unbounded body under tension across the crack plane (Figure 5.21). Introducing the polar coordinates r, φ, we describe the crack boundary with the function $r = R(\varphi)$. We may consider $R(\varphi)$ as a continuous set of G-coordinates when φ varies in $[0, 2\pi]$. Then small increments $\delta R(\varphi) \geq 0$ take the part of G-variations. A different approach is to consider displacements $\delta n(s)$ of the boundary in the direction of the external normal. Here s is an arc measured along the boundary. Then we treat $\delta n(s)$ as G-variations. However, $\delta R(\varphi)$ and $\delta n(s)$ are in a simple geometrical connection when the equation of the boundary is

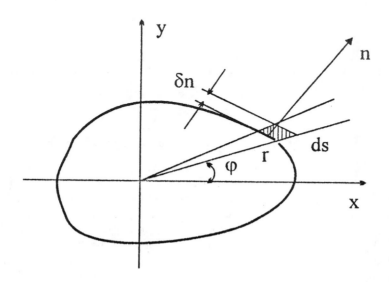

FIGURE 5.21
Planar crack of arbitrary shape and variation of its boundary.

known.

The virtual work performed by external and internal forces on G-variations is

$$\delta W_e + \delta W_i = \int_S G(s)\delta n(s)ds \qquad (5.19)$$

Here $G(s)$ is the driving generalized force, and the integral is taken over all the crack boundary S. The right-hand side in (5.19) can be presented also as follows:

$$\delta W_e + \delta W_i = \int_0^{2\pi} G(\varphi)R(\varphi)\delta R(\varphi)d\varphi \qquad (5.20)$$

It is evident that $G(s)$ and $G(\varphi)$ are the energy fluxes through a unit arc of the crack front. They can be evaluated using Irwin's equation.

Similar to (5.19), the virtual fracture work is determined as follows:

$$\delta W_f = -\int_S \gamma(s)\delta n(s)ds \qquad (5.21)$$

Here $\gamma(s)$ is the specific fracture work. In fatigue the influence of microdamage accumulated ahead of the crack must be taken into account. The typical formula for $\gamma(s)$ is

$$\gamma(s) = \gamma_0[1 - \psi^\alpha(s)] \qquad (5.22)$$

where γ_0 is the specific fracture work for the nondamaged material, $\psi(s)$ is the damage measure at the boundary point given with coordinate s, and $\alpha > 0$. Similar to (5.19), one may replace equation (5.21) for the following:

$$\delta W_f = -\int_0^{2\pi} \gamma(\varphi)R(\varphi)\delta R(\varphi)d\varphi \qquad (5.23)$$

The system cracked body-loading will be in a sub-equilibrium, equilibrium or non-equilibrium state depending on the sign in the relationship

$$\delta W_e + \delta W_i + \delta W_f \gtrless 0$$

for every $\delta n(s) \geq 0$ or $\delta R(\varphi) \geq 0$. Using equations (5.20) and (5.23), we obtain

$$G(\varphi) \gtrless \Gamma(\varphi) \qquad (5.24)$$

There $\Gamma(\varphi) \equiv \gamma(\varphi)$ and $\gamma(\varphi)$ are defined similarly to (5.22).

From the first look, equation (5.24) seems to be a local criterion because it connects generalized forces at each point of the crack boundary. But, in fact, it is not so because both $G(\varphi)$ and $\Gamma(\varphi)$ are functionals of the crack shape. In particular, the local strain energy release rate and the local damage measure depend on the stress-strain field in the whole body. On the other side, equation (5.24) describes the local behavior of cracks. At the points where $G(\varphi) < \Gamma(\varphi)$ the crack does not propagate, and at the points where $G(\varphi) \geq \Gamma(\varphi)$ it grows (in a continuous or jump-wise way).

Unfortunately, the above described scheme needs information on stress-strain field around a crack of arbitrary shape, and it is a three-dimensional problem of elasticity theory. The number of available analytical solutions is limited. Perturbation technique allows finding solutions in the vicinity of cracks which slightly deviate from a regular shape. A number of problems concerning, in particular, the initially planar cracks with small kinking and branching are discussed in the literature on fracture mechanics. However, the aim of those studies is to evaluate the stress intensity factors while in the mechanics of fatigue we are interested in the stress distribution around cracks with finite tip curvature as well as in the stress in the far field. Some results of fracture mechanics will be used later in the analysis of kinking and branching of fatigue cracks; but such an approach requires additional assumptions.

5.8   Mixed-Mode Fatigue Cracks

Under arbitrary loading and/or in the case of arbitrary geometry, fatigue cracks propagate in mixed modes. Putting aside three-dimensional problems, let us consider fatigue crack growth in the two-dimensional case. The typical problems are plane-stress or plane-strain states with applied cyclic stresses $\sigma_x^\infty(t), \sigma_y^\infty(t), \tau_{xy}^\infty(t)$. Even a plate under remote uni-axial cyclic tension enters this class when a crack is inclined with respect to the direction of tension.

Two kinds of complications are met in connection with mixed-mode fatigue cracks. First, such cracks generally do not propagate in their initial plane. Their trajectories are not known beforehand and must be found in the process of solution. Second, the microdamage associated with mixed-mode fatigue cracks has a more complicated nature than that in the case of planar mode I cracks. In particular, both tensile and shear microcracking ought to be taken into account. These types of microdamage, generally, interact; that requires a widespread generalization of common models of continuum damage mechanics.

The simplest problem is when a mixed-mode crack propagates in its initial plane, i.e., without kinking or branching. It is, for example, the case of cracking between two bodies whose plane surfaces are glued or welded with a layer that is weaker than the materials of both bodies. The properties of these materials may be similar as well as dissimilar. Another example is provided by cracks in strongly anisotropic, e.g., orthotropic, materials. Then the crack configuration is prescribed by the material properties. In both types of problems, we consider an arbitrary two-dimensional cyclic loading. It results in the necessity of introducing, along with microcracking in tensile mode, microcracking in shear mode also.

To demonstrate the main features of this kind of problem, consider a planar crack between two bodies of the same material with Young's modulus E and Poisson's ratio ν. Let the crack be single-parametric, and its size given with length $2a$ (Figure 5.22). The first step is to evaluate the generalized forces G and Γ as functions of the current loading as well as functionals of loading and crack growth histories.

The stress field in the vicinity of the crack tip is characterized by stress intensity

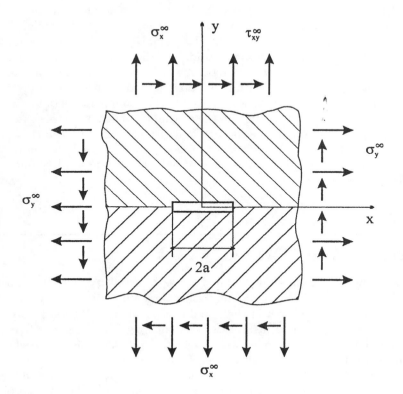

FIGURE 5.22
Mixed mode planar crack.

factors

$$K_I = Y_I \sigma_y^\infty (\pi a)^{1/2}, K_{II} = Y_{II} \tau_{xy}^\infty (\pi a)^{1/2} \tag{5.25}$$

with geometric correction factors Y_I and Y_{II} of the order of unity. Let us neglect the contribution of microdamage in the generalized driving force. Then Irwin's equation may be applied

$$G = \frac{K_I^2 + K_{II}^2}{\eta E} \tag{5.26}$$

Here $\eta = 1$ for the plane-stress state and $\eta = (1 - \nu^2)^{-1}$ for the plain-strain state.

The dominating factors affecting the microdamage accumulation near the tips are stresses σ_y and τ_{xy} on the prolongation of the crack, i.e., at $|x| \geq a, y = 0$. The normal stress σ_y produces opening mode microcracks, and the tangential stress τ_{xy} produces transverse shear mode microcracks.

Without going into further interpretation, let us introduce two microdamage measures, ω_σ and ω_τ. Measure ω_σ characterizes the level of mode I microcracking, and measure ω_τ that of mode II microcracking. It is natural to attribute ω_σ to the range $\Delta\sigma_y$ of $\sigma_y(t)$ within a cycle, and ω_τ to the range $\Delta\tau_{xy}$ of $\tau_{xy}(t)$. Later on, equations similar to (4.2) are used

$$\frac{\partial\omega_\sigma}{\partial N} = \left(\frac{\Delta\sigma_y - \Delta\sigma_{th}}{\sigma_d}\right)^{m_\sigma}, \qquad \frac{\partial\omega_\tau}{\partial N} = \left(\frac{|\Delta\tau_{xy}| - \Delta\tau_{th}}{\tau_d}\right)^{m_\tau} \qquad (5.27)$$

with material parameters σ_d, τ_d, $\Delta\sigma_{th}$, $\Delta\tau_{th}$, m_σ and m_τ which have a transparent meaning. In general, they depend on the stress ratios $R_\sigma = \sigma_y^{\min}/\sigma_y^{\max}$ and $R_\tau = \tau_{xy}^{\min}/\tau_{xy}^{\max}$. The right-hand sides in these equations must be put to zero at $\Delta\sigma_y < \Delta\sigma_{th}$ or $|\Delta\tau_{xy}| < \Delta\tau_{th}$, respectively. For the generalized resistance force we use the equation

$$\Gamma = \Gamma_0[1 - (\psi_\sigma + \psi_\tau)^\alpha] \qquad (5.28)$$

similar to (4.11).

The assumption is used in equations (5.27) and (5.28) that microdamage is originated from two different sources. Generally, they are mutually connected. One might replace such a model in various ways. For example, one might introduce a single microdamage measure ω. The rate of this measure will depend on a certain effective stress σ_{eff} such as a linear combination of acting stresses (similar to those in Mohr's criterion in the elementary strength analysis). It might also be a combination of the mean normal stress responsible for microporosity and the deviatoric stress responsible for shear yielding. Then instead of (5.27) we obtain

$$\frac{\partial\omega}{\partial N} = \left(\frac{\Delta\sigma_{eff} - \Delta\sigma_{th}}{\sigma_d}\right)^m$$

In the general case, all parameters entering the right-hand side depend on extremal values of σ_x, σ_y and τ_{xy}, as well as on the current value ω. However, such a generalization meets a number of difficulties. Apart from the absence of sufficient experimental data, we have to assume that all stresses vary in time synchronously, or to introduce a number of other ad hoc hypotheses.

The next move is to estimate the stresses acting at the crack tips and their prolongation. The easiest way is to follow equations similar to (5.5) or (5.6). Then the information on stress intensity factors can be used for the approximate evaluation of the stress field. There is some doubt concerning the validity of equation (5.3) for shear stresses. In any case, in most situations we may present the stresses at the tips as

$$\sigma_y \approx Z_\sigma \sigma_y^\infty (a/\rho)^{1/2}, \tau_{xy} \approx Z_\tau \tau_{xy}^\infty (a/\rho)^{1/2}$$

The coefficients Z_σ and Z_τ do not necessarily coincide with the geometry correction factors Y_σ and Y_τ entering equations (5.25).

Let us discuss briefly how to apply the quasistationary approximation approach to this problem. For example, when certain characteristic scales λ_σ and λ_τ exist for tensile and shear process zones, this approach results in equation:

$$\frac{da}{dN} = \frac{\lambda_\sigma\left(\dfrac{\Delta K_I - \Delta K_{Ith}}{K_{Id}}\right)^{m_\sigma} + \lambda_\tau\left(\dfrac{|\Delta K_{II}| - \Delta K_{IIth}}{K_{IId}}\right)^{m_\tau}}{\left[1 - \dfrac{(K_I^2 + K_{II}^2)_{\max}}{K_{IC}^2}\right]^{1/\alpha} - \omega_{ff}(N)} \qquad (5.29)$$

Similar to equation (4.46), the following notation is used in (5.29): ΔK_{Id} and ΔK_{IId} are material parameters characterizing the resistance to microdamage accumulation, ΔK_{Ith} and ΔK_{IIth} are threshold resistance parameters, and K_{IC} is the quasi-brittle fracture toughness.

Equation (5.29) is a natural expansion of equation (4.46) over two-dimensional stress or strain states. The nominator of the right-hand side accounts for microdamage in the process zone. Positiveness of the denominator signifies stability of the system with respect to the quasi-brittle fracture taking into account the microdamage $\omega_{ff}(N)$ accumulated in the far field. Equation (5.29) includes all three stages of fatigue crack growth: slow growth at the ranges of stress intensity factors close to the threshold magnitudes; stationary (in logarithmic scale) growth at moderate ranges; accelerated growth when the crack size approaches its critical magnitude.

When $\omega_{ff}(N) \ll 1$, $\lambda_\sigma/a \ll 1$, $\lambda_\tau/a \ll 1$, the loss of stability takes place at $K_I^2 + K_{II}^2 = K_{IC}^2$. Here we meet a contradiction with experimental data on fracture toughness for most materials. This contradiction remains even for macroscopically isotropic materials. It shows that one has to distinguish the magnitudes of specific fracture work in different modes. A way to account for this difference is discussed in Chapter 10 in an application to laminate composite materials. The leading idea is to distinguish, even within the same process zone, the crack fronts corresponding to tensile and shear modes. It allows introduction of two different magnitudes for specific fracture work, $\gamma_I \neq \gamma_{II}$ and for fracture toughness, $K_{IC} \neq K_{IIC}$. Then instead of (5.29) we obtain

$$\frac{da}{dN} = \frac{\lambda_\sigma \left(\dfrac{\Delta K_I - \Delta K_{Ith}}{K_{Id}} \right)^{m_\sigma} + \lambda_\tau \left(\dfrac{|\Delta K_{II}| - \Delta K_{IIth}}{K_{IId}} \right)^{m_\tau}}{\left[1 - \left(\dfrac{K_I^2}{K_{IC}^2} + \dfrac{K_{II}^2}{K_{IIC}^2} \right)_{\max} \right]^{1/\alpha} - \omega_{ff}(N)}$$

The final failure occurs when the equality

$$\frac{K_I^2}{K_{IC}^2} + \frac{K_{II}^2}{K_{IIC}^2} = 1 - \omega_{ff}^\alpha$$

is primarily attained.

5.9   Nonplanar Crack Propagation

Even under monotonous quasistatic loading, cracks in real macroscopic homogeneous materials do not propagate in a straight, planar way. It is well known that when a plate under tension contains an initially inclined crack or an artificial cut, the trajectory of the further quasi-brittle crack propagation significantly deviates from the initial plane. Similarly, the direction of crack trajectory changes when the applied stresses vary during the process of crack propagation. The change also takes place near a boundary of two dissimilar materials. In addition, small local deviations from the initial crack plane may be caused by local material nonhomogeneities resulting into kinking and branching of cracks. In this case we say on meandering cracks [33].

Several groups of criteria have been suggested in fracture mechanics to predict the start of quasi-brittle fracture under mixed mode monotonous loading. First, an intuitively transparent criterion of maximal circumferential tensile stresses around the crack tip ought to be listed. The equivalent or very close criteria postulate that the angle of the first crack path corresponds to zero circumferential tangential stresses, to maximal stress intensity factor in mode I, or to zero stress intensity factor in mode II for a kinked (slightly curved, branched) crack tip. The second group uses, as criteria variables, energy conditioned parameters. Among them is the maximal strain energy release rate criterion or, what is frequently the same, the maximal J-integral criterion for a crack with a small kink or branch. The third group of criteria does not follow so clearly from physical considerations. Among them the minimum strain energy density criterion proposed by Sih [136] ought to be mentioned. This criterion postulates that a crack begins to propagate in the direction of minimal density of strain energy. All the listed criteria predict, at least for the plane-stress and plane-strain states and the first step of crack growth, rather close numerical results that are in satisfactory agreement with experiments. A survey of the literature can be found in the book by Gdoutos [60] and the paper of Ramulu and Kobayashi [121].

Consider an initially oblique planar crack under uni-axial tension (Figure 5.23). The counter-clockwise angles θ and φ are treated as positive, and in Figure 5.23 at $\theta > 0$ we have $\varphi < 0$. Some predicted results are presented in Figure 5.24. There the first angle of kinking φ is plotted against the initial angle θ with the use of various criteria. The range of available experimental data is also shown in Figure 5.24. The number of criteria might be increased. In particular, criteria can be formulated for initial (planar) cracks as well as for slightly kinked cracks.

To estimate the crack growth rate under cyclic loading, the well-known semi-empirical equations are widely used. For example, using the Paris-Erdogan equation and maximal circumferential stress criterion, we obtain that the fatigue crack growth rate is

$$\frac{da}{dN} = c \left(\Delta K_{eff} \right)^m \tag{5.30}$$

Here c and m are empirical constants, and ΔK_{eff} is the range of the mode I stress intensity factor within a cycle for the kinked crack, or a certain equivalent value. For example, the following effective stress intensity range has been used

$$\Delta K_{eff} = \left(\Delta K_I^4 + 8 \Delta K_{II}^4 \right)^{1/4}$$

with the ranges ΔK_I and ΔK_{II} in modes I and II. When the maximum strain energy release criterion is used, we obtain instead of (5.30)

$$\frac{da}{dN} = c_1 \left(\Delta G_{eff} \right)^{m_1}$$

where ΔG_{eff} is the range of the strain energy release rate for the kinked crack, c_1 and m_1 are empirical constants. In the case of the minimum strain energy density criterion, similarly, we have

$$\frac{da}{dN} = c_2 \left(\Delta S \right)^{m_2}$$

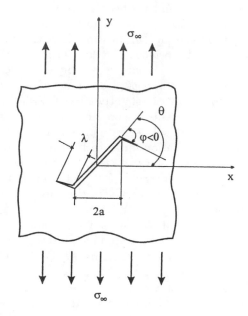

FIGURE 5.23
Initially oblique crack under uni-axial tension.

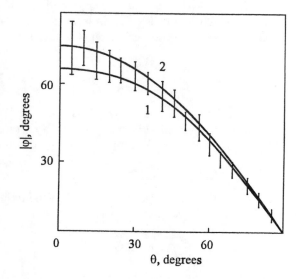

FIGURE 5.24
Angle of first kinking of oblique cracks under monotonous tension according to criteria: (1) maximal normal stress; (2) maximal strain energy release rate.

with the range ΔS of the strain energy density in the direction of expected crack growth. Analogous equations can be written using Forman's equation and its generalizations. A recent survey of models for mixed mode fatigue crack propagation has been performed by Ramulu and Kobayashi [121]. That paper also contains results of experimental analysis and the comparison with various theoretical predictions.

A theory of fatigue crack propagation based on analytical fracture mechanics with application to mixed mode cracks has been suggested by Bolotin [14]. Additional details and references to earlier publications can be found in the book [19] and in the survey paper [21].

From the viewpoint of analytical fracture mechanics, the statement of the problem involves a continuous set of G-coordinates. For example, in the two-dimensional case the tip advancement $da(\varphi)$ and the corresponding variation $\delta a(\varphi)$ must be considered as functions of the kinking angle φ with magnitudes from $[-\pi, \pi]$. This angle plays the same part as the subscripts $1, \ldots, m$ at G-coordinates $a_1, \ldots a_m$ when the number of G-coordinates is bounded. If a problem is solved numerically, which is practically the only way, the first tip advancement is a finite step Δa with a certain kinking angle φ. After that a new problem arises to find the new angle and the new tip advancement. As to the first kinking or branching, there are a number of approximate solutions of the corresponding problems of elasticity theory (see, e.g., the survey by Gdoutos [60]).

It is understandable that, staying in the framework of the theory of fatigue developed in this book and considering the material as elastic, we will follow the path that is related to the criterion of maximal strain energy release rate. The essential difference is accounting for microdamage accumulated along the expected crack trajectories.

5.10 First Kinking of Fatigue Cracks

A typical fatigue crack trajectory initiated from an oblique edge notch is depicted in Figure 5.25a. The path of the crack is distinctly curvilinear; it usually begins from the kinking and then the crack gradually approaches its dominant direction (when no change of loading regime is involved). The primary problem is to evaluate the direction of the first kinking (Figure 5.25b). We can use results obtained in linear fracture mechanics where kinking and branching of cracks have been studied extensively. Most of the studies concern the first small deviations from the planar shape of a crack. Beginning from the paper by Banichuk [4], a number of papers have been published where small parameter and related perturbation techniques were used to evaluate stress intensity factors and strain energy release rates for cracks with small kinks [55, 98, 156].

Difficulties arise when we approach the second, third, etc. kinkings, moreover when a crack propagates in a meandering way. In principle, the techniques of computational mechanics [2, 104] allow calculating of stress intensity factors for cracks and crack-like defects of arbitrary shape. But in fatigue the problem is much more complicated due to the necessity for studying stress-strain fields in a wide domain ahead of the crack and for evaluating damage produced by cyclic

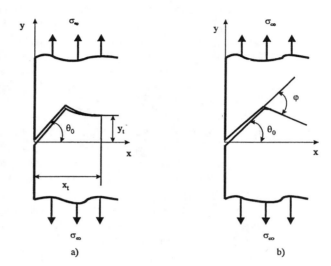

FIGURE 5.25
Oblique crack and its growth: (a) crack trajectory; (b) first kinking.

loading. In this section we briefly discuss the problem of the first kinking which will be used later to develop an approximate semi-analytical approach to more complicated problems.

Consider an initially oblique crack in the plane-stress state, e.g., an edge crack shown in Figure 5.25b. Let the applied stress $\sigma_\infty(t)$ vary in time cyclically. The maximal stress level is sufficiently low to provide the inequality $G < \Gamma_0$. Here G is the generalized driving force corresponding to maximal applied stress within a cycle, and Γ_0 is the generalized resistance force for the nondamaged material. At a certain cycle number N_* the equality $G = \Gamma$ is primarily attained due to decreasing resistance produced by damage accumulation near the crack tip.

Denote by a_0 the projection of the initial crack size on the x-axis, the initial angle of crack inclination θ_0, and the angle of first kinking φ. All angles are considered as positive when counter-clockwise. Then the angle φ can be found from condition

$$\max\left[G(a_0, \theta_0, \varphi, N) - \Gamma(a_0, \theta_0, \varphi, N)\right] = 0 \qquad (5.31)$$

The condition (5.31) is illustrated in Figure 5.26, where G and Γ are schematically plotted against φ for three various situations, notated 1, 2 and 3. When $\max\{G - \Gamma\} < 0$ (case 1), the system is in a sub-equilibrium state, and the crack does not propagate in any direction. At $\max\{G - \Gamma\} = 0$ that corresponds to the condition given in (5.31), the state is an equilibrium one with respect to G-coordinate a. This state is attained at $\varphi = \varphi_*$ (case 2). When $\max\{G - \Gamma\} > 0$ (case 3), there is an interval of angles (φ_1, φ_2) of unstable crack propagation.

Equation (5.31) is in fact the condition of equilibrium with respect to G-coordinate $a(\varphi)$ where φ is also subjected to variation. The equilibrium state is

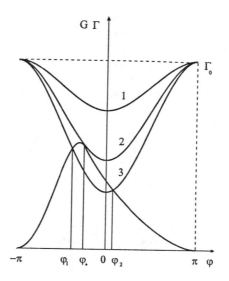

FIGURE 5.26
Relationship between generalized forces for fatigue cracks subjected to kinking; states of the system: (1) sub-equilibrium; (2) equilibrium; (3) non-equilibrium.

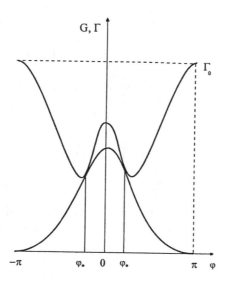

FIGURE 5.27
Crack branching due to entering the higher resistance domain.

stable with respect to $a(\varphi_*)$ if $\partial G/\partial a << \partial\Gamma/\partial a$ at $\varphi = \varphi_*$. Stability with respect to $a(\varphi)$ at other angles is provided by the inequality $G < \Gamma$ at $\varphi \neq \varphi_*$. Equation (5.31) allows prediction of both the angle φ_* of the kinking and the cycle number N_* at this event, i.e., at the first advancement of the crack tip. This equation is also valid for crack branching. A typical situation is illustrated in Figure 5.27 drawn for a transverse crack $(\theta_0 = 0)$ meeting in its propagation a domain with higher resistance to damage accumulation. Two tangent points at $\varphi = \pm\varphi_*$ correspond to the angles of symmetric branching. When the crack propagates along the boundary, we have $\varphi_* = \pi/2$.

The next step is to evaluate the generalized forces, G and Γ. In the case of linear elastic material, the generalized driving force G is just equal to the strain energy release rate for a crack with a small kink. Before kinking the stress intensity factors are

$$K_I = Y_I(\theta)\sigma_\infty(\pi a)^{1/2}, \qquad K_{II} = Y_{II}(\theta)\sigma_\infty(\pi a)^{1/2} \qquad (5.32)$$

with correction factors Y_I and Y_{II} (subscript at θ_0 is omitted). These factors depend on the initial angle θ and, maybe, on the normalized lengths, such as the ratio a/w for a plate of finite width w.

When an oblique crack receives a small kink with length $\lambda << a$, the stress intensity factors may be presented as follows:

$$k_I(a,\theta,\varphi) = K_I(a,\theta)f_{11}(\varphi) + K_{II}(a,\theta)f_{12}(\varphi)$$
$$(5.33)$$
$$k_{II}(a,\theta,\varphi) = K_I(a,\theta)f_{21}(\varphi) + K_{II}(a,\theta)f_{22}(\varphi)$$

Here $K_I(a,\theta)$ and $K_{II}(a,\theta)$ are the stress intensity factors defined in (5.32). Functions $f_{jk}(\varphi)$ depend on the angle of kinking from the initial plane (Figure 5.28). These functions have been estimated by a number of authors cited above. Some functions $f_{jk}(\varphi)$ can take negative values. Therefore, the right-hand side of the first of equations (5.33) may become negative. It means crack closure, and in that case the right-hand side must be put to zero. The sign of $K_{II}(a,\theta,\varphi)$, evidently, does not matter.

The functions $f_{jk}(\varphi)$ appear to be related to the Inglis-Williams functions entering the asymptotic equations for circumferential stresses σ_n, τ_n in the vicinity of tips:

$$\sigma_n(a,\theta,\varphi) = \frac{1}{(2\pi r)^{1/2}}[K_I(a,\theta)g_{11}(\varphi) + K_{II}(a,\theta)g_{12}(\varphi)]$$
$$(5.34)$$
$$\tau_n(a,\theta,\varphi) = \frac{1}{(2\pi r)^{1/2}}[K_I(a,\theta)g_{21}(\varphi) + K_{II}(a,\theta)g_{22}(\varphi)]$$

The angular functions g_{jk} are even numerically close to the corresponding functions in (5.33). The comparison of numerical results given by equations (5.33) and (5.34) is presented in [156].

Using Irwin's equation for the energy release rate, we may assume that similar to (5.26)

$$G = \frac{k_I^2(a,\theta,\varphi) + k_{II}^2(a,\theta,\varphi)}{\eta E} \qquad (5.35)$$

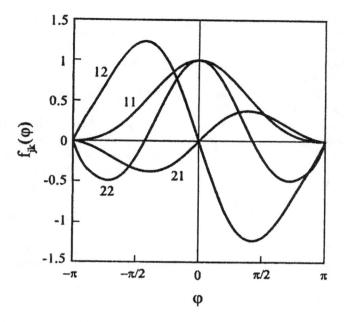

FIGURE 5.28
Angular functions for cracks with small kinking; figures at curves correspond to the subscripts at $f_{jk}(\varphi)$.

Here, compared with (5.26), the stress intensity factors $k_I(a, \theta, \varphi)$ and $k_{II}(a, \theta, \varphi)$ for the kinked crack are used.

Rigorously, Irwin's equation is invalid for kinked, branched and curvilinear cracks. However, the direct numerical analysis shows that equation (5.35) gives paradoxically good results even for large angles of kinking. In the paper by Wu [156] it was stated that the error of equation (5.35) does not exceed 0.2% even for cracks with kinking up to $\varphi = 90°$. A similar situation takes place in many other cases showing surprisingly high robustness of the tools of linear fracture mechanics even if they are used beyond the initially stated boundaries.

In any case, we assume that equation (5.35) is valid, at least as a good approximation, for fatigue cracks when the material is linear elastic and microdamage affects fracture toughness only. As to the generalized resistance force, we assume that it is given in equation (5.28). Thus, this force depends on the two damage measures, ω_σ and ω_τ. These measures are to be evaluated by equations similar to (5.27) where $\sigma_y(x)$ and $\tau_{xy}(x)$ are replaced by $\sigma_n(\varphi)$ and $\tau_n(\varphi)$, respectively:

$$\frac{\partial \omega_\sigma}{\partial N} = \left(\frac{\Delta \sigma_n - \Delta \sigma_{th}}{\sigma_d} \right)^{m_\sigma} , \qquad \frac{\partial \omega_\tau}{\partial N} = \left(\frac{|\Delta \tau_n| - \Delta \tau_{th}}{\tau_d} \right)^{m_\tau} \qquad (5.36)$$

As in the case of planar crack propagation, equations (5.27) and (5.36) allow various, among them far-reaching, generalizations. We will not go into this domain now.

To evaluate the stress field in the vicinity of the tip of a kinking crack, several approaches can be suggested. In [14] a combination has been used of a Neuber's type formula for the stress concentration factors near a notch and the Inglis–Williams equations. Another way is to treat the crack as a narrow elliptical slit with a given curvature radius ρ at the tip. Then the well-known solutions of elasticity theory can be used. Here we follow the latter approach.

Equations that calculate the circumferential stresses $\sigma_n(\varphi)$ and $\tau_n(\varphi)$ at arbitrary angles θ and φ are rather cumbersome and are omitted here. They can be found in many manuals on elasticity theory.

One of the most interesting points in question is the comparative contribution of tensile and shear microcracking in the first kinking. Evidently crack behavior strongly depends on the ratio of the resistance to shear and tensile microdamage. Hence, the ratio τ_d/σ_d is of the main importance here. The ratios equal to 0.5 and $0.573 \approx 3^{-1/2}$ correspond to common ratios of yield stresses in shear and uniaxial tension assumed in classical plasticity theory.

As an illustrative example let us consider a central crack with the initial half-length (in the projection on x-axis) $a_0 = 1$ mm inclined to this axis at $\theta = 45^0$. The plate is subjected to the cyclic tensile stress $\sigma_\infty^{\max} = 150$ MPa, $R = 0.2$. The width of the plate is large compared with $2a_0$. The material parameters are: $E = 200$ GPa, $\nu = 0.3, \gamma_0 = 30$ kJ/m^2, $\sigma_d = 10$ GPa, $\Delta\sigma_{th} = \Delta\tau_{th} = 100$ MPa, $m_\sigma = m_\tau = 4$, $\rho = 50$ μm, $\alpha = 1$.

Before going into a detailed numerical analysis, consider briefly the influence

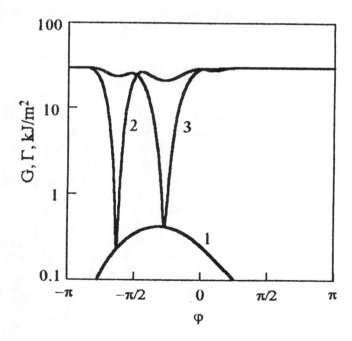

FIGURE 5.29
Angular distribution of generalized forces at first kinking curves 2 and 3 correspond to $\tau_d/\sigma_d = 0.25, 0.5$.

of τ_d/σ_d. The angular distribution of generalized forces $G(\varphi)$ and $\Gamma(\varphi)$ at the moment of crack growth start is shown in Figure 5.29. The driving force $G(\varphi)$ is drawn with line 1. Lines 2 and 3 correspond to the ratios τ_d/σ_d equal to 0.25 and 0.5. When this ratio is equal to 0.25, the crack begins to propagate at the angle φ_* that is near $\varphi = -60°$. This magnitude is very close to the angle of initial crack propagation in quasibrittle fracture under monotonous loading. When $\tau_d/\sigma_d = 0.5$, i.e., the resistance of the material to shear microcracking is high, the crack begins to propagate at the angle which is closest to $\varphi = 0$ (line 3). Note that the two cases demonstrated in Figure 5.29, in general, correspond to materials with various mechanical properties.

5.11 Prediction of Fatigue Crack Trajectories

In the general case, trajectories of mixed mode fatigue cracks are not easy to predict. In principle, the current fields of stresses, strains and microdamage must be found at each small increment of crack size. This requires a very large amount of laborious computations. To simplify the problem, a set of additional assumptions is expedient. First, we might try to replace the curvilinear crack (Figure 5.30a) by a planar crack (Figure 5.30b) or piece-planar one, e.g., a two-link crack (Figure 5.30c). Then the available information concerning stress intensity factors can be

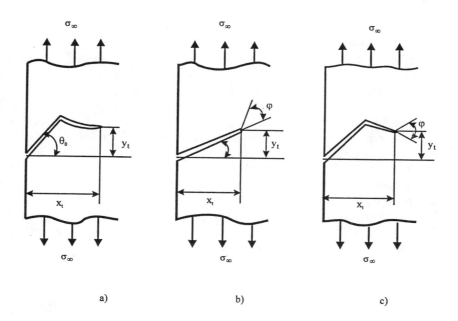

FIGURE 5.30
Meandering crack (a) and its one-link (b) and two-links (c) approximations.

used to estimate the generalized driving force. Second, a special assumption might be introduced about the stress distribution in the vicinity of the crack tip. Similar to the problem of first kinking, various ad hoc assumptions are to be used. We are interested mostly in the near field which is controlled, in the first place, by stress intensity factors. Of course, all opportunities are open for the application of computational mechanics, up-to-date software for stress analysis, etc. In this book we avoid, as far as possible, the use of direct computational techniques.

Later on, let us apply the approximation shown in Figures 5.30b. Namely, we replace the curvilinear crack with a plane oblique crack having small additional increments arbitrarily inclined with respect to the main part of the crack. This replacement must be made at each step of computation. Therefore, the main crack length and the angle of inclination depend on the current cycle number N. In particular, for an edge crack with tip coordinates x_t, y_t we obtain that the length of the main crack a and its inclination angle θ are

$$a = (x_t^2 + y_t^2)^{1/2}, \qquad \theta = \tan^{-1}(y_t/x_t)$$

Additional increments are inclined to the plane of the main crack under the angle φ (Figure 5.30b). For a central crack with the tip coordinates x_t^+, y_t^+ and x_t^-, y_t^- we obtain the half-length a and the angle θ as follows:

$$a = \tfrac{1}{2}\left[(x_t^+ - x_t^-)^2 + (y_t^+ - y_t^-)^2\right]^{1/2}$$

$$\theta = \tan^{-1}\left[(y_t^+ - y_t^-)/(x_t^+ - x_t^-)\right]$$

A similar "straight line approximation" has been used by Sih and Barthelemy [136] in the framework of the S-approach to predict the path of quasi-brittle fracture under monotonous loading; however, the crack has been presented as a mathematical cut. Comparison to experiments can be found in [60] that demonstrate a rather good agreement. In this study, we schematize fatigue cracks as inclined narrow elliptical slits with fixed tip curvatures. The contribution of far field damage is assumed negligible. To facilitate computations, we assume the size λ of the process zone and the effective tip radius ρ constants. Therefore, minimal numerical data are required to close the system of governing equations and to perform computations.

The numerical procedure consists of the following steps. Using equations (5.32) and (5.33) for the stress intensity factors $k_I(a, \theta, \varphi), k_{II}(a, \theta, \varphi)$, the generalized driving force (5.35) is computed for each position of the crack tip. The acting circumferential stresses $\sigma_n(r, \varphi)$ and $\tau_n(r, \varphi)$ are calculated within the process zone for each magnitude of the local polar radius r and the polar angle φ. The next step is to evaluate the damage measures $\omega_\sigma(r, \varphi)$ and $\omega_\tau(r, \varphi)$ using equations (5.36). Then equation (5.28) is applied to calculate the generalized resistance force. The process of computations proceeds at increasing cycle number N until the condition

$$\max\left[G(a, \theta, \varphi, N) - \Gamma(a, \theta, \varphi, N)\right] = 0 \tag{5.37}$$

is satisfied. The corresponding cycle number N and the angle φ are recorded; the crack is prolongated in the angle φ at the distance $\Delta r = \lambda$. Then coordinates of the new tip position are found, and the procedure based on equations (5.32), (5.33) and (5.35)-(5.37) must be repeated for a newly schematized crack.

The numerical results for an initially inclined central crack in a wide plate are presented in Figures 5.31–5.34. The material parameters are taken as follows: $E = 200$ GPa, $\nu = 0.3$, $\rho = 50$ μm, $\gamma_0 = 30$ kJ/m^2. The latter magnitude corresponds to the fracture toughness $K_{IC} \approx 77$ MPa $\cdot$ m$^{1/2}$. In equations (5.28) and (5.36) we assume that $\sigma_d = 10$ GPa, $\tau_d = 5$ GPa, $\Delta\sigma_{th} = \Delta\tau_{th} = 100$ MPa, $m_\sigma = m_\tau = 4, \alpha = 1$. The plate with the initial crack is subjected to tensile stress with the range $\Delta\sigma_\infty = 150$ MPa at $R = 0.2$. The initial crack size is $a_0 = 1$ mm, and the initial inclination angle $\theta_0 = 45°$.

Figure 5.31 shows the relationship between the generalized forces at three states. The crack growth start takes place at the first contact of curves corresponding to G and Γ (Figure 5.31a). The angle of first kinking is negative which means that the crack begins to grow in its primary direction. Then the angle of kinking is approaching zero (Figure 5.31b) until the final failure (Figure 5.31c).

The crack trajectory is shown in Figure 5.32 when x_t and y_t are current tip coordinates. Note that the different scales are used along the horizontal and vertical axes. It is seen that the crack changes direction of growth almost immediately after the growth starts.

Additional information is given in Figure 5.33 where the tip coordinates are plotted against the cycle number. The crack starts to grow at $N_* \approx 4.5 \cdot 10^6$ cycles and the final failure takes place at $N_* \approx 2.5 \cdot 10^6$ cycles. Looking at Figures 5.32 and 5.33, one can observe that the oblique crack rapidly transforms in the process of growth in a crack close to the mode I.

The influence of the initial angle θ_0 is illustrated in Figure 5.34. The initial conditions are $x_t = 1$ mm at four angles $\theta_0 = 15$, 30, 45, and 60° (lines 1, 2, 3, 4, respectively). Figure 5.34a shows the crack trajectories which shape changes considerably when the initial crack slope increases. The corresponding growth rate diagrams are shown in Figure 5.34b, where dx_t/dN is used as a measure of growth rate. The stress intensity factor range is taken as $\Delta K = \Delta\sigma_\infty(\pi a)^{1/2}$ where $a(N) = x_t(N)$. This means that the diagram as a whole is referred to as the x-projection of the current crack size. Obviously, the obtained diagram is similar to those for the mode I cracks (see, for example, Figures 5.3-5.5). The lines in Figure 5.34b differ only at their initial sections.

Edge oblique cracks, in general, are more interesting from the viewpoint of applications than central cracks. Actually, damage due to fatigue usually originates from edge cracks that are frequently inclined to the surface as well as to the direction of applied stresses. We use the same simplified model for cracks as for central cracks including the assumption that the circumferential stresses are distributed in the process zone as in the vicinity of a narrow elliptical slit. Without going into further detail about computations, let us consider some numerical results. The same data are used as for central cracks.

Some numerical results are presented in Figures 5.35-5.38. The initial crack length is $a_0 = 1$ mm and the initial angle $\theta_0 = 45°$. The relationship between the generalized forces is shown in Figure 5.35 for the start state, intermediate state, and final failure. The initial angle of kinking is strongly negative (Figure 5.35a) which means that kinking occurs just at the beginning of crack growth. Later the slope of the crack is approaching zero (Figures 5.35a and b).

The trajectory is shown in Figure 5.36, and the tip coordinates in function of the cycle number in Figure 5.37. It is evident that an edge oblique crack is

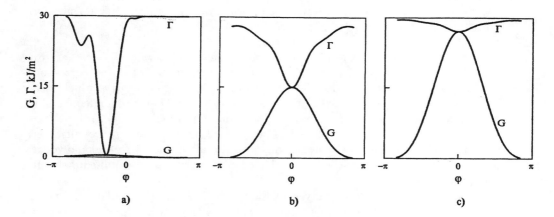

FIGURE 5.31
Relationship between generalized forces for central oblique crack: (a) start of growth; (b) intermediate state; (c) final failure.

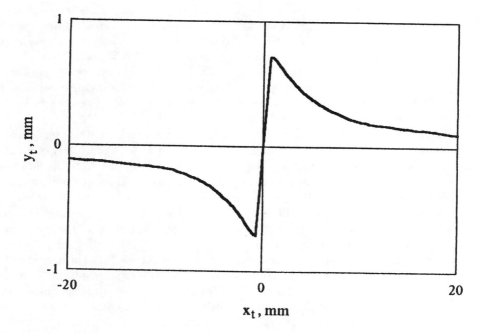

FIGURE 5.32
Crack trajectory of central oblique crack.

05-33

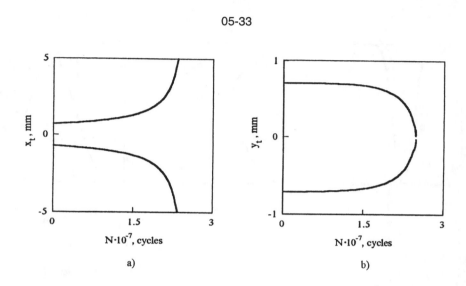

FIGURE 5.33
Crack tip coordinates versus cycle number: (a) abscissa, (b) ordinate.

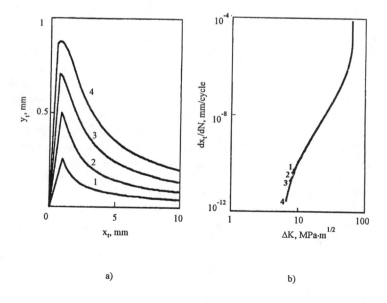

FIGURE 5.34
Growth rate diagram for central oblique crack.

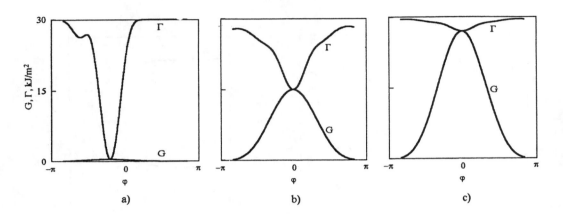

a)   b)   c)

FIGURE 5.35
Relationship between generalized forces for edge oblique crack: (a) start of growth; (b) intermediate state; (c) final failure.

becoming a mode I crack rather rapidly.

The growth rate diagram (Figure 5.38) is very similar to that for the central crack. Here dx_t/dN is plotted against $\Delta K = \Delta\sigma_\infty(\pi a)^{1/2}$ at $a(N) = x_t(N)$ and $Y = 1.12$.

The above numerical results relate to the fixed initial angle of crack inclination, namely to $\theta_0 = 45°$. Figure 5.39a shows the crack trajectories for edge cracks inclined in various angles. Lines 1, 2, 3, and 4 are drawn for $\theta_0 = 15°, 30°, 45°$, and $60°$, respectively. Note again that different scales are used along the horizontal and vertical axes. It is seen from Figure 5.39a that the cracks change direction immediately after the beginning of the loading process.

The effect of the initial crack inclination on the growth rate diagram is illustrated in Figure 5.39b. The same notation is used there as in Figure 5.34. The divergence of curves corresponding to different initial angles is significant only at the earlier stage of crack growth. Then all the curves merge, which is natural because all the cracks are approaching the mode I.

Though there are many similarities in patterns of growth for central and edge cracks, there are some differences. The most obvious difference is in the crack trajectories. This is illustrated in Figure 5.40 where the crack trajectories for the same numerical data are plotted for a central crack (1) and edge crack (2). Only the right-hand side of the central crack is shown. The difference is observed both at the crack growth start and at the approaching mode I features.

Frankly, there is a feeling of doubt and dissatisfaction concerning the single-link approximation (Figure 5.30b). To validate, at least partially, this approximation, a more consistent two-link model (Figure 5.30c) was used. The first link was iden-

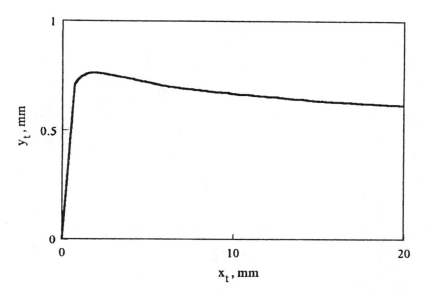

FIGURE 5.36
Crack trajectory of edge oblique crack.

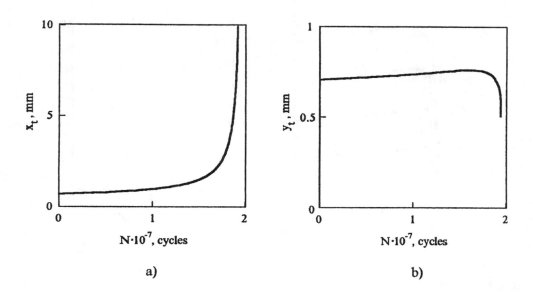

a)   b)

FIGURE 5.37
Crack tip coordinates versus cycle number: (a) abscissa; (b) ordinate.

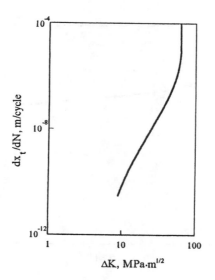

FIGURE 5.38
Growth rate diagrams for edge oblique crack.

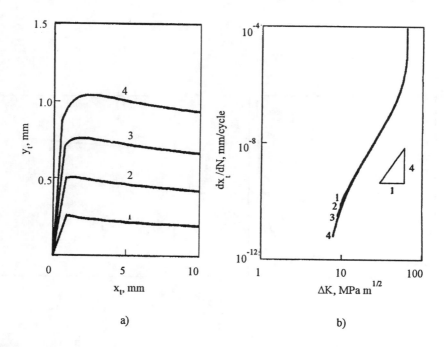

a)   b)

FIGURE 5.39
Crack trajectories (a) and growth rate diagrams (b) for edge cracks at various
initial inclination angles; lines 1, 2, 3, and 4 correspond to $\theta_0 = 15°$, $30°$, $45°$,
and $60°$.

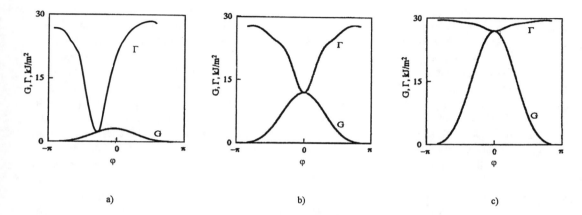

a)   b)   c)

FIGURE 5.40
Crack trajectories for central (1) and edge (2) cracks.

tified with the initial crack, and the second link with the crack increment at the considered time moment. The results of the stress intensity factors for two-link cracks [106] were used along with the interpolation upon available numerical data. The computations show that the agreement between the two approaches is good at the beginning of crack growth as well as at the later stage when the crack propagates practically in mode I. However, the discrepancy is not significant even at the intermediate stage. This may be interpreted as evidence that the single-link approximation gives a sufficiently reasonable prediction for nonplanar fatigue cracks. It should be noted, by the way, that the prediction of crack trajectories in linear fracture mechanics, i.e., for the quasi-brittle fracture under monotonous quasistatic loading, is also usually based, at least implicitly, upon a single-link approximation.

5.12 Fatigue Crack Propagation in Random Media

Random factors influencing fatigue crack growth can be classified in several ways. In particular, one could distinguish randomness of material properties inherent to a material's structure, random variability of properties within a sample of specimens or structural components, and random variability of properties of commercial materials for which suppliers are responsible. In the paper by Bolotin [26] names are suggested for these kinds of variability as within-specimen, specimen-to-specimen, and batch-to-batch scatters, respectively. Some aspects of this classification have been already discussed in Chapter 2 in the context of fatigue crack nucleation and early growth. Hereafter we turn to the crack propagation in the presence of various random factors. It is not easy to draw a border between some groups of random factors. In particular, instabilities of manufacturing enter, perhaps, all three groups. Nevertheless, from the viewpoint of analytical modeling, the difference is rather distinctive. To describe randomness on the within-specimen scale, random fields must be considered. The fields are continuous for amorphous polymers, and piece-continuous for multi-phase materials or materials with multiple microcracking. The batch-to-batch and specimen-to-specimen scatters are described by random variables. Some of these variables are parameters characterizing random fields; they enter probability distributions and spectral densities of local material properties.

It is more difficult to draw the border between the specimen-to-specimen and batch-to-batch scatters. However, there is a distinction there also. For example, initial crack sizes and damage parameters at a crack tip refer to the first type of scatter while the scatter of bulk parameters such as Young's modulus or fracture toughness may be attributed both to the specimen-to-specimen and batch-to-batch randomness. The decision depends primary upon the context. Laboratory results are usually discussed in terms of specimen-to-specimen randomness. On the design stage of a structure, one might attribute most of the uncertainties to the batch-to-batch randomness. Some additional comments concerning random factors in fatigue have already been given in Chapter 2. Systematic study of the influence of random material properties on fatigue crack propagation began at the turn of the 60/70's. A comprehensive survey of earlier publications in this area has been presented in the books [19, 137]. Two kinds of mathematical models are being discussed currently in the literature. The first way is to introduce randomness into a model with the direct randomization of deterministic equations. Consideration of the batch-to-batch and specimen-to-specimen scatter of material properties requires only a slight modification of deterministic equations by replacing some material parameters with their stochastic equivalents. To take into consideration random variability of properties within a specimen or a structural component, these parameters must be replaced with random functions of coordinates measured along the cracks trajectories and, sometimes, randomly varying in time. The second way is to use available mathematical models which seem to be suitable to describe irreversible processes of any physical nature. There are a lot of such models beginning from elementary Markov chains and including more sophisticated stochastic diffusion processes. A survey of both kinds of models can be found in the book by Spencer and Sobczyk [137]. The most popular object for randomization is the Paris–Erdogan equation (4.25) whose right-hand side is replaced either with a random variable or a random function. The latter might depend on a, N_*, or both. Some of these approaches lead to models of a second kind. For example, assuming that the right-hand side of (4.25) contains an additional multiplier which is a Gaussian stationary white noise, we arrive to a stochastic differential equation corresponding to the simplest diffusion process. Obviously, this model is not sound from the physical viewpoint; but it shows the direction of how to develop more consistent models.

The approach presented hereafter is based on the randomized version of the theory developed in this book. In the beginning we discuss various ways for randomization of material parameters entering this theory. After the discussion of results concerning plane crack propagation, the case is considered of cracks that are meandering due to random distribution of material properties at the crack tip. Samples of crack trajectories and crack growth rates are presented as well as results of the primary statistical treatment of numerical data. Some qualitative conclusions are made concerning the meandering propagation of fatigue cracks through bodies with randomly distributed local properties. The content of this and the next section is based on the papers [26, 33].

Material properties within a body are to be described with random functions of coordinates and, sometimes, of time or cycle number. The main characteristics of spatial distributions are correlation scales and variability measures. Evidently, the correlation scale of nonhomogeneities which originate from manufacturing instabilities is rather large. It means that in the study of within-specimen randomness only nonhomogeneities inherent to material microstructure are of significance. The corresponding correlation scale is of the order of the grain size if polycrystalline or concrete-like materials are concerned. In composite materials, the correlation scale depends on orientation. For example, in fiber composites, it is of the order of the fiber cross-section size in one direction, and of the ineffective fiber length in another direction. In any case, the correlation scales may be treated as small compared with characteristic sizes of a specimen.

Variability of material bulk parameters depends on the ratio between the local scatter scale and the thickness of a specimen (or, generally, on the width of the crack front). During its propagation, the crack front crosses a number of grains and other elements of microstructure acting as a kind of an averager. The wider the crack front is compared with the microlevel correlation scale, the less is the scatter on the macrolevel. This scale effect may be described with a comparatively simple stochastic model. Let the grain properties be given with the specific fracture work γ_g that varies randomly and independently among grains when one moves along the crack front. The grain size λ_g is small compared with the crack front width w (Figure 5.41a). Then the corresponding bulk parameter γ may be estimated as a random variable with the expected value close to that of γ_g. The variance of γ is the lower as the larger is the ratio w/λ_g. Hence, the size effect takes place that is illustrated in Figure 5.41b. There the probability density functions are plotted schematically: $p(\gamma_g)$ for the local (grain-to-grain) specific fracture work, and $p(\gamma; w)$ for the result of averaging along the crack front.

When the crack front is wide, the above type of scattering becomes small. On the other side, instabilities of manufacturing and assembly, peculiarities of the service life, etc. produce nonhomogeneities whose scale is large compared with, say, the side of the cross-section. This type of randomness is described with the same analytic modes of small-scale randomness. The only difference is in scale parameters. However, the large-scale within-specimen randomness may result in significant scattering of fatigue life and sometimes must be included in the analysis.

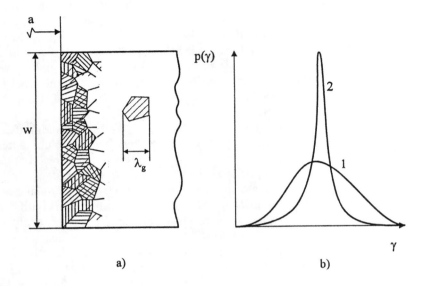

a)   b)

FIGURE 5.41
Crack front as averager of material properties (a) and probability density functions (b) of the specific fracture work: (1) for individual grains; (2) averaged upon the crack front.

5.13 Randomization of Material Resistance to Fatigue Crack Growth

Hereafter, we are going to discuss the ways to randomize the equations of fatigue crack growth given in Section 4.1. These equations contain a number of parameters that characterize local material properties along the crack trajectory. Generally, almost all the parameters entering equations (4.2)–(4.12) and (4.17) must be treated as random. To avoid unnecessary complications at this stage of study, we assume that one of the material parameters is responsible for scattering of other random parameters. For example, one may take the threshold resistance stress $\Delta\sigma_{th}$ proportional to σ_d.

For simplicity, let ρ be treated as a deterministic constant. Then all material parameters with the dimension of stress intensity factors may be taken proportional to σ_d. In particular, $K_C \sim K_{IC} = \text{const} \cdot \sigma_d \lambda_\gamma^{1/2}$, where λ_γ is a material scale parameter, say a characteristic grain size. Similarly, when γ is considered as a leading material parameter, σ_d, $\Delta\sigma_{th}$ will be deterministic functions of the local magnitude of γ. In particular, the specific fracture work for nondamaged material is connected with σ_d as

$$\gamma \sim \text{const} \cdot \sigma_d^2 \lambda_\gamma / E \tag{5.38}$$

Treating Young's modulus E as a deterministic constant, we obtain that γ is proportional to σ_d^2, etc. Details will be considered later, in connection with special problems.

To describe both specimen-to-specimen (batch-to-batch) and within-specimen random scatter, the random functions $\sigma_d(x)$ and/or $\gamma(x)$ must depend on random variables with given distributions. For example, $\sigma_d(x)$ might be taken as a homogeneous function of x that is acceptable if the fluctuations of $\sigma_d(x)$ are sufficiently small. The mean value and the variance of this function must be considered as random ones. This comment concerns also, in the general case, all parameters entering the covariance function and power spectral density of $\sigma_d(x)$. If a Markov model is used for $\sigma_d(x)$, the transition probabilities and the initial probability distribution, in general, depend on random parameters that describe the scatter of properties among specimens or structural components.

Another approach is to present functions $\sigma_d(x)$ and/or $\gamma(x)$ as a sum of random variables describing the specimen-to-specimen scatter, and random functions which describe local nonhomogeneities. There are also a number of ways to fit experimental data and microstructural aspects. For instance, $\gamma(x)$ may be both a continuous or discrete random function. Although the microstructure of most structural materials is discontinuous (from the viewpoint of continuum mechanics), a crack, as a rule, propagates rather slowly. The crack front acts here as a kind of an averager, smoothing irregularities of the microstructure. On the other hand, the computerized version of equations (4.2) or (4.17) is discrete. For example, if we assume a continuous Markov model for $\gamma(x)$, after discretization it becomes a kind of Markov or semi–Markov discrete (chain) model.

A special situation appears at the stage when elasto-plastic behavior has to be taken into account. The jump-wise propagation of cracks is also covered by the mechanics of fatigue. To simulate this process, we must return to the alternating relationships $G < \gamma$ and $G > \gamma$ between the generalized forces. The transition from one sub-equilibrium state to another is governed by these relationships under variation of the sign between the left- and right-hand sides. It is easy to simulate this discrete process numerically, but it is not at all a Markov process.

In this analysis, we present the local specific fracture work in the form

$$\gamma(x) = \gamma_0 + (\gamma_+ - \gamma_-)v + \gamma_1 u(x). \tag{5.39}$$

Here γ_0, γ_+, γ_- and γ_1 are deterministic values, v is a random variable, and $u(x)$ is a normalized homogeneous random function of the coordinate x measured along the crack path. Values γ_+, γ_- and v as well as parameters of the power spectral density $S_u(k)$ of the function $u(x)$ characterize the specimen-to-specimen scatter; meantime the within-specimen randomness enters with $u(x)$. Similarly, one may present the resistance stress $\sigma_d(x)$ as

$$\sigma_d(x) = \sigma_d^0 + (\sigma_d^+ - \sigma_d^-)v + \sigma_{d1}u(x). \tag{5.40}$$

Evidently the random variable v and the random function $u(x)$ in (5.40) differ from those entering (5.39).

The right-hand sides of equations (5.39) and (5.40) are, in general, in strong correlation. This follows, at least by the order of magnitude, from the relationship given in (5.38). We cannot change independently either deterministic material parameters or probabilistic functions characterizing $\sigma_d(x)$ and $\gamma(x)$. To satisfy the relationship (5.38), we will treat $\gamma(x)$ as the leading random function and determine $\sigma(x)$ as follows:

$$\sigma_d = \sigma_d^0 \left(\frac{\gamma}{\gamma_0} \right)^{1/2} \tag{5.41}$$

The threshold resistance stress $\Delta\sigma_{th}(x)$ will be taken as a portion of the local stress $\sigma_d(x)$, and other material parameters as deterministic constants. Further specialization of random properties is needed to predict fatigue crack growth with material nonhomogeneities. For example, the function $u(x)$ is taken in [26] as a nonlinear transform of the normalized stationary Gaussian process with the power spectral density

$$S_u(k) = \frac{2k_1 k_0}{\pi} \frac{1}{(k^2 - k_0^2)^2 + 4k_1^2 k^2}. \tag{5.42}$$

Here k_0 is an equivalent of the frequency (the wave number connected with the characteristic length of nonhomogeneities, $k_0 = 2\pi/\lambda_0$), and λ_1 describes the space correlation of nonhomogeneities. We consider, for simplification, k_0 and k_1 as fixed. As the specimen-to-specimen scatter is bounded, it is natural to assume for v the beta-distribution: Then

$$p(v) = \frac{\Gamma(\mu + \nu)v^{\mu-1}(1 - v)^{\nu-1}}{\Gamma(\mu)\Gamma(\nu)} \tag{5.43}$$

where $\mu > 0, \nu > 0$, and $\Gamma(\cdot)$ is a gamma function.

Samples of functions $\gamma(x)$ and $\sigma_d(x)$ are presented in Figure 5 42. The material parameters are taken as follows: $\gamma_0 = 20$ kJ/m^2, $\gamma_1 = 0.5$ kJ/m^2, $\gamma_+ - \gamma_- = 4$ kJ/m^2. The function $u(x)$ in equation (5.39) is taken as $u(x) = [n_1^2(x) + n_2^2(x)]^{1/2}$. Here $n_1(x)$ and $n_2(x)$ are two independent Gaussian processes with zero mean and unity variance. It means that $u(x)$ is a homogeneous (stationary with respect to x) function following Rayleigh's distribution. The power spectral density for $n_1(x)$ and $n_2(x)$ is taken according to equation (5.42). Corresponding characteristic lengths are taken as $\lambda_0 = \lambda_2 = 100$ μm. Figure 5.42a shows a sample $\gamma(x)$ of the distribution of the specific fracture work along the crack prolongation, and Figure 5.42b a corresponding sample of the resistance stress $\sigma_d(x)$. The numerical values of $\gamma_d(x)$ and $\sigma_d(x)$ are connected as assumed in (5.41), and it is assumed that $\sigma_d^0 = 10$ GPa in equation (5.41). The magnitude of v is distributed according to (5.43).

When a Markov model is used, the initial probability density function $p_0(u)$ for $|x| = a_0$ and the transition probability function $p(u_k, x_k|u_{k-1}, x_{k-1})$ for all $|x| > a_0$ must be given (or estimated from experimental data).

Let $\gamma(x)$ follow equation (5.40), and $u(x)$ be a homogeneous Markov function with the transition probability density

$$p(u_k, x_k|u_{k-1}, x_{k-1}) = \frac{\Gamma(\mu_1 + \nu_1)u_k^{\mu_1-1}(1 - u_k)^{\nu_1-1}}{\Gamma(\mu_1)\Gamma(\nu_1)} \tag{5.44}$$

The power exponents $m_1 > 0$ and $m_2 > 0$ depend on the relationship between two consequent magnitudes u_k and u_{k-1} of the random variable u as well as on the

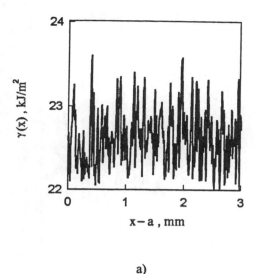

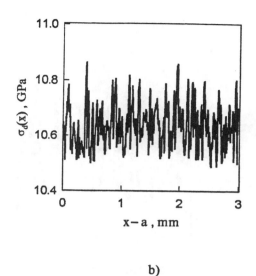

a)   b)

FIGURE 5.42
Distribution of specific fracture work (a) and resistance stress (b) along the crack path as samples of the Rayleigh random process.

relationship between coordinates x_k and x_{k-1}. This dependence is to satisfy some conditions arising, generally speaking, from heuristic considerations. In particular, if in the preceding point x_{k-1} the magnitude of u_{k-1} has been close to the median of the unconditional distribution, one expects that the probability of transition to a distant value u_k is comparatively small. When u_{k-1} is near the lower or upper boundary of the interval $(0,1)$, the probability of moving to the central part of the distribution should be higher.

Let the stationary distribution be symmetric, i.e., $\mu_1 = \nu_1 = \mu_0$, and $\lambda = x - x_{k-1} = $ const. Then equations

$$\mu_1 = \mu_0[1 + C|u_{k-1} - 1/2|^n \operatorname{sign}(u_{k-1} - 1/2)]$$

$$\nu_1 = \mu_0[1 - C|u_{k-1} - 1/2|^n \operatorname{sign}(u_{k-1} - 1/2)]$$

(5.45)

with constant parameters $C > 0, n > 0$ satisfy the stated above assumptions. For numerical simulation we divide the interval $(0, 1)$ into equal parts. Each point within this interval corresponds to a certain magnitude of $\gamma(x)$ and other related parameters. Instead of the probability density given in equations (5.44) and (5.45), we come to the transition probability matrix P. For example, setting in (5.44) and

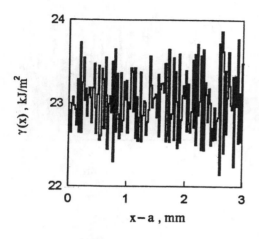 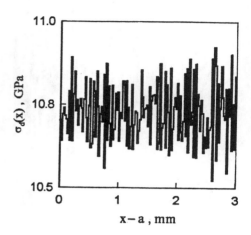

FIGURE 5.43
The same as in Figure 5.42 as samples of Markov random process.

(5.45) $\mu_0 = 4, n = 2, C = 0.1, v = 0.1, \ 0.2, \ldots, 0.9$, the matrix takes the form

$$P = \begin{bmatrix} 0.04 & 0.12 & 0.20 & 0.22 & 0.20 & 0.13 & 0.07 & 0.02 & 0.00 \\ 0.02 & 0.09 & 0.17 & 0.22 & 0.21 & 0.16 & 0.09 & 0.03 & 0.00 \\ 0.01 & 0.07 & 0.15 & 0.21 & 0.22 & 0.18 & 0.11 & 0.05 & 0.00 \\ 0.01 & 0.06 & 0.13 & 0.20 & 0.22 & 0.19 & 0.13 & 0.05 & 0.01 \\ 0.01 & 0.06 & 0.13 & 0.20 & 0.22 & 0.20 & 0.13 & 0.06 & 0.01 \\ 0.01 & 0.05 & 0.13 & 0.19 & 0.22 & 0.20 & 0.13 & 0.06 & 0.01 \\ 0.00 & 0.05 & 0.11 & 0.18 & 0.22 & 0.21 & 0.15 & 0.07 & 0.01 \\ 0.00 & 0.03 & 0.09 & 0.16 & 0.21 & 0.22 & 0.17 & 0.09 & 0.02 \\ 0.00 & 0.02 & 0.07 & 0.13 & 0.20 & 0.22 & 0.20 & 0.12 & 0.04 \end{bmatrix}$$

Only the first two decimal figures of the transition probabilities are given here. This matrix looks quite different from the birth-and-death Markov chain matrix, and other classical Markov models. The samples of functions $\gamma(x)$ and $\sigma_d(x)$ generated by this matrix are shown in Figure 5.43 where the path $\lambda = 20 \ \mu$m is chosen to obtain a better correspondence to the case of the Rayleigh distribution (Figure 5.42).

5.14 Monte Carlo Simulation

Evidently the numerical simulation is the only way to obtain quantitative results in the framework of the discussed model. It ought to be stressed that not only statistical estimates, but also separate samples are of interest. Actually, the primary information obtained by experimenters consists of the samples of crack behavior. In the author's experience, practicing engineers responsible for decisions often do not believe in the numerical results predicted by probabilistic models. They are more interested in how an engineering system may response to various actions in the presence of scatter and uncertainty in the initial data. The "worst" response among a family of 10^2 or 10^3 samples might look for them more convincing than the discussion of an event that happens with the probability 10^{-2} or 10^{-3}.

In the context of the fatigue theory, one of the most interesting questions is how the local, small-scale nonhomogeneities affect the crack growth. Another question concerns the comparative contribution of local nonhomogeneities and specimen-to-specimen scatter into the fatigue life.

We begin with the first question considering the planar crack propagation. In this case we put $v = 0.5$ in (5.39) and (5.40). We consider a central mode I crack in a wide plate under remote cyclic tension, i.e., the Griffith problem in fatigue.

The following numerical data are used in numerical simulation of crack growth: $E = 200$ GPa, $\nu = 0.3$, $\gamma_0 = 20$ kJ/m^2, $\gamma_1 = 0.5$ kJ/m^2, $\gamma_+ - \gamma_- = 4$ kJ/m^2, $\sigma_d^0 = 10$ GPa, $\Delta\sigma_{th} = 0$, $m = 4$, $\alpha = 1$, $\rho_s = 10$ μm, $\rho_b = \lambda_\rho = 100$ μm, $\lambda_0 = \lambda_1 = 40$ μm. The loading regime is considered at $\Delta\sigma_\infty = 150$ MPa and $R = 0.2$. A special question arises concerning the initial conditions. For simplification, the initial crack size a_0, initial tip radius ρ_0 and the microdamage along the crack path are supposed to be deterministic. In particular, $\omega(x, 0) = 0$ at $|x| \geq a_0$. The initial specific fracture work $\gamma(x)$ is given with its probability density function. For example, one can assume that $\gamma(a_0)$ follows the same distribution as $\gamma(x)$ at $|x| > a_0$. It means that material properties are the same both near the tip of the initial crack and along the crack path in the bulk of the material. In fact, it is not true even for laboratory specimens with initial cracks specially implanted before testing. There is a much more complicated situation in service due to the various origins of cracks, environmental actions, etc. The randomness of initial conditions that does not coincide with the randomness of material properties along the crack path may be interpreted as the specimen-to-specimen scatter and treated correspondingly.

Typical samples of fatigue crack growth history $a(N)$ at $a_0 = 1$ mm, $\rho_0 = 50$ μm are presented in Figure 5.44a. Three sample curves from a family of 25 curves were chosen: the "worst" curve corresponding to a shortest fatigue life, the "best" one corresponding to a longest fatigue life, and an "average" one. It is necessary to note that the divergence of the curves begins at the earlier stages of fatigue life, and no intersection of curves is observed at later stages.

A set of crack growth rate diagram is shown in Figure 5.44b. The scatter of da/dN is half of the order of magnitude. It ought to be noted that Figure 5.44b is plotted without any smoothing. Therefore, part of the fluctuations may be attributed to discretization, though a rather fine mesh has been used. Time histories of the effective tip radius ρ and the tip damage measure ψ are shown in Figure 5.45. As in the deterministic case, three stages of fatigue life are quite distinct.

Similar results are obtained when the distribution of material properties is modeled as a Markov process. As an example, a set of samples of crack growth histories and growth rate diagrams is presented in Figure 5.46 for the same numerical data as in Figure 5.45.

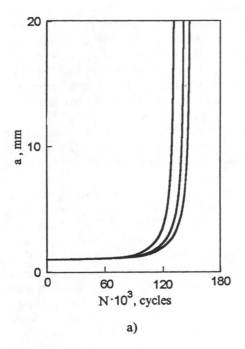
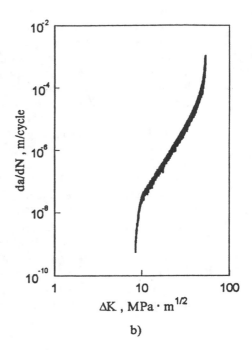

a)

b)

FIGURE 5.44
Three samples of crack growth history (a) and growth rate diagram (b) for random media with continuous distribution of material properties.

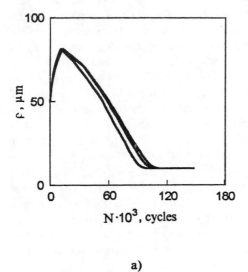
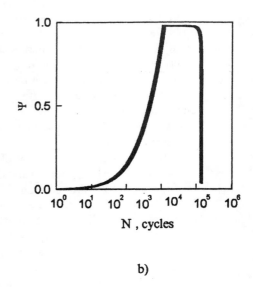

a)

b)

FIGURE 5.45
Three samples of effective tip radius (a) and tip damage diagrams (b) for the same data as in Figure 5.44.

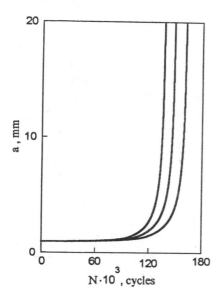 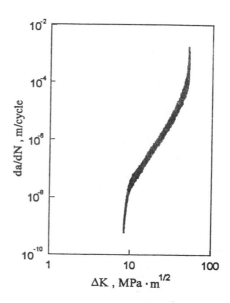

FIGURE 5.46
The same as in Figure 5.44 for material properties modeled as Markov process.

The main conclusion from the analysis of computations is as follows. Fluctuations of $\gamma(x)$ and $\sigma_d(x)$ along the path of the crack result in variability of the crack growth rate and the crack size as function of the cycle number N. The scatter of fatigue life is not significant even when only within-specimen randomness of material properties is taken into account. Some details of the samples from the same family are also of interest. It appears that only a pair of intersecting samples happens to be observed among 15-25 trials. It is essential that the divergence of sample functions begins at the earlier stage of the crack growth. It signifies that the conditions near the initial crack tip play an important part in the within-specimen variability of the crack growth history and the total fatigue life.

The survey and systematic statistical treatment of published experimental data are given by Ditlevsen and Sobczyk [56], Spencer and Sobczyk [137], Rocha, Schuëller and Okamura [127]. The most popular are data by Virkler, Hillberry and Goel [151]. These data were obtained for the specimens of aluminum alloy cut from the same sheet and having the same initial crack size. Looking at the experimental diagrams where the crack size a is plotted against the cycle number N, one cannot miss a striking point: the curves that correspond to various specimens intersect very rarely. This is in agreement with the results of numerical simulation (compare with Figures 5.44a and 5.46a). In the meantime, most models of fatigue damage based on Markov chains or diffusion processes do not exhibit this effect. Moreover, one can see an evident difference in the behavior of sample functions in an experiment and during simulation: in contrast to experimental samples, simulated ones intersect too frequently. Despite a satisfactory agreement achieved on the level of means and mean squares, one cannot consider such models consistent because they do not describe the actual behavior of individual fatigue histories. This point has been discussed in more details in the paper by Bolotin [26].

It seems significant that the scatter that originated from the local nonhomogeneity is qualitatively similar to the specimen-to-specimen scatter. However, the presence of both types of randomness increases the scatter of fatigue life. A similar effect produces large-scale local nonhomogeneity of material properties. Let, for example, the function $\gamma(x)$ be given as for Figures 5.42, 5.44 and 5.45 while the characteristic lengths for the power spectral density (5.42) are $\lambda_0 = \lambda_1 = 5$ mm, and the fluctuating component of $\gamma(x)$ given by $\gamma_1 = 0.8$ kJ/m^2. Other numerial data are the same as in the above example. The samples of $\gamma(x)$ and $\sigma_d(x)$ are presented in Figure 5.47. The samples of $a(N)$ and da/dN plotted as functions of ΔK are shown in Figure 5.48.

The qualitative character of Figure 5.47 is similar to that in Figure 5.44. Though the scatter of properties is less than above, the scatter of results is more significant. In fact, the crack size before final failure is about 20 mm which is equal to several characteristic scales λ_0 and λ_1. The averaging effect is not significant for such a short distance.

A large number of problems remain open for further research. Among them are: more accurate approach to "natural" averaging of material properties, the use of various models for microdamage accumulation and crack tip conditions, the assessment of material parameters from available experimental data. Even in the above scope, the performed analysis shows that the statistical scatter of fatigue crack growth rates and cycle numbers at final failure originate at the initial stage of crack propagation. Therefore, more attention must be paid to initial conditions and specimen-to-specimen scatter of mechanical properties to obtain a more reliable prediction of the fatigue life of structural components.

5.15 Meandering of Fatigue Cracks

As has been shown in Section 5.11, the problem of nonplanar fatigue propagation is difficult even in the deterministic case. Difficulties arise when we approach the second, third, etc. kinkings, and moreover when a crack propagates in a meandering way. In principle, the techniques of computational mechanics allow calculation of stress intensity factors and strain energy release rates for cracks of arbitrary shape. But fatigue problems are more complicated due to the necessity of studying the random stress-strain field in a wide domain ahead of the crack to evaluate the damage produced by cyclic loading. To avoid computational difficulties which could be overcome only with the use of powerful computers, we follow the idea of the single-link approximation. Namely, we replace the actual meandering crack (Figure 5.30a) with a planar oblique crack whose tips coincide with the tips of the actual crack (Figure 5.30b). In this case we consider the crack propagation through a body with randomly distributed properties as a sequence of first small kinkings and branchings. Evidently, it is a coarse approximation to actual fatigue cracks,

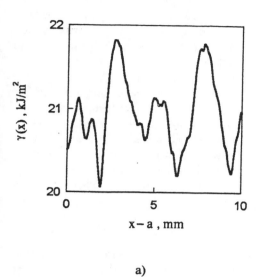

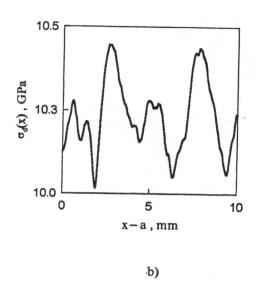

a)

.b)

FIGURE 5.47
The same as in Figure 5.42 for large-scale heterogeneity.

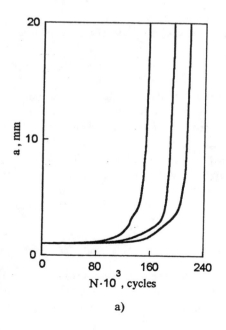

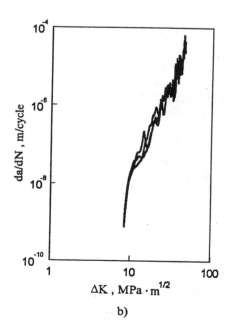

a)

b)

FIGURE 5.48
The same as in Figure 5.44 for large-scale heterogeneity.

especially from the viewpoint of stress distribution in the vicinity of crack tips. To get a better approximation one must replace the actual meandering crack with a zig-zag crack. But there are no reliable data on the stress distribution around such cracks. On the other hand, a planar crack with small kinks is acceptable on the earlier growth stage of a primarily planar crack as well as at the later stage when a crack propagates with small deviations from a planar shape.

Let us choose the specific fracture work as a leading material parameter. Similar to (5.41), we connect the local magnitudes γ with a certain ultimate stress σ_U as $\gamma \approx \sigma_U^2 \lambda_\gamma / E$ where λ_γ is a scale parameter. The stresses σ_d and τ_d are to be in strong correlation with σ_U. For simplification, we assume that all these stresses are connected deterministically. Then we write the following relationships: $\sigma_d / \sigma_d^0 = \sigma_U / \sigma_U^0$, $\tau_d / \tau_d^0 = \sigma_U / \sigma_U^0$. Here σ_d^0, τ_d^0, and σ_U^0 are some representative (deterministic) magnitudes of random variables σ_d, τ_d, and σ_U. This results in equations

$$\sigma_d = \sigma_d^0 (\gamma/\gamma_0)^{1/2}, \qquad \tau_d = \tau_d^0 (\gamma/\gamma_0)^{1/2}$$

where $\gamma_0 = (\sigma_U^0)^2 \lambda / E$ is the representative specific fracture work. As in Section 5.14, we assume the threshold resistance stresses $\Delta\sigma_{th}$ and $\Delta\tau_{th}$ to be proportional to the corresponding stresses σ_d and τ_d. The exponents m_σ, m_τ, and α entering equations (5.27) and (5.28) we consider as deterministic. As a result, all relevant random variables in the constitutive equations are presented as functions of a single variable, namely, the local magnitude of the specific fracture work $\gamma(x, y, \varphi)$. Here x and y are coordinates of the point in consideration, and φ is the angle of expected crack propagation. The field $\gamma(x, y, \varphi)$ is homogeneous with respect to x, y in body volume, and homogeneous with respect to φ in the interval $(-\pi, \pi)$.

In the further specification of the model, we assume that the magnitudes of $\gamma(x, y, \varphi)$ are independent at points separated by the characteristic distance λ. Thus the crack trajectories are schematized as sequences of links with the length λ (Figure 5.48). Each connection point is, in general, the point of kinking. Some of these points, at least in principle, can be the points of branching. However, in general, crack branchings are rather rare phenomena, especially when quasistatic fracture and plane-stress or plane-strain states are considered. Even after branching, both branches do not necessarily proceed to propagate.

One of the ways to describe the field $\gamma(x, y, \varphi)$ is to introduce the probability density function $p(\varphi)$ at each point of kinking. This angular distribution, due to homogeneity and isotropy, is the same at each point. Hence, we have a homogeneous random function of φ given in the interval $(-\pi, \pi)$. For example, the angular distribution

$$\gamma(u) = \gamma_0 + \gamma_1 u(\varphi) \tag{5.46}$$

is appropriate. Here γ_0 and γ_1 are positive deterministic constants; $u(\varphi)$ is a normalized random function of φ. When this function is defined in the segment [0,1], the minimum specific fracture work for the nondamaged material is equal to γ_0.

Another model, especially suitable for cracks meandering in polycrystalline materials, is as follows. All the body consists of a number of small domains whose size λ is of the order of material grains. The material parameters λ, σ_d, τ_d, $\Delta\sigma_{th}$

and $\Delta\tau_{th}$ are constant within each domain, varying independently from one domain to neighboring ones. This makes the length λ both the characteristic length of nonhomogeneity and correlation. Last, we schematize each elementary domain as an isosceles triangle with the angle $\Delta\varphi$ at its apex (Figure 5.48).

For further computation, we use the beta-distribution for the magnitudes of u within each three-angular domain. The probability function for $u(\varphi)$ is similar to that for v in equation (5.43):

$$p(u) = \frac{\Gamma(\mu + \nu)u^{(\mu-1)}(1 - u)^{(\nu-1)}}{\Gamma(\mu)\Gamma(\nu)} \tag{5.47}$$

Equations (5.46) and (5.47) cover a wide variety of special cases, from the specific fracture work that is uniformly distributed in $[\gamma_0, \gamma_0 + \gamma_1]$, $\mu = \nu = 1$ up to the strongly centered one at $\mu \gg 1$, $\nu \gg 1$.

The following data are used for the numerical simulation: $E = 200$ GPa, $\nu = 0.3$, $\gamma_0 = 30$ kJ/m^2, $\gamma_1 = 5$ kJ/m^2, $m_\sigma = m_\tau = 4$, $\alpha = 1$, $\sigma_d^0 = 10$ GPa, $\tau_d^0 = 5$ GPa, $\Delta\sigma_{th} = \Delta\tau_{th} = 100$ MPa, $\rho = 50$ μm. The parameters of angular distribution (5.47) are assumed $\mu = \nu = 4$. An edge mode I crack is considered at the loading conditions $\Delta\sigma_\infty = 150$ MPa, $R = 0.2$. Initial conditions are $a_0 = 1$ mm, $\rho_0 = 50$ μm, $\theta_0 = 45^0$.

Some numerical results are presented in Figures 5.49– 5.52. Figure 5.49 is in some aspects similar to Figure 5.35 for a related purely deterministic problem. Figure 5.49a corresponds to the crack growth start. The first contact of G- and Γ-curves exhibit the tendency of a crack to branching. Figure 5.49b is drawn for an intermediate stage, and Figure 5.49c for the final failure. As in Figure 5.35, meandering cracks are gradually approaching mode I. The last fact is illustrated in Figure 5.50 where the samples of crack trajectories are shown. The solid line is obtained by the averaging of 25 sample trajectories. Two "extreme" trajectories and a "moderate" one chosen from 25 samples are presented in Figure 5.50. The scales along the vertical and horizontal axes differ ten times. As in the deterministic case (see, e.g., Figure 5.36), all samples exhibit the tendency to approach mode I cracks. However, the initial growth is frequently oriented along the primary crack direction. One of the reasons is a comparatively low resistance to shear microcracking.

Additional information is given in Figure 5.51, similar to Figure 5.37 in the deterministic case. Fluctuations in the y-direction are significant; they are increasing when we approach the final failure. By the way, the scatter of fatigue life is moderate. The duration of the initiation stage varies from $N_* = 4.95 \cdot 10^6$ to $N_* = 5.93 \cdot 10^6$ among 25 samples. Total fatigue life varies from $N_{**} = 1.95 \cdot 10^7$ to $N_{**} = 2.20 \cdot 10^7$.

To obtain a crack growth rate diagram, one must perform smoothing of the recorded numerical results. The result is presented in Figure 5.52 where dx_t/dN is plotted versus $\Delta K = 1.12\Delta\sigma_\infty(\pi a)^{1/2}$, where $a \equiv x_t$. This relationship is similar to ordinary diagrams, both experimental and theoretical. The scatter of crack growth is significant even in the $\log - \log$ scale. It seems that the level of scattering depends on the numerical procedure (smoothing included).

A central, initially oblique crack under uni-axial remote tension can be studied by the same approach. At each step of computation the actual crack is replaced with a planar crack whose tips coincide with the tips of the actual crack. The

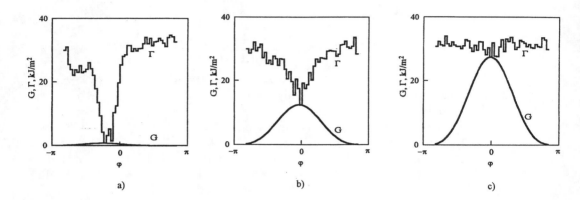

FIGURE 5.49
Relationship between generalized forces for an edge crack in random media: (a) start of growth; (b) intermediate state; (c) final failure.

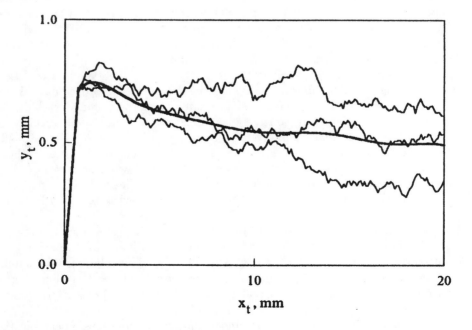

FIGURE 5.50
Three samples of crack trajectory and average trajectory.

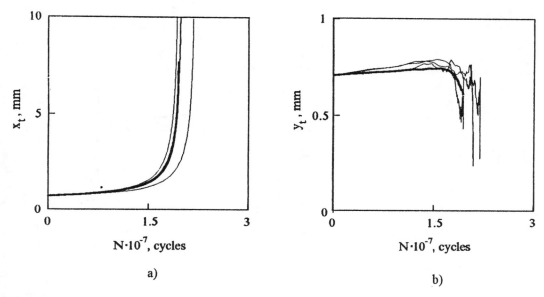

FIGURE 5.51
Crack tip coordinates versus cycle number: (a) abscissa; (b) ordinate.

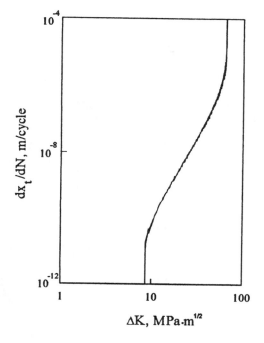

FIGURE 5.52
Crack growth rate diagram for an edge, initially oblique crack in random media.

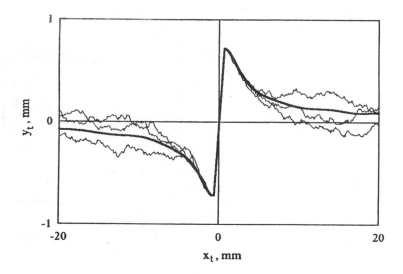

FIGURE 5.53
Three samples of trajectory of an initially central oblique crack and an averaged trajectory.

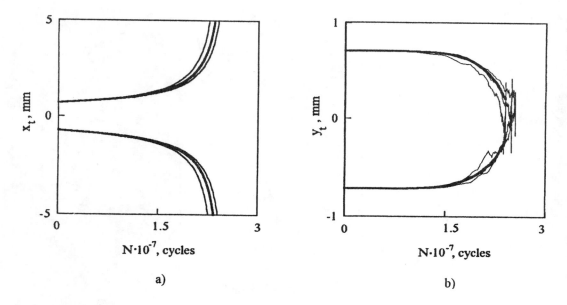

FIGURE 5.54
Crack tip coordinates for an initially central, oblique crack: (a) abscissas; (b) ordinates.

schematized crack is given with length parameters a_1 and a_2 and the angle θ. Two angles of kinking φ_1 and φ_2 satisfy conditions

$$\max\left[G_1(a_1, a_2, \theta, \varphi_1, N) - \Gamma_1(a_1, \theta, \varphi_1, N)\right] = 0$$

$$\max\left[G_2(a_1, a_2, \theta, \varphi_2, N) - \Gamma_2(a_2, \theta, \varphi_2, N)\right] = 0$$

similar to (5.37).

The generalized forces G_1 and G_2 depend on both G-coordinates which, in turn, depend on the expected angles of kinking, φ_1 and φ_2. In general, a crack propagates in both sides nonuniformly even if the initial crack and the applied loading are centered.

The following data are used for numerical simulation: $a_1(0) = a_2(0) = 1$ mm, $\theta_0 = 45^0$. Other parameters are as for Figures 5.49- 5.52. Two typical samples of crack trajectories and the result of averaging upon 25 samples are given in Figure 5.53. They are similar to those for the corresponding deterministic problem (Figure 5.31). The remote parts of the diagrams may be interpreted as relating to "almost" mode I cracks. A slight deviation from symmetry with respect to the y-axis is a result of local nonhomogeneities.

Abscissas x_t and ordinates y_t of the crack tips are plotted in the function of the cycle number in Figure 5.54. The above mentioned loss of symmetry as well as the higher scattering near the final failure are observed. The scatter of fatigue life is seen in Figure 5.54 also. In the "worst" case (among 25 samples) we have $N_* = 3.99 \cdot 10^6$ and $N_{**} = 2.40 \cdot 10^7$ while in the "best" case $N_* = 4.00 \cdot 10^6$ and $N_{**} = 2.54 \cdot 10^7$. As to the crack growth rate diagram obtained after smoothing of samples $x_t(N)$, it looks very similar to that given in Figure 5.52.

Some additional comments concerning the random factors in fatigue were given in Chapter 2 in the context of fatigue crack nucleation and early growth.

5.16 Summary

A wide scope of fatigue crack propagation problems were considered in this chapter, including planar cracks in plates under various loading, circular and elliptical penny-shaped cracks, initially oblique cracks as well as cracks propagating in materials with properties randomly varying along the crack path. All these problems were treated in the framework of a simplified model (linear elastic material with the power-threshold damage law, and a simple equation describing conditions at the crack tip). Nevertheless, the numerical results presented in the form of crack growth rate diagrams, crack trajectory samples, etc. exhibit qualitative agreement with experimental data. In particular, numerical simulation of crack growth in the presence of statistical scatter of local properties predicts rather well the observed behavior of fatigue cracks in real materials. The obvious drawback of the models discussed in this chapter, namely neglect of plastic deformations (especially in the process zone), is partially compensated for by the explicit introduction of crack tip blunting. An alternative approach will be considered in the next chapter, dedicated to the application of the thin plastic zone model to fatigue problems.

Chapter **6**

FATIGUE CRACKS IN ELASTO-PLASTIC BODIES

6.1 Effects of Plastic Straining on Fatigue Crack Propagation

The model of fatigue crack growth developed in the two previous chapters is related to classical models of linear fracture mechanics supplemented with effects of modeling of microdamage accumulation.

One of the most vulnerable features of linear fracture mechanics, namely singularity at the crack tips, is removed in this study. However, the feeling of dissatisfaction remains because of sharp stress concentration at the tips. One could manipulate parameters of the model to decrease both acting tip stresses and corresponding resistance stresses. However, the principal way to overcome conceptual difficulties is to introduce plastic straining in the model.

One distinguishes small-scale (confined) and full-scale plastic deformations. In the first case, the size of the plastic zone is small compared with the characteristic body sizes and the length of the ligament between the plastic zone and the opposite body border. These conditions are not satisfied in the case when a general elasto-plastic behavior of the cracked body is considered. In low-cycle fatigue, when the life until final failure is measured in thousands of cycles, we deal with both small-scale and full-scale plastic straining. In the latter case, fatigue crack growth is certainly strain-controlled or, in terms of global loading, displacement-controlled.

A brief introduction into low-cycle fatigue was presented in Section 1.3. Compared with classical, high-cycle fatigue, low-cycle fatigue is accompanied by plastic straining whose effects are significant and must be taken into account in the prediction of fatigue life. Elasto-plastic fracture is the subject of nonlinear fracture mechanics. In fatigue, the models of nonlinear fracture mechanics are complicated not only by the necessity to include microdamage accumulation. Considering cyclic elasto-plastic straining, one must take into account hysteretic phenomena as well as cyclic softening and/or hardening. However, all general notions of the theory of fatigue crack propagation remain valid for low-cycle fatigue. Difficulties arise in the evaluation of driving and resistance generalized forces that depend on loading and crack propagation history.

Three main ways may be listed in nonlinear fracture mechanics [71, 114, 118, 153]. The first one is based on the deformation theory of plasticity which is in fact a special kind of nonlinear theory of elasticity. The direct application of the deformation theory of plasticity to fatigue is certainly out of the question because this theory does not cover hysteresis and cyclic hardening/softening. The second way is connected with the thin plastic zone and related models applicable in the case of confined yielding. Using these models, we have a rather wide freedom to choose equations of microdamage accumulation. In addition, these models contain, partially as asymptotic and partially as border cases, models of linear fracture mechanics associated with microdamage. The third way is based on the application of the theory in full scope, in any form such as plastic flow, incremental, or endochronic versions supplemented with damage accumulation laws. The latter approach is necessary in the case of fully plastic yielding, and we do not discuss this approach here.

Plastic strains appear at crack tips at already low applied stresses. Nevertheless, the domain of applicability of linear fracture mechanics is rather wide. Linear fracture mechanics covers even such situations where good agreement with experimental data seems paradoxical. The explanation of this paradox is rather obvious. Actually, a certain domain near the crack tips is plastically deformed and strongly damaged. The material in this domain is more or less unloaded, and the stress concentration is diminished compared with the prediction of elasticity theory. As a result, the behavior of a cracked body primarily depends on the state a little ahead of the tip. Stress distribution in this area is mostly governed by stress intensity factors. This makes stress intensity factors (or strain energy release rates) the leading parameters in quasi-brittle fracture when confined yielding takes place.

But if the plastically strained zone is not small, the state in this zone becomes an important factor. The approximation of linear fracture mechanics is asymptotic in the sense that it is more accurate the less applied stresses are compared with the yield stress or similar mechanical characteristics or, which is almost equivalent, the smaller the size of the plastic zone compared with the crack length.

According to the thin plastic mode model, the length of the plastic zone in the case of confined yielding is defined by equation (1.20)

$$\lambda = a \left[\sec \left(\frac{\pi \sigma_\infty}{2\sigma_0} \right) - 1 \right] \tag{6.1}$$

Here σ_∞ is the applied stress and σ_0 is a material parameter related to the yield tensile stress σ_Y. The stress σ_0 could be equal to σ_Y or $\sqrt{3}\sigma_Y$ (the latter in the plane-strain state). In engineering practice it is sometimes assumed $\sigma_0 = (\sigma_Y + \sigma_U)/2$, where σ_U is the ultimate tensile stress. To keep some generality, we denote the tip zone stress σ_0 (instead of the common notation σ_Y) and name it the limit stress. At the first stage of study we consider σ_0 as a material constant. Later on, the effects of hardening and softening, as a result of plastic deformation or microdamage accumulation (or both), will be discussed.

At sufficiently small σ_∞/σ_0, equation (6.1) can be replaced with the following:

$$\lambda = \frac{\pi^2 a}{8} \left(\frac{\sigma_\infty}{\sigma_0} \right)^2 \tag{6.2}$$

Linear fracture mechanics is applicable when $\lambda \ll a$ or, as follows from (6.1) and (6.2), at $\sigma_\infty^2/\sigma_0^2 \ll 1$. The generalized driving force taking into account the plastic zone is

$$G = G_e \left[1 + O(\lambda/a) \right] \tag{6.3}$$

where G_e is the driving force in the linear elastic case. Hence, at moderate σ_∞/σ_0 the generalized driving forces are not very sensitive to plastic phenomena. The reason is that the generalized driving forces are connected to the general balance of energy and forces in the system cracked body-loading. The generalized resistance forces depend on plastic deformations in a larger degree. In particular, the so-called R-curves (see Section 1.7) take into account the influence of the history of crack growth on the resistance to further crack propagation.

Until now we have discussed the situation in classical fracture mechanics, i.e., in the context of fracture under monotonous loading. In fatigue the picture becomes complicated due to the cyclic character of stress-strain fields. In particular, hysteretic phenomena, residual stresses and strains, cyclic hardening and softening are to be taken into account. The first step is to introduce, along with the plastic zone λ whose size is given in (6.1), a cyclic process zone with the size λ_p (Figure 6.1). Within the process zone the stress-strain state varies essentially during each cycle of loading.

The origin of the cyclic process zone (later on, just process zone) is illustrated in Figure 6.2. Let the material be ideal elasto-plastic with the yield stress σ_0 both in tension and compression. During the first loading half-cycle the yield limit is equal to σ_0 and the tensile stress to $\sigma_\infty^{\max}$. During the first down-loading half-cycle,

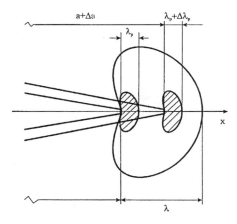

FIGURE 6.1
Fatigue crack growth in elasto-plastic material.

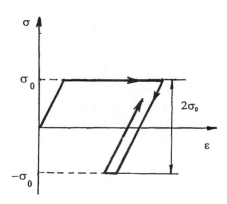

FIGURE 6.2
One-dimensional cyclic loading of ideal elasto-plastic material.

the material is deformed as one with the yield limit $2\sigma_0$ under the compression stress $\sigma_\infty^{max} - \sigma_\infty^{min}$. If the applied stress range $\Delta\sigma_\infty = \sigma_\infty^{max} - \sigma_\infty^{min}$ is sufficiently large, further cyclic straining near the tip follows the same pattern. Therefore, to evaluate the size λ_p of the process zone, we have to replace in (6.1) σ_∞ with $\Delta\sigma_\infty$ and σ_0 with $2\sigma_0$. The size λ_p within the framework of the thin plastic zone model is

$$\lambda_p = a \left[\sec\left(\frac{\pi\Delta\sigma_\infty}{4\sigma_0} \right) - 1 \right] \tag{6.4}$$

The corresponding approximate evaluation similar to equation (6.2) takes the form

$$\lambda_p = \frac{\pi^2 a}{8} \left(\frac{\Delta\sigma_\infty}{2\sigma_0} \right)^2 \tag{6.5}$$

For example, at $R = 0$, i.e., $\sigma_\infty^{min} = 0$, $\Delta\sigma_\infty = \sigma_\infty^{max}$, equation (6.5) predicts the size λ_p as one fourth of the size λ. At $R > 0$ the size of the process zone may be much smaller than the size of the plastic zone (Figure 6.3). However, at each crack tip advancement both zones, plastic and process ones, advance, too.

Stress distribution near the crack tip during a loading cycle is schematically depicted in Figure 6.4. Line 1 shows the stresses corresponding to the first up-loading. The length of the plastic zone is equal to λ. Line 2 is drawn under the assumption that the material behaves as elasto-plastic with the yield limit $2\sigma_0$ under the compression stress $\sigma_\infty^{max} - \sigma_\infty^{min}$. The length of the plastic zone becomes equal to λ_p. Line 3 depicts the residual stresses. This picture is valid for the first cycle only and, morever, only for fixed crack tips. During crack propagation, stress distribution changes; however, the process zone length remains one of the leading parameters for fatigue growth in elasto-plastic materials.

When we discuss the problem from the viewpoint of low-cycle and high-cycle fatigue, we have difficulties drawing the border between them. In practice, the border is usually stated in terms of cycle number N_{**} counted until the final failure.

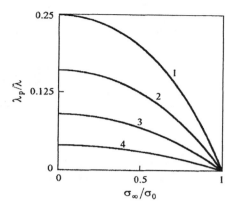

FIGURE 6.3
The size of a cyclic process zone as a portion of the size of a plastic zone in monotonous loading in the function of the applied stress and cyclic stress ratio.

If $N_{**} = 10^2 \ldots 10^4$, one says low-cycle fatigue. At $N_{**} = 10^5$ and more we say that high-cycle fatigue takes place. In other terms, one could say low-cycle fatigue when the crack growth rate is sufficiently high, e.g., $da/dN > 10^{-6}$ m/cycle. The above criteria are of the a posteriori character. Low-cycle fatigue is expected when the maximal applied stresses are sufficiently close to the yield limit. For example, it could be the criterion $\sigma_\infty^{\max} > 0.5\sigma_Y$, etc. In theoretical analysis, it is expedient to formulate the transition conditions in terms of variables directly entering the models of fatigue crack growth.

We might draw the border between low- and high-cycle fatigue comparing the

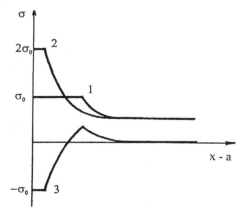

FIGURE 6.4
Stress distribution during a loading cycle: (1) at up-loading; (2) at down-loading; (3) residual stresses.

maximum stresses at the tip with the yield stress σ_Y or an equivalent limit stress σ_0. Then the border condition is

$$\sigma_{\max}(a) \approx \sigma_0$$

According to equation (4.16), $\sigma_{\max}(a) \sim \sigma_\infty (a/\rho)^{1/2}$, where ρ is the effective tip radius in high-cycle fatigue. Transition from high-cycle to low-cycle fatigue takes place in the vicinity of the stress intensity factor

$$K_{tr} \sim \sigma_0 \left(\pi \rho_b\right)^{1/2} \tag{6.6}$$

where ρ_b is the "blunt" tip radius (Section 4.2). The same estimate follows when we draw the border comparing the plastic zone size λ with ρ_b.

However, the estimate (6.6) seems conservative. For example, for low-carbon steels when $\sigma_0 \sim 500$ MPa, $\rho_b \sim 100$ μm, equation (6.6) gives $K_{tr} \sim 10$ MPa·m$^{1/2}$. It means that the crack growth rate diagram, except for a comparatively small threshold region, is covered by the low-cycle fatigue model, at least in its thin plastic zone version.

As an alternative, we might assume that the transition is governed by the crack tip opening displacement. It seems that, when the plastic tip opening displacement

$$\delta \approx \frac{\pi a \sigma_\infty^2}{E \sigma_0} \tag{6.7}$$

(E is Young's modulus) is less than the "blunt" tip radius ρ_b, we may apply a high-cycle fatigue model. Then transition to the high-cycle fatigue model becomes necessary at $K_{\max} > K_{tr}$ where

$$K_{tr} \sim (E \sigma_0 \rho_b)^{1/2} \tag{6.8}$$

Substitution in (6.8) of numerical values $E \sim 200$ GPa, $\sigma_0 \sim 500$ MPa, $\rho_b \sim 100\mu$m gives $K_{tr} \sim 100$ MPa·m$^{1/2}$. Obviously, this estimate is too high. Perhaps the real border is situated somewhere between the estimates given in (6.6) and (6.8).

In this chapter, the thin plastic model is applied systematically to predict the crack propagation in low-cycle fatigue as well as in the case of nonsteady loading when both low-cycle and high-cycle fatigue features are present. We follow here mainly the papers by Bolotin and Lebedev [36] and Bolotin and Kovekh [35].

6.2   Stress and Displacement Distribution for Cracks with Loaded Faces

The thin plastic zone model is discussed in most manuals on fracture mechanics [1, 41, 50, 62, 66]. The well-known final results such as the formulas for plastic zone length and for crack tip opening displacement are presented in equations (1.20), (1.21), (6.1) and (6.2). For further study, we need more detailed information on stress, strain and displacements fields for a mathematical slit with loaded faces.

These fields were evaluated by a number of authors. Using various analytical techniques, they obtained results which, although identical, differ in the final form. In later research, we will choose a suitable form every time.

The general solution for plane problems of elasticity theory is given by Kolosov (1909) and developed by Muskhelishvili (1933). This solution is based on the Goursat (1889) general solution of the biharmonical equation $\Delta \Delta F = 0$ with the use of two analytical functions $\varphi(z)$ and $\chi(z)$ of the complex variable $z = x + iy$:

$$F(z) = \frac{1}{2}\left[\bar{z}\varphi(z) + z\overline{\varphi(z)} + \chi(z) + \overline{\chi(z)}\right]$$

Here the bar denotes the complex conjugate. Since the Airy function in plane elasticity problems satisfies the biharmonic equation, the complex presentation for stresses σ_x, σ_y, τ_{xy} and displacement u, v follows:

$$\sigma_x + \sigma_y = 2[\Phi(z) + \overline{\Phi(z)}]$$

$$\sigma_y - \sigma_x + 2i\tau_{xy} = 2[\bar{z}\Phi'(z) + \Psi(z)]$$

$$2\mu(u + iv) = \kappa\varphi(z) - z\overline{\varphi'(z)} - \overline{\psi(z)}$$

(6.9)

Here the notation used is $\Phi(z) = \varphi'(z)$, $\Psi(z) = \psi'(z)$, $\psi(z) = \chi'(z)$; μ is shear modulus; $\kappa = 3 - 4\nu$ for plane-strain state, and $\kappa = (3 - \nu)/(1 + \nu)$ for plane-stress state.

Let the faces of the slit $y = 0^+$, $|x| \leq b$ and $y = 0^-$, $|x| \leq b$ be loaded with the given normal stresses $\sigma_y(x, 0)$. The tangential stresses along the x-axis are absent, i.e., $\tau(x, 0) = 0$. Then the second equation (6.9) gives

$$\text{Im}\left[\bar{z}\Phi'(z) + \Psi(z)\right] = 0.$$

It results in the following presentation of functions $\Phi(z)$ and $\Psi(z)$:

$$\Phi(z) = \frac{1}{2}Z(z), \qquad \Psi(z) = -\frac{1}{2}zZ'(z)$$

Here $Z(z)$ is an analytical function except, maybe, the bifurcation points. This function can be chosen as

$$Z(z) = \frac{g(z)}{\{(z + b)(z - b)\}^{1/2}}$$

(6.10)

with the new analytical function $g(z)$. The latter is to be found from boundary conditions. The function $Z(z)$ is commonly called the potential of Westergaard (1937). Instead of (6.9) we obtain

$$\sigma_x = \text{Re}Z - y\text{Im}Z', \qquad \sigma_y = \text{Re}Z + y\text{Im}Z', \qquad \tau_{xy} = -y\text{Re}Z'$$

$$\mu u = \frac{\kappa - 1}{4}\text{Re}W - \frac{1}{2}y\text{Im}Z$$

$$\mu v = \frac{\kappa - 1}{4}\text{Im}W - \frac{1}{2}y\text{Re}Z$$

(6.11)

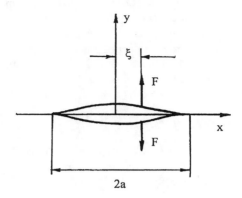

FIGURE 6.5
Axillary elasticity problem for the thin plastic zone model.

where $W(z)$ is the analytical function defined with the relationship $Z = W'$.

The solution of the boundary problem is provided by the appropriate choice of the function $g(z)$ in equation (6.10). The condition $\sigma_y(x, 0) = p(x)$ transforms into the following:

$$\mathrm{Re} Z(z) = p(x), \qquad |x| \le b, \quad y = 0$$

At infinity $Z(z)$ decreases not slower than z^{-2}. In particular, under the remotely applied stresses σ_∞ we have

$$Z(z) = \frac{\sigma_\infty z}{(z^2 - b^2)^{1/2}} \tag{6.12}$$

When two concentrated forces F are applied at $x = \xi$, $y = 0^+$ and $y = 0^-$ (Figure 6.5), we come to the function

$$Z(z) = \frac{F(b^2 - \xi^2)^{1/2}}{\pi(z - \xi)(z^2 - b^2)^{1/2}}. \tag{6.13}$$

For the thin plastic zone model we have

$$p(x) = \begin{cases} 0, & |x| < a \\ -\sigma_0, & a \le |x| \le b \end{cases}$$

where for brevity the notation is used

$$b = a + \lambda$$

Substituting $F = p(\xi)d\xi$ in equation (6.13) and integrating, we obtain

$$Z = -\int\limits_a^b \frac{2\sigma_0 z(b^2 - \xi^2)^{1/2} d\xi}{\pi(z^2 - \xi^2)(z^2 - b^2)^{1/2}} dx$$

or, after the integration at $\sigma_0 = \text{const}$:

$$Z(z) = -\frac{2\sigma_0}{\pi} \left\{ \frac{z}{(z^2 - b^2)^{1/2}} \cos^{-1}\left(\frac{a}{b}\right) - \right.$$

$$\left. - \cot^{-1}\left[\frac{a}{z}\left(\frac{z^2 - b^2}{b^2 - a^2}\right)^{1/2}\right]\right\} \tag{6.14}$$

The superposition of equations (6.12) and (6.14) gives

$$Z(z) = \frac{\sigma_0 z}{(z^2 - b^2)^{1/2}} \left\{ \frac{\sigma_\infty}{\sigma_0} - \frac{2}{\pi}\cos^{-1}\left(\frac{a}{b}\right)\right\} +$$

$$+ \frac{2\sigma_0}{\pi}\cot^{-1}\left[\frac{a}{z}\left(\frac{z^2 - b^2}{b^2 - a^2}\right)^{1/2}\right] \tag{6.15}$$

The first summand in the braces has a singularity at $|x| = b$. To remove this singularity, we have to put this term to zero. Therefore, it is

$$\sigma_\infty = \frac{2\sigma_0}{\pi}\cos^{-1}\left(\frac{a}{a + \lambda}\right)$$

As a result, we come again to equation (6.1) for the size of the plastic zone. The final form of Westergaard's potential for the thin plastic zone model is

$$Z(z) = \frac{2\sigma_0}{\pi}\cot^{-1}\left[\frac{a}{z}\left(\frac{z^2 - b^2}{b^2 - a^2}\right)^{1/2}\right] \tag{6.16}$$

It ought to be mentioned that $b = a + \lambda$ with λ determined by equation (6.1). Hence the applied stress σ_∞ implicitly enters equation (6.16) with the plastic zone length λ.

The stress and displacement fields can be now found with the use of equations (6.11) and (6.16). We are interested in the stress $\sigma_y(x, 0)$, notated hereafter simply $\sigma(x)$:

$$\sigma(x) = \begin{cases} 0, & |x| < a \\ \sigma_0, & a \le |x| \le b \\ \dfrac{2\sigma_0}{\pi}\cot^{-1}\left[\dfrac{a}{x}\left(\dfrac{x^2 - b^2}{b^2 - a^2}\right)^{1/2}\right], & |x| > b \end{cases} \tag{6.17}$$

As to displacements, the final form of the equations depends on the method of analytical computations. The direct application of equations (6.11) and (6.16) is performed by Burdekin and Stone (1966). The displacement $v(x, 0)$ at $a \le x \le b$ notated later on $v(x)$ is

$$v(x) = \frac{4\eta a \sigma_\infty}{\pi E} \left\{ \coth^{-1} \left[\left(\frac{b^2 - x^2}{b^2 - a^2} \right)^{1/2} \right] - \right.$$

$$\left. - \frac{x}{a} \coth^{-1} \left[\frac{a}{x} \left(\frac{b^2 + x^2}{b^2 - a^2} \right)^{1/2} \right] \right\}$$

Here $\eta = 1$ for the plane-stress state and $\eta = 1 - \nu^2$ for the plane-strain state. Another version is obtained by Goodier and Field (1963):

$$v(x) = \frac{\eta \sigma_\infty b}{\pi E} \left[\cos\theta \log \left(\frac{\sin^2(\beta - \theta)}{\sin^2(\beta + \theta)} \right) + \cos\beta \log \left(\frac{[\sin\beta + \sin\theta]^2}{[\sin\beta - \sin\theta]^2} \right) \right],$$

$$\theta = \cos^{-1} \left(\frac{x}{b} \right), \qquad \beta = \frac{\pi \sigma_\infty}{2 \sigma_0}$$

(6.18)

In further analysis, we use also the presentation given by Panasyuk (1960) without using Westergaard's technique:

$$v(x) = \frac{\eta \sigma_0}{\pi E} \left[(x - a) M(b, x, a) - (x + a) M(b, x, -a) \right]$$

(6.19)

Here the notation used is

$$M(b, x, \xi) = \log \frac{b^2 - x\xi - [(b^2 - x^2)(b^2 - \xi^2)]^{1/2}}{b^2 - x\xi + [(b^2 - x^2)(b^2 - \xi^2)]^{1/2}}$$

(6.20)

The function given in (6.20) corresponds to loading of the crack faces with two unity normal forces (Figure 6.5). Therefore, it can be considered as a kernel for more complicated situations when the stress $\sigma_0(x)$ within the plastic zone is a function of x (Section 6.10).

6.3   Generalized Driving Forces

The generalized driving force in the thin plastic zone model is commonly associated with the J-integral taken along the contour closely enveloping the plastic zone. The evaluation of the J-integral is illustrated in Figure 6.6a. The countour of integration at the crack tip is chosen in such a way that equation (1.24) takes the form

$$J = - \int_a^{a+\lambda} \sigma \frac{\partial v_-}{\partial x} dx - \int_{a+\lambda}^a \sigma \frac{\partial v_+}{\partial x} dx$$

where $v_+(x)$ and $v_-(x)$ are magnitudes of $v(x)$ at $y = 0^+$ and $y = 0^-$, respectively. As $\sigma(x, 0) = \sigma_0$, $v_+(x) = v_-(x)$, we obtain

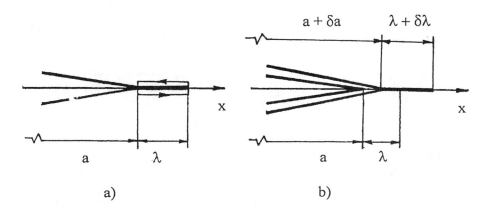

FIGURE 6.6
Two approaches to the thin plastic zone model: (a) evaluation of J-integral; (b) variation of the crack size together with the plastic zone size.

$$J = -2 \int_{a}^{a+\lambda} \sigma_0 \frac{\partial v}{\partial x} dx. \qquad (6.21)$$

At $\sigma_0 = \text{const}$, equation (6.21) gives

$$J = \sigma_0 \delta \qquad (6.22)$$

Here $\delta = 2v(a)$ is the crack tip opening displacement (1.21)

$$\delta = \frac{8\eta \sigma_0 a}{\pi E} \log \left[\sec \left(\frac{\pi \sigma_\infty}{2\sigma_0} \right) \right] \qquad (6.23)$$

The final result is

$$J = \frac{8\eta \sigma_0^2 a}{\pi E} \log \left[\sec \left(\frac{\pi \sigma_\infty}{2\sigma_0} \right) \right] \qquad (6.24)$$

However, the true generalized force must be derived from the condition that the summed virtual work of external and internal forces is $\delta W_e + \delta W_i = G \delta a$. The parameter given in (6.24) does not satisfy this condition. Following the general approach, we must use the relationship

$$G = \frac{\delta W_e + \delta W_i}{\delta a} \qquad (6.25)$$

It is essential that both the crack size a and the plastic zone size λ be subjected to variation (Figure 6.6b). Actually, the size λ increases, as it follows from equation (6.1), with the crack size a. The thickness of the plastic zone is infinitely small (more accurately, this is a discontinuity of displacements in an elastic body). Therefore, the virtual work of internal forces within the integration domain can be put to zero. Hence

$$G\,da = 2 \int\limits_{a}^{a+\lambda+\delta a+\delta\lambda} \sigma\delta v dx = 2\delta a \int\limits_{a}^{a+\lambda} \sigma\frac{\partial v}{\partial a}dx + O(\delta a^2)$$

The final formula for the generalized driving force takes the form

$$G = 2 \int\limits_{a}^{a+\lambda} \sigma_0 \frac{\partial v}{\partial a} dx \qquad (6.26)$$

Being very similar to (6.21), equation (6.26) differs in the derivatives of $v(x)$. These equations coincide if we might assume

$$\frac{\partial}{\partial a} = -\frac{\partial}{\partial x}$$

It is valid in the case of elastic solids but fails when a plastic zone is present. Coming to calculations, we assume that $\sigma_0 = $ const and $v(x,a)$ are determined according to (6.18). These calculations are rather cumbersome though elementary. The final result is

$$G = \frac{8\eta\sigma_0^2 a}{\pi E}\left[\log\cos\left(\frac{\pi\sigma_\infty}{2\sigma_0}\right) + \frac{\pi\sigma_\infty}{2\sigma_0}\tan\left(\frac{\pi\sigma_\infty}{2\sigma_0}\right)\right] \qquad (6.27)$$

The details can be found in publications [50, 114, 152] where they are performed in a somewhat different context, namely, for evaluation of the work spent on fractures in the thin plastic zone model. In terms of this study, the authors have dealt with the generalized resistance force Γ. As in the equilibrium state $G = \Gamma$, the final result can be interpreted as the generalized driving force without any reference to crack (non)propagation.

When $\sigma_\infty \ll \sigma_0$, equations (6.24) and (6.27) lead to the result that is in agreement with Irwin's equation (1.15) for the strain energy release rate:

$$G = \frac{\pi\eta\sigma_\infty^2 a}{E} \qquad (6.28)$$

At $\sigma_\infty \sim \sigma_0$ the discrepancy between the magnitudes given in equations (6.24), (6.27), and (6.28) becomes significant. It is illustrated in Figure 6.7 where the normalized force G/G_0 is plotted versus the normalized applied stress σ_∞/σ_0. Here $G_0 = 8\eta\sigma_0^2 a/(\pi E)$. The lines 1, 2 and 3 correspond to equations (6.28), (6.24) and (6.27), respectively. Fracture criterion $\delta = \delta_c$ is often used in applications where δ_c is the critical crack tip opening displacement. In that case, the use of equation (6.24) is justified because the critical condition $J = J_c$ according to (6.22) is equivalent to the condition $\delta = \delta_c$.

In fatigue we have to take into account the influence of dispersed damage, crack tip blunting, strain softening and hardening, etc. Then the driving force must be evaluated by equation (6.26) and, in the special case of the classical thin plastic zone model and $\sigma_0 = $ const, by equation (6.27). On the other hand, the limit stress in fatigue may depend on the damage level. Then σ_0 varies both in time and as a function of x. In this case we must apply equations (6.26) without any further simplifications.

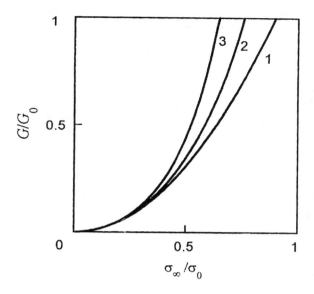

FIGURE 6.7
Normalized load parameters: (1) Irwin's energy release rate; (2) *J*-integral for thin plastic zone model; (3) generalized driving force.

6.4   Models of Damage Accumulation

The crucial point of the presented study is certainly the choice of the damage accumulation law. When material is deformed plastically, it is not clear which field parameters are responsible for damage: cyclic stresses, cyclic strains, or both. Considering strain-controlled damage, we have to choose between total and plastic strains, to analyze the contribution of mean and deviatoric strains, etc. In the framework of the thin plastic zone model, the problem is not so complicated due to the assumed one-dimensionality of the field within the plastic zone. In any case, the adequate choice of control parameters in damage accumulation law is an open question.

Damage accumulation is described by the equation with respect to the damage rate $\partial \omega / \partial N$ whose right-hand side is the subject for discussion. The first idea is to consider the maximum tensile stress as the control parameter. The simplest equation that might fit the experimental data is similar to equation (4.2):

$$\frac{\partial \omega}{\partial N} = \left(\frac{\Delta \sigma - \Delta \sigma_{th}}{\sigma_d} \right)^{m_\sigma} \tag{6.29}$$

Here, as in (4.2), σ_d, $\Delta \sigma_{th}$, and m_σ are material parameters, and the subscript at m_σ indicates that we consider stress-controlled damage. At $\Delta \sigma < \Delta \sigma_{th}$ we must put $\partial \omega / \partial N = 0$.

However, it is reasonable to expect that the damage in the plastic zone is controlled not by stresses but by deformations. In particular, assuming that the control parameter is a relative cyclic displacement $2\Delta v$ of the plastic zone faces, we replace equation (6.29) as follows:

$$\frac{\partial \omega}{\partial N} = \left(\frac{2\Delta v - \Delta \delta_{th}}{\delta_d} \right)^{m_\delta} \tag{6.30}$$

Here $2\Delta v$ is the range of the opening displacement within the process zone. A set of material parameters enters equation (6.30), δ_d, $\Delta \delta_{th}$, and m_δ. In general, $m_\delta \neq m_\sigma$. Moreover, to provide a continuous transition to the purely elastic case, when equation (6.29) is valid, one must assume $m_\delta \approx 1/2m_\sigma$. Actually, Δv is approximately proportional to $(\Delta \sigma_\infty)^2$, and in most situations $\Delta \sigma \gg \Delta \sigma_{th}$, $2\Delta v \gg \Delta \delta_{th}$.

Instead of displacements, plastic strains may be treated as control parameters. Then we come to the equation

$$\frac{\partial \omega}{\partial N} = \left(\frac{\Delta \varepsilon - \Delta \varepsilon_{th}}{\varepsilon_d} \right)^{m_\varepsilon} \tag{6.31}$$

with material parameters ε_d, $\Delta \varepsilon_{th}$, and m_ε. It is natural to expect that $m_\varepsilon \sim m_\delta$. To evaluate the strain range $\Delta \varepsilon$ in the framework of the thin plastic zone model, we have to "smooth" the cyclic displacement $2\Delta v$ upon the total transverse size of the process zone.

To make a sensible choice between the models given in equations (6.29)-(6.31), we have to compare the results based on each equation with experimental data. The quasistationary approximation (Sections 3.8-3.9) opens the possibility to obtain a coarse approximation for the middle part of the crack propagation history. In particular, the prediction may be made on power exponent m_p in the Paris equation (4.25).

Consider equation (6.29). Let the crack tip be propagating in a slow, continuous way. For an elasto-plastic material with the ideal Bauschinger's effect, the stress range $\Delta \sigma$ in plastic straining is equal to $2\sigma_0$ (Figure 6.2). Neglecting the far field damage and assuming that $2\sigma_0 \gg \Delta \sigma_{th}$, it is easy to obtain the estimate for the damage measure ψ within the process zone

$$\psi \sim \lambda_p \left(\frac{da}{dN} \right)^{-1} \left(\frac{2\sigma_0}{\sigma_d} \right)^{m_\sigma}$$

Here λ_p is the length of the process zone. The crack growth at the Paris stage takes place at the damage measure level close to the critical one, $\psi_* = 1$. Assuming $\psi \sim \psi_*$ and using equation (6.5) for λ_p, we conclude that the crack growth rate follows the relationship

$$\frac{da}{dN} \sim \text{const} \, (\Delta K)^2 \tag{6.32}$$

Equation (6.32) shows that when we follow equation (6.29), the estimate of the Paris exponent is $m_p \sim 2$.

Turning to equation (6.30), we obtain in a similar way that

$$\psi \sim \lambda_p \left(\frac{da}{dN} \right)^{-1} \left(\frac{\Delta \delta}{\delta_d} \right)^{m_\delta}$$

The range $\Delta\delta$ of the crack tip opening displacement, following equation (6.23), is approximately proportional to $a(\Delta\sigma_\infty)^2$. This results in the estimate

$$\frac{da}{dN} \sim \text{const}\,(\Delta K)^{2(1+m_\delta)}$$

and the Paris exponent $m_p \sim 2(1 + m_\delta)$.

At last, consider equation (6.31). Let us relate the range of the crack tip opening displacement to the transverse dimension of the process zone in "real" modeling of the elasto-plastic material. This dimension is of the same order as the length of the process zone. Then we obtain the estimate $\Delta\varepsilon \sim \Delta\delta/\lambda_p \sim \sigma_0/E$. Therefore, the plastic strain smoothed across the process zone does not depend (more exactly, weakly depends) on the applied stress. Such a paradoxical conclusion may be explained as a result of constraints put on plastic straining within a small-scale zone confined in the bulk of the elastically deformed material. As a result, the rate equation in the case of strain-controlled damage is the same as equation (6.32), i.e., $m_p \sim 2$.

Before making the choice between equations (6.29)-(6.31), we must decide which of them gives agreeable magnitudes of the Paris exponent. The choice is between $m_p \sim 2$ and $m_p \sim 2(1+m_\delta)$. In the latter case, we obtain $m_p \sim 4$ even at $m_\delta = 1$. The intuitively more acceptable exponent $m_\delta = 2$ gives $m_p \sim 6$, etc.

It seems that the magnitude $m_p \sim 2$ is more acceptable: the numerical analysis in the framework of the suggested theory gives higher magnitudes such as $m_p = 2.5$ and even more. By the way, the magnitude $m_p = 2$ has been predicted by many simple models which use, as a characteristic length scale, either the crack tip opening displacement or the length of the plastic process zone. This is not surprising as it follows even from a transparent dimension consideration. However, the decisive point is in experimentation.

Unfortunately, experimental data concerning the Paris exponents show a wide scatter even for the same material and the same test conditions. Most tests distinctly correspond to high-cycle fatigue. In the case when low-cycle fatigue is expected, there is not sufficient information to make a conclusive decision concerning the numerical value of the Paris exponent. As a rule, the lower the yield stress and the higher the ductility, the lower the Paris exponent, and vice versa. For mild carbon steels and aluminum alloys the magnitudes $m_p = 2 \ldots 3$ are frequently observed. Higher exponents are typical for low-ductile materials fractured without significant plastic deformations. On the other side, even for the same material, steady test procedure and environmental conditions, the exponent m_p varies in a wide range. The low magnitudes such as $m_p = 1.5 \ldots 2.5$ are met rather frequently. It is evident that the result also depends on the choice of the interval used for estimation of the slope exponent.

The above analysis has been done in the terms of standard crack growth rate diagrams presenting da/dN in the function of ΔK. Rigorously, such an approach is not valid for low-cycle fatigue. Even the notion of stress intensity factor has no physical meaning when plastic straining is of significance. However, if it is desirable to study the relationship between the high-cycle and low-cycle fatigue laws, their joint presentations may be useful.

Let us use equation (6.29) in both cases. To achieve the matching of crack growth rate curves, we must take the exponent m_σ for high-cycle fatigue equal to

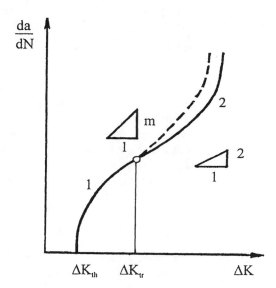

FIGURE 6.8
Schematic presentation of high-cycle fatigue (1) and low-cycle fatigue (2) in the crack growth rate diagram.

2 or 3 as well, and choose appropriate magnitudes for some geometry parameters such as the "blunt" tip radius ρ_b. However, the perfect matching of crack growth laws is not physically sound.

A typical situation is demonstrated in Figure 6.8. Curve 1 corresponds to low applied stress and, therefore, to high-cycle fatigue. The slope of the middle part of the curve is, as usual, close to the power exponent m in the damage accumulation equation (4.21). Curve 2 is drawn for a higher load level. The left-hand part corresponding to near-the-threshold crack growth is absent. The curves intersect at $\Delta K = \Delta K_{tr}$ where ΔK_{tr} is the transition stress intensity factor range, $\Delta K_{tr} = (1 - R)K_{tr}$.

If we accept the scheme presented in Figure 6.8, the crack growth history looks as follows. At the low stress level the crack begins to propagate in the high cycle fatigue pattern (curve 1).When ΔK attains the transition magnitude ΔK_{tr}, the growth switches to the low-cycle fatigue pattern (curve 2).

When the initial stress level is high, and from the beginning the condition $\Delta K \geq \Delta K_{tr}$ holds, the process of crack propagation follows the low-cycle fatigue mechanism. There is some evidence that the scheme presented in Figure 6.8 is acceptable for materials whose rate diagrams in the $\log - \log$ scale may be interpreted as bilinear ones. In general, the real picture is more complicated due to the influence of stress ratio, load, frequency, crack closure and size effect, not mentioning the natural scatter of test data.

In concluding this section, it is expedient to emphasize again that the above analysis concerns small-scale, confined plastic straining. In the case of full-scale straining, the models of plasticity theory complemented with damage models must be used. Moreover, when the plastic straining is significant, its direct contribution into damage must be taken into account. The simplest one-dimensional equation that includes both cyclic and purely plastic components of damage is

$$d\omega = f(\Delta\sigma, \omega, \varepsilon_p)dN + g(\omega, \varepsilon_p)d\sigma$$

The second term in the right-hand side corresponds to the plastic, instantaneous component of damage ($g = 0$ at $d\sigma < 0$). The acquired plastic strain ε_p enters both in $f(\Delta\sigma, \omega, \varepsilon_p)$ and $g(\omega, \varepsilon_p)$. In the author's opinion, it is possible to build up a more consistent theory of plasticity considering plastic straining as a kind of damage. Then, of course, damage measures must be treated as tensorial variables, namely, as the fields of second and fourth rank to take into account the acquired anisotropy of compliance as well as residual stress and stress fields, both in macro- and microscale.

6.5   Generalized Resistance Forces

Following the general theory (Section 3.4), let us apply generalized resistance forces to specific fracture work, i.e., to the amount of work spent on the formation of a unit of new cracked area or, in two-dimensional problems, a unit of new crack length. There are some additional features concerning the specific fracture work when plastic straining is involved. Not only microdamage, but also deformations affect the amount of this work, and these factors, in general, enter into the interaction.

In the framework of the thin plastic zone model we may assume that

$$\gamma = \gamma_0 f(\omega, 2v) \tag{6.33}$$

where γ_0 is the specific fracture work for nondamaged material. The function $f(\omega, 2v)$ takes into account both effects mentioned above. For fatigue cracks the generalized resistance force is equal to the specific fracture work at the tip. Therefore,

$$\Gamma = \Gamma_0 f(\psi, \delta) \tag{6.34}$$

with the same function as in (6.33) depending on the tip damage $\psi(N) \equiv \omega[a(N), N]$ and crack tip opening displacement $\delta(N) \equiv 2v[a(N), N]$.

The right-hand sides of equations (6.33) and (6.34) describe a number of different and, sometimes, controversial mechanisms. Apart from the direct influence of microcracking and plastic straining, the influence of plastic tip blunting on the fracture work must be taken into consideration. As no sufficient experimental data are available to describe the details of these effects, we use in equation (6.33) the multiplicative model:

$$\gamma = \gamma_0 \left[1 + \left(\frac{2v - \delta_{th}}{\delta_p} \right)^p \right] \left[1 - \left(\frac{\omega}{\omega_*} \right)^\alpha \right] \tag{6.35}$$

The first bracket with material parameters δ_p, δ_{th}, and p describes purely plastic effects, the second one the effect of microdamage. At $\delta < \delta_{th}$ the first bracket must

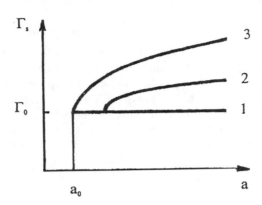

FIGURE 6.9
Generalized resistance force under monotonous loading; lines 1, 2, 3 correspond to three increasing load levels.

be put to unity. Then we come to the model (4.11) used for linear elastic materials. When the loading is monotonous, and cracking at the tip is absent, equation (6.35) gives the generalized resistance force

$$\Gamma_s = \Gamma_0 \left[1 + \left(\frac{\delta - \delta_{th}}{\delta_p} \right)^p \right] \tag{6.36}$$

This force may be interpreted in terms of R-curves (see Section 1.7). As to the cyclic effects in plasticity, they may be attributed to microdamage and taken into account by the second bracket in (6.35).

Equation (6.36) is illustrated in Figure 6.9 where the resistance force Γ_s is plotted against the crack size a. Line 1 corresponds to a crack propagating in linear elastic material under the condition that $\delta < \delta_{th}$ and $\gamma_0 = $ const. Line 2 is drawn for a crack that begins to grow in elastic material while the plastic effects appear later. At last, line 3 is drawn for a crack that propagates in an elasto-plastic pattern from the beginning. In terms of applied stress σ_∞, lines 1, 2, 3 correspond to three increasing load levels.

The picture changes in the case of cyclic loading. In Figure 6.10 the distribution of γ is schematically shown ahead of the fixed crack tip under monotonous loading (line 1) and cyclic loading (lines 2 and 3). Influence of monotonous straining covers all the plastic zone $a \leq |x| \leq a + \lambda$, while the influence of cyclic straining is concentrated within the process zone $a \leq |x| \leq a + \lambda_p$.

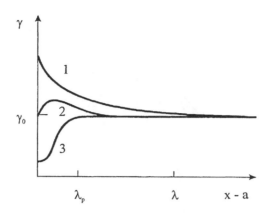

FIGURE 6.10
Specific fracture work distribution ahead of the crack tip: (1) at the first up-loading; (2) after moderate cyclic damage; (3) after significant cyclic damage.

6.6   Constitutive Relationships and Computational Procedure

As in the case of linear elastic material, the governing relationship for a single-parameter fatigue crack is

$$G \gtrless \Gamma \tag{6.37}$$

Here G is the generalized driving force, and Γ is the generalized resistance force. A crack does not propagate at $G < \Gamma$, propagates in a stable, continuous pattern at $G = \Gamma, \partial G/\partial a < \partial \Gamma/\partial a$, and becomes unstable at $G > \Gamma$ or $G = \Gamma, \partial G/\partial a > \partial \Gamma/\partial a$.

The process of crack growth is illustrated in Figure 6.11 in terms of the relationship (6.37) between the generalized forces. There the graphs for the maximum (within a cycle) driving force G, the monotonous resistance force Γ_s, and the current resistance force Γ are plotted schematically. At $N = 0$ we have $G < \Gamma = \Gamma_s$, i.e., subequilibrium state takes place, and the crack tip is fixed. Due to damage accumulation, the resistance force is decreasing until the equality $G = \Gamma$ becomes satisfied. The attained equilibrium state is unstable because $\partial G/\partial a > \partial \Gamma/\partial a$. The first crack advancement is jump-wise with the step covering the distance of the order of the process zone. It is a crack growth pattern depicted in Figure 3.10. The next and, perhaps, a few following advancements also may be jump-wise. Later the process zone becomes less distinctive. This results in the continuous crack growth at $G = \Gamma$, $\partial G/\partial a < \partial \Gamma/\partial a$. It is also shown in Figure 3.11. At the conclusive stage, when each new cycle results in the significant increasing of the driving force, the process may again turn to be jump-wise (compare with Figure 3.11).

For calculations, we assess the left-hand side in (6.37) using equation (6.27).

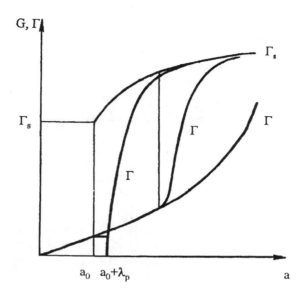

FIGURE 6.11
Relationship between generalized forces during crack propagation in elasto-plastic material.

The right-hand side is taken, according to (6.35), as follows:

$$\Gamma = \Gamma_s \left[1 - \left(\frac{\psi}{\omega_*} \right)^\alpha \right] \tag{6.38}$$

Here Γ_s is determined in (6.36). However, to check the stability conditions, we must use equation (6.35) for the fracture toughness of damaged material ahead of the tip.

Damage accumulation is assumed to be governed by equation (6.29) where $\Delta\sigma(x, N)$ is the range of the tensile stress at $|x| \geq a$, $y = 0$. This range is equal to the stress corresponding to the down-loading part of the cycle, i.e., to $\sigma_\infty^{\max} - \sigma_\infty^{\min}$. Plastic straining must be taken into account both in up-loading and down-loading parts. In particular, when σ_0 is constant, equation (6.17) gives

$$\Delta\sigma = \begin{cases} 2\sigma_0, & a \leq |x| \leq a + \lambda_p \\[2ex] \dfrac{4\sigma_0}{\pi} \cot^{-1}\left[\dfrac{a}{x} \left(\dfrac{x^2 - (a + \lambda_p)^2}{(a + \lambda_p)^2 - a^2} \right)^{1/2} \right], & |x| > a + \lambda_p \end{cases} \tag{6.39}$$

The computational procedure is similar to that in the case of linear elastic material. The principal difference is that we do not deal with the effective tip radius ρ. Its role is transferred to the tip opening displacement δ. Until the crack growth initiation, the steps ΔN of the cycle number N are used. Later they are replaced by the steps Δa of the size a. To diminish numerical errors, the step Δa

is taken as a small portion of the process zone size λ_p estimated in (6.4). In both cases the cycle number N corresponding to equilibrium states is estimated from the condition

$$\psi(N) = \omega_* \left[1 - \frac{G(N)}{\Gamma_s(N)} \right]^{1/\alpha}$$

where $\psi(N)$ is found by integration of equation (6.29) for damage accumulation. To check the stability conditions we must know the distribution of ω ahead of the tip. Therefore, equation (6.29) must be integrated along the future crack path. In the case of instability, the new position of the tip may be found using equation (3.43). However, in most numerical examples presented below, the continuous process of crack growth occurs during the major part of the fatigue life. This circumstance makes the amount of computations comparatively moderate.

6.7   Discussion of Numerical Results

Typical numerical results are presented in Figures 6.12-6.18. The following data are used: $E = 200$ GPa, $\sigma_0 = 500$ MPa, $\gamma_0 = 15$ kJ/m^2, $\sigma_d = 2.5$ GPa, $\Delta\sigma_{th} = 250$ MPa, $m_\sigma = 4$, $\delta_p = 100$ μm, $\delta_{th} = 0$, $\alpha = p = 1$. Computations are performed for $\Delta\sigma_\infty = 200$ MPa, $R = 0.2$, $a_0 = 2$ mm.

The relationship between the generalized forces is given in Figure 6.12a, and

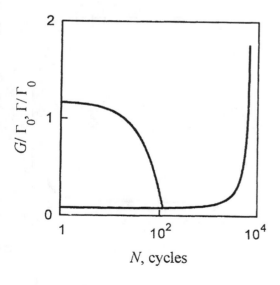
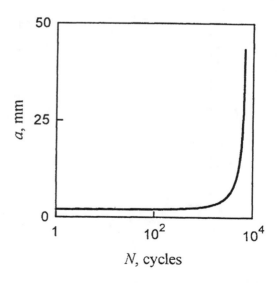

FIGURE 6.12
Relationship between generalized forces (a) and crack growth history (b) in low-cycle fatigue.

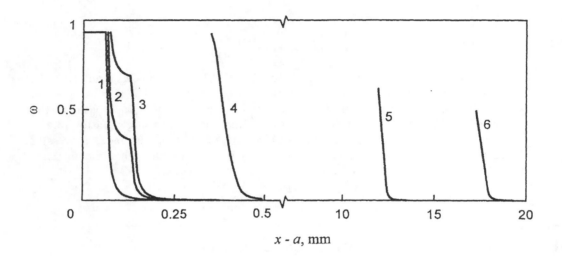

FIGURE 6.13
Damage distribution ahead of the crack tip at crack growth initiation (line 1),
at the earlier stage of growth (lines 2, 3, 4), and at the final stage (lines 5, 6).

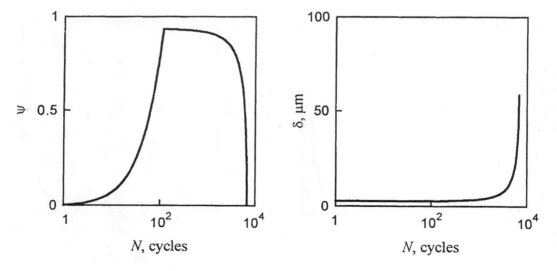

FIGURE 6.14
Crack tip damage measure (a) and reversible crack tip displacement (b) in
low-cycle fatigue.

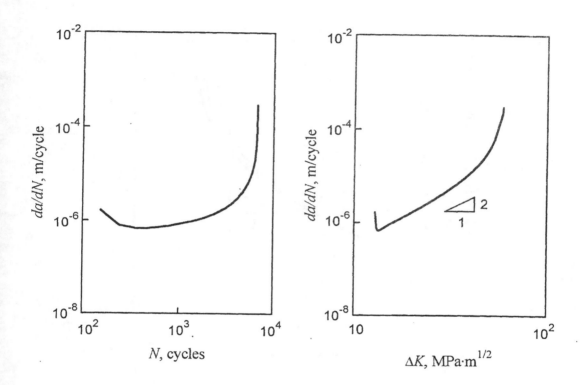

FIGURE 6.15
Crack growth rate history (a) and rate diagram (b) in low-cycle fatigue.

the crack growth history in Figure 6.12b. At $N = 0$ we have $G < \Gamma$. The start of crack growth occurs at $N = N_* = 116$ when the equality $G = \Gamma$ primarily becomes valid. This equality holds later, until the final failure at $N = N_{**} = 6758$. The final crack size at that moment is approaching $a = 50$ mm.

Crack growth is accompanied with gradual decreasing of the damage measure ahead of the crack. It is illustrated in Figure 6.13 where the damage measure ω is plotted versus the coordinate $x - a$. At $N = N_*$ (line 1) we have a uniform distribution of damage within the process zone and a rapid decrease of damage when we go in depth. Very soon (line 2) the picture changes. The plateau corresponding to the process zone vanishes during the further crack propagation, though one can distinguish the domain of more intensive damage accumulation (lines 3, 4). The right-hand part of Figure 6.13 (lines 5 and 6) exhibits an essential decrease of damage everywhere including the tip vicinity. The final failure takes place at

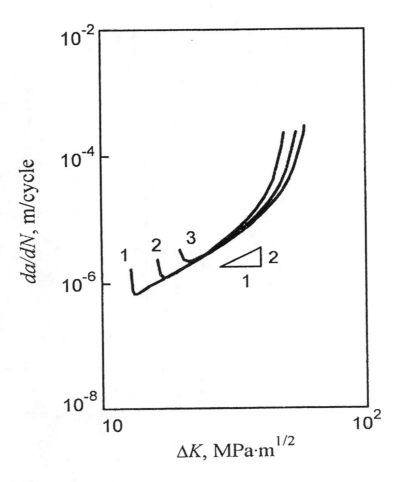

FIGURE 6.16
Crack growth rate versus stress intensity factor range for load levels $\Delta\sigma_\infty$ = 200, 250, and 300 MPa (lines 1, 2, 3, respectively).

comparatively low damage. In this aspect the pattern is similar to that in the linear elastic case.

The damage measure history $\omega(N)$ is presented in Figure 6.14a. The picture is similar to that for the case of linear elastic material. However, in contrast to the latter case where we have observed the crack tip sharpening at the final stage, here the monotonous tip blunting takes place (Figure 6.14b).

There are also some differences in the crack growth rate history as well as in the crack growth rate diagram. It is seen in Figure 6.15a where the ratio da/dN is plotted versus the cycle number and in Figure 6.15b where the rate is presented in the function of the stress intensity factor range $\Delta K = \Delta\sigma_\infty(\pi a)^{1/2}$. First, the typical numerical values of da/dN are of the order of 10^{-5} m/cycle and more. Second, the initial parts of the curves in Figure 6.15a and 6.15b have a "tail up."

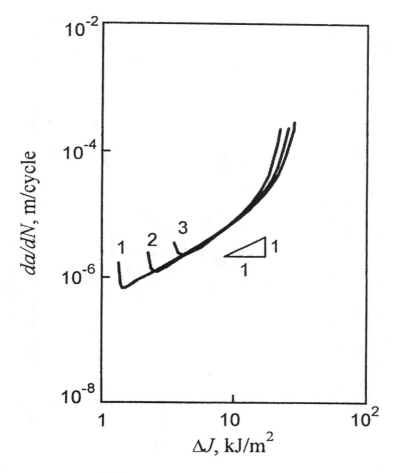

FIGURE 6.17
Crack growth rate versus *J*-integral range for three load levels (notation as in Figure 6.16).

In other words, the crack growth rate has a tendency to decrease at the initial stage. The cause is the practically uniform damage distribution within the process zone during the first stage of the fatigue life. In particular, the first step of the crack tip is a jump covering the initial process zone. Later crack growth becomes more regular and similar to that in the linear elastic case.

The use of standard crack growth rate diagrams for low-cycle fatigue is, rigorously, invalid. Actually, the stress intensity factor range ΔK is not an intrinsic loading parameter for elasto-plastic materials. Perhaps, it is not the best (at least not the only one) parameter that may be treated as correlating with the crack growth rate. This issue is discussed below in terms of diagrams where various load parameters are used.

A comparison of three types of diagrams is given in Figures 6.16-6.18. The

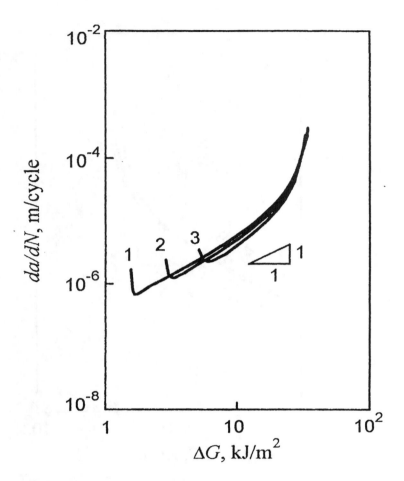

FIGURE 6.18
Crack growth rate versus generalized driving force range for three load levels (notation as in Figure 6.16).

diagrams are obtained for load conditions $\Delta\sigma_\infty = 200$, 250 and 300 MPa (lines 1, 2 and 3 respectively) at load ratio $R = 0.2$. In Figure 6.16 the rate da/dN is plotted, as usual, in ΔK function. The lines diverge at the initial stage as well as close to the final failure. The Paris stage is rather short. However, it is possible to estimate the power exponent m_p at this stage. In particular, the exponent estimated within the interval $\Delta K = 15 \ldots 30 \text{MPa} \cdot \text{m}^{1/2}$ is close to $m_p = 2.6$.

The range ΔJ of the J-integral is used as a load parameter in Figure 6.17. It is seen from equation (6.22) that at $\sigma_0 = \text{const}$ it actually means that the crack tip opening displacement is considered a parameter correlating with the crack growth rate. The divergence of curves is almost of the same order of magnitude as in Figure 6.16. The slope of the middle stage is approximately twice less.

The range ΔG of the generalized stress G given in equation (6.27) is used in

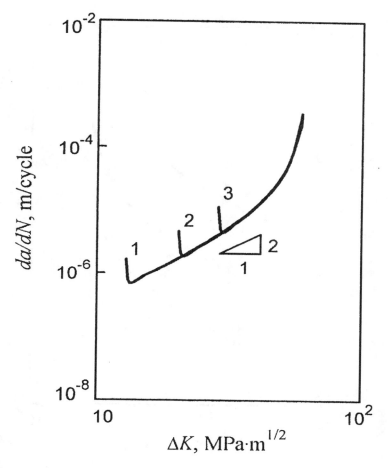

FIGURE 6.19
Crack growth rate diagrams for different initial crack length $a_0 = 2$, 5, 10 mm (lines 1, 2, 3, respectively).

Figure 6.18 as a correlating parameter. Compared to Figures 6.16 and 6.17, the divergence of curves is minimal.

The common feature of the diagrams presented in Figures 6.16–6.18 is that the Paris exponent $m_p \approx 2.5$ when da/dN is plotted against ΔK, and twice less when da/dN is plotted against ΔJ and ΔG. Meantime, the damage accumulation exponent in equation (6.29) is $m = 4$. The order of magnitude $m_p \sim 2$ has been predicted in Section 6.4 with the use of the quasistationary approximation.

It is of interest to vary the exponent m_σ in equation (6.29) and observe how this change affects the crack growth rate. Some numerical results are given in Figure 6.19 for three magnitudes $m_\sigma = 2$, 4, 6 of the exponent in the equation of damage accumulation. Computations are performed for the same data as in Figures 6.12–6.18. The stress range is $\Delta \sigma_\infty = 200$ MPa at $R = 0.2$. All the

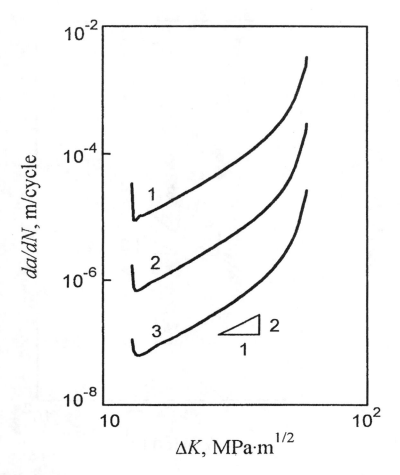

FIGURE 6.20
Influence of the damage accumulation exponent on crack growth rate; lines 1, 2, 3 are drawn for m_p = 2, 4, 6, respectively.

curves exhibit the Paris exponent near $m_p \sim 2.5$ in their middle parts. However, the crack growth rate varies with m_σ to a large degree. Different magnitudes of m_σ correspond to essentially different resistance to damage accumulation, i.e., to different materials. The conclusion is that, assuming damage is stress-controlled, we come to a comparatively low slope in the middle part of the fatigue crack growth rate diagram.

The above analysis has been performed for a Griffith-type crack in an ideal elasto-plastic material. In reality, plasic hardening and/or softening are involved, as well as the influence of damage on material compliance. Some attempts at a more sophisticated analysis will be given at the end of this chapter. In particular, the effect of damage on the yield stress will be discussed in Section 6.10.

6.8   Penny-Shaped Fatigue Crack

A circular planar crack in an unbounded body under cyclic tensile stresses was considered by Bolotin, Lebedev, Murzakhanov, and Nefedov in the paper [37]. The model of thin plastic zone was used there as well as analytical results on the propagation of a similar crack under monotonous loading [114, 152]. Here, following the paper [37], we briefly discuss this problem (Figure 6.21a).

Let the loading be given by the applied stress $\sigma_\infty(t)$ acting along the axis z that is orthogonal to the x, y plane. The circular crack is situated in the x, y plane and propagates, by assumption, remaining circular and situated in this plane. The material is ideal elasto-plastic with the limit stress $\sigma_0 = \text{const}$ within the plastic zone $a \geq r \geq a + \lambda$ under monotonous loading. Here r is the polar radius, a is the crack radius, and λ is the width of the plastic zone (Figure 6.21b). The size λ is given by equation

$$\lambda = a \left[\left(1 - \frac{\sigma_\infty^2}{\sigma_0^2} \right)^{-1/2} - 1 \right] \tag{6.40}$$

and the tip displacement by equation

$$\delta = \frac{8(1 - \nu^2)\sigma_0 a}{\pi E} \left[1 - \left(1 - \frac{\sigma_\infty^2}{\sigma_0^2} \right)^{1/2} \right] \tag{6.41}$$

The tensile stress σ_z ahead of the crack tip is

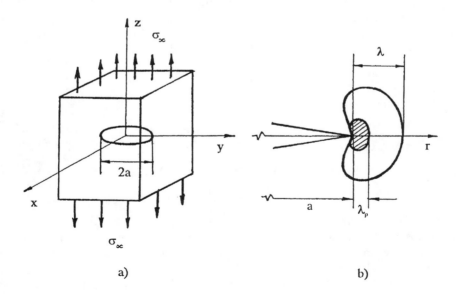

a)   b)

FIGURE 6.21
Penny-shaped crack in elasto-plastic material.

$$\sigma_z = \begin{cases} \sigma_0, & a \le r \le a + \lambda \\ \dfrac{2\sigma_0}{\pi}\left[\sin^{-1}\left(\dfrac{1 - \rho_0^2}{\rho^2 - \rho_0^2}\right)^{1/2} - \zeta \sin^{-1}\dfrac{1}{\rho} + \dfrac{\pi\zeta}{2}\right], & r > a + \lambda \end{cases} \tag{6.42}$$

Hereafter, for brevity the notation used is

$$\zeta = \sigma_\infty/\sigma_0, \quad \rho = r/(a + \lambda), \quad \rho_0 = a/(a + \lambda). \tag{6.43}$$

When the applied stresses vary in time cyclically with extremal magnitudes $\sigma_\infty^{\max}$ and $\sigma_\infty^{\min}$, the reversible plastic zone width λ_p may be calculated replacing in (6.40) σ_∞ with $\Delta\sigma_\infty = \sigma_\infty^{\max} - \sigma_\infty^{\min}$ and the limit stress σ_0 with $2\sigma_0$. This results in the process zone width

$$\lambda_p = a\left[\left(1 - \frac{\Delta\sigma_\infty^2}{4\sigma_0^2}\right)^{-1/2} - 1\right] \tag{6.44}$$

The maximal tensile stress distribution within a cycle is given by equations (6.42) and (6.43) at $\sigma_\infty = \sigma_\infty^{\max}$. The stress range $\Delta\sigma_z$ is also given by these equations when we replace σ_∞ with $\Delta\sigma_\infty$, σ_0 with $2\sigma_0$, and λ with λ_p. In general, the situation is very similar to that for Griffith's crack under cyclic loading (Section 6.7).

As before, the condition of fatigue crack (non)propagation is given by the relationship $G \lessgtr \Gamma$. The generalized driving force G is taken at the moment when $\sigma_\infty(t)$ attains the maximum. The generalized resistance force Γ is to be calculated taking into account microdamage and plastic straining. Neglecting the influence of plastic deformations and microdamage on the generalized driving force, we identify this force with the strain energy release rate around all of the crack boundary. The stress intensity factor for mode I circular cracks is

$$K_I = 2\sigma_\infty(a/\pi)^{1/2}$$

The application of Irwin's equation gives

$$G = \frac{8\sigma_\infty^2 a^2(1 - \nu^2)}{E} \tag{6.45}$$

The J-integral technique results in

$$J = 2\pi a\sigma_0\delta \tag{6.46}$$

where δ is determined in (6.41). However, as in the case of Griffith's crack, equation (6.46) does not take into account the variation of the plastic zone size and, therefore, cannot be interpreted as the generalized force in the sense of the virtual work principle. The "true" generalized driving force is given in equation (6.25).

Let us determine the virtual work $\delta W_e + \delta W_i$ as the amount of work performed by the tensile stress $\sigma_z(r, 0)$ in the case of returning sizes $a + \delta a$ and $\lambda + \delta\lambda$ to the former sizes a and λ. Then

$$G = \frac{\delta W_e + \delta W_i}{\delta a} = \frac{\partial}{\partial a}\left[\int_0^{2\pi}\int_0^a \sigma_\infty w r\,dr\,d\varphi + \int_0^{2\pi}\int_a^{a+\lambda}(\sigma_0 - \sigma_\infty)w r\,dr\,d\varphi\right]$$

Here $w(r)$ is the displacement in z-direction at $z = 0$:

$$w(r) = \begin{cases} \dfrac{4(1-\nu^2)\sigma_0 a}{\pi E}\left[E(\mu_1,\eta) - \dfrac{\zeta}{\rho_0}\dfrac{\rho^2}{(1-\rho^2)^{1/2}}\right], & 0 \le r \le a \\[3mm] \dfrac{4(1-\nu^2)\sigma_0 a}{\pi E}\left[\dfrac{\rho}{\rho_0}E(\mu_2,\eta^{-1}) - \dfrac{(1-\zeta^2)(1-\rho^2)^{1/2}}{\zeta\rho_0} - \right. \\[3mm] \left. - \dfrac{\rho^2 - \rho_0^2}{\rho\rho_0}F(\mu_2,\eta^{-1})\right], & a \le r \le a+\lambda \end{cases}$$

Apart from notation (6.43), the following notation is used here:

$$\mu_1 = \sin^{-1}\left(\frac{1-\rho_0^2}{1-\rho^2}\right)^{1/2}, \quad \mu_2 = \sin^{-1}\left(\frac{1-\rho^2}{1-\rho_0^2}\right)^{1/2}, \quad \eta = \rho/\rho_0$$

as well as the notation for noncomplete elliptical integrals

$$E(\mu,\eta) = \int_0^\mu (1 - \eta^2\sin^2\xi)^{1/2}d\xi$$

$$F(\mu,\eta) = \int_0^\mu (1 - \eta^2\sin^2\xi)^{-1/2}d\xi$$

Further calculation results into the formula

$$G = \frac{16(1-\nu^2)\sigma_0^2 a^2}{E\left[(1-\zeta^2)^{-1/2} - 1\right]} \tag{6.47}$$

The details can be found in [152] where the energy dissipation in the vicinity of the tip is estimated in the framework of the thin plastic zone model. In equilibrium states the virtual fracture work $\delta W_f = -(\delta W_e + \delta A_i)$; hence the final result is valid for the generalized driving force.

The comparison of numerical results with use of equations (6.45), (6.46), and (6.47) is presented in Figure 6.22. The curves 1, 2 and 3 correspond to that sequence of equations. The driving force is normalized with respect to its characteristic magnitude

$$G_0 = \frac{8(1-\nu^2)\sigma_0^2 a^2}{E}$$

Similar to the case of the central crack in a plate, the divergence of curves is of the same order, and this divergence grows when σ_∞/σ_0 increases (see also Figure 6.7).

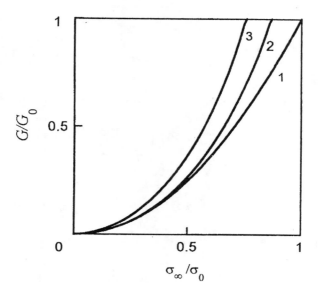

FIGURE 6.22
Normalized load parameters for circular crack: (1) energy release rate;
(2) *J*-integral; (3) generalized driving force.

The generalized resistance force depends on the specific fracture work γ_0 for nondamaged and elastically deformed material as well as on the conditions in the vicinity of the crack tip. Two groups of factors are to be taken into account: plastic straining in the plastic zone and microdamage accumulation. Each factor can produce both hardening and softening effects. In a typical situation, the resistance to crack growth increases due to plastic straining. On the contrary, microdamage usually results in decreasing resistance. For the further analysis we assume the model for the specific fracture work similar to (6.35):

$$\gamma = \gamma_0 \left[1 + \left(\frac{2w - \delta_{th}}{\delta_f} \right)^\beta \right] \left[1 - \left(\frac{\omega}{\omega_*} \right)^\alpha \right], \quad a \le r \le a + \lambda$$

Here δ_f, δ_{th}, ω_*, α and β are material parameters, and $2w(r)$ is the crack opening displacement within the plastic zone. The corresponding resistance force Γ satisfies the condition $\delta W_f = -\Gamma_0 \delta a$. Hence $\Gamma = 2\pi a \gamma$ where γ follows equation (6.46) with $\Gamma_0 = 2\pi a \gamma_0$. At $2w < \delta_{th}$ the first bracket must be put to unity, and we return to equation (4.6).

To close the system of equations, we need an equation governing damage accumulation. We take equation (4.3) which in the notation of this section takes the form ($\Delta\sigma_z > \Delta\sigma_{th}$):

$$\frac{\partial \omega}{\partial N} = \left(\frac{\Delta\sigma_z - \Delta\sigma_{th}}{\sigma_d} \right)^m (1 - \omega)^{-n} \tag{6.48}$$

The predicted crack growth rate diagrams are shown in Figures 6.23-6.25 for

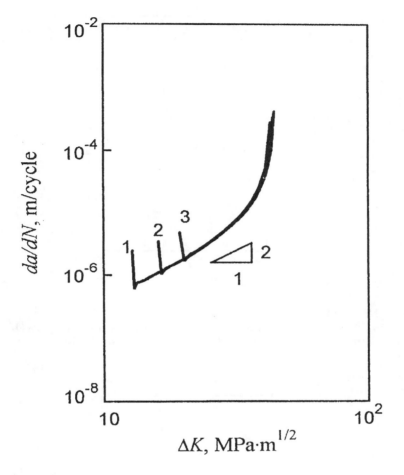

FIGURE 6.23
Influence of load level on the penny-shaped crack growth rate; lines 1,
2, 3 correspond to $\Delta\sigma_\infty = 200, 250, 300$ MPa.

$E = 200$ GPa, $\nu = 0.3$, $\sigma_0 = 500$ MPa, $\sigma_d = 25$ GPa, $\Delta\sigma_{th} = 250$ MPa, $\gamma_0 = 20$ kJ/m^2, $\alpha = \beta = n = 1$. Figure 6.23 illustrates the influence of load level on the crack growth rate. Lines 1, 2, 3 are drawn for $\Delta\sigma_\infty = 200, 250, 300$ MPa at $R = 0.2$. General character of the diagram is similar to that in Figure 6.16. The slope exponent m_p on the middle stage as in Griffith's type cracks takes the value between 2 and 3.

The influence of the initial crack size a_0 is demonstrated in Figure 6.24. Lines 1, 2, 3 are drawn for $a_0 = 2, 10, 15$ mm. The initial size affects only the early part of the diagrams. However, the influence upon total fatigue life may be drastic.

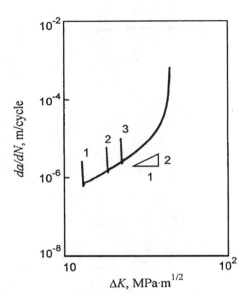

FIGURE 6.24
Influence of initial size of penny-shaped crack on the growth rate; lines
1, 2, 3 correspond to $a_0 = 2, 10, 15$ mm, respectively.

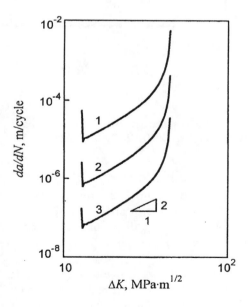

FIGURE 6.25
Crack growth diagram for power exponents $m = 2, 4, 6$ in equation of
damage accumulation (lines 1, 2, 3, respectively).

At last, the influence of exponent m in equation (6.48) of damage accumulation is illustrated in Figure 6.25. Lines 1, 2, 3 correspond to $m = 2, 4, 6$. The effect of m on the rate da/dN is significant, though various m, in general, refer to various materials. Some additional examples can be found in the paper [37].

6.9 Crack Propagation Under Nonsteady Loading

Fatigue crack growth, in general, is strongly history dependent. This dependence enters into analytical models not only with the current crack length and damage accumulated during the early stage, but also with the conditions at the crack tip. If the material is deformed plastically, the tip may be schematized as a blunt one. Then the crack tip opening displacement becomes one of the main variables in the model of fatigue crack growth.

Three representative load histories are schematically depicted in Figure 6.26. Two applied stress levels are considered, moderate and high. A moderate cyclic loading with a single overloading is shown in Figure 6.26a. Switching from one load regime to another is shown in Figures 6.26b and 6.26c. The corresponding crack growth histories are presented in Figure 6.27. Lines 1 and 2 in Figure 6.27 are drawn for moderate and high stress levels, respectively. In the first case the crack tip is comparatively sharp; in the second case blunting is significant. Line 3 shows crack growth in the case of single overloading (Figure 6.26a). Under overloading the crack tip is subjected to significant blunting, and a domain of essential compression residual stresses appears at the tip. Both factors produce a temporary arrest of the crack growth. A certain number of cycles is needed to form tip conditions corresponding to the lower stress level, and an additional number is required to cross the

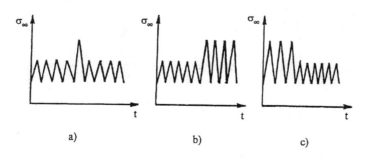

FIGURE 6.26
Nonsteady loading histories: (a) single overloading; (b) switch to higher level; (c) switch to lower level.

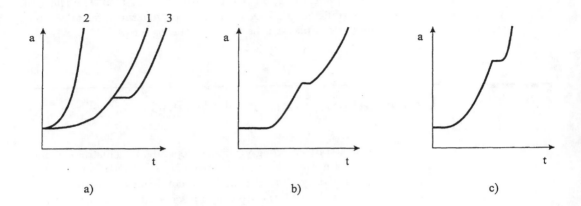

FIGURE 6.27
Crack growth histories: (a) at single overloading; (b) and (c) at switching to other load levels.

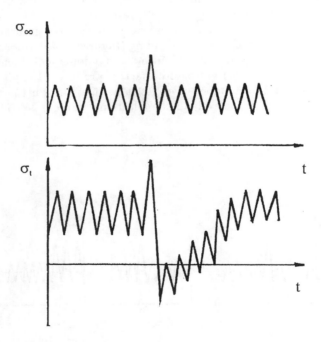

FIGURE 6.28
Schematic stress history in case of single overloading: (a) applied stress; (b) crack tip stress.

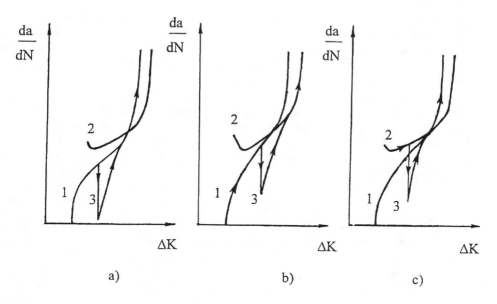

FIGURE 6.29
Schematic crack growth rate diagrams for cases a, b, and c in Figure 6.26.

domain of residual compression stresses. It is illustrated in Figure 6.28 where applied $\sigma_\infty(t)$ and tip $\sigma_t(t)$ stresses are presented schematically. After overloading the tip stress becomes compressive and remains such until the new regime is restored. Because of this, the total fatigue life becomes longer than that in the absence of overloading (Figure 6.27a). This effect has been observed by many experimenters and discussed in literature.

The influence of load history is especially significant when we have alternating high-cycle and low-cycle fatigue crack propagation. It is of interest to consider the interaction of two mechanisms, just to show the application of different analytical models for special numerical examples. The load histories depicted in Figure 6.26 are illustrated in Figure 6.29 in terms of fatigue crack growth diagrams. For simplification, we do not include into consideration the threshold and closure effects which, in general, expose themselves in different ways in high- and low-cycle fatigue.

Line 1 in Figure 6.29 corresponds to high-cycle fatigue, line 2 to low-cycle fatigue. Figure 6.29a presents the case when the high-cycle fatigue crack growth is interrupted by overloading that is accompanied with distinct plastic straining. In the beginning we follow line 1 until the overloading occurs. Then the arrest of crack growth takes place with the consequent slow return to line 1. Figure 6.29b corresponds to the loading regime depicted in Figure 6.26b. We follow line 1. Then, due to the switch to higher applied stress, we transfer to line 2. Figure 6.29c is drawn for the case shown in Figure 6.26c. The initial stage of crack growth follows the low-cycle pattern (line 2). Then we observe a temporary crack arrest until appropriate conditions are attained for the further crack propagation in the high-cycle pattern (line 1).

The border between the two types of fatigue is rather vague. Moreover, presenting the crack growth rate in the function of the stress intensity factor range and using the transition estimates (6.6) or (6.8), we must change the model when ΔK crosses the transition border $\Delta K_{tr} = K_{tr}(1 - R)$. In any case, the schematic diagrams presented in Figures 6.27 and 6.29 offer the general picture of the influence of the load history on fatigue crack growth.

A more detailed analysis of the influence of overloadings is given in the paper by Bolotin and Lebedev [36]. The analytical model developed in Chapters 4 and 5 has been applied for the regular loading regime, and the thin plastic zone model for overloadings. Equation (4.2) has been used for damage accumulation. The load ratio has been assumed constant, while the resistance threshold stress has been taken in the form

$$\Delta \sigma_{th} = \Delta \sigma_{th}^0 - k\sigma_{res} \tag{6.49}$$

Here $\Delta \sigma_{th}^0$ and k are material parameters, and σ_{res} is the residual stress appearing as a result of overloading (positive in tension). Equation (6.49) is similar to equations (4.5), (4.28) etc. describing threshold effects in fatigue. Compared with them, equation (6.49) allows the effect of residual compression within the plastic zone, as well as the recovery of the regular stress distribution with the further crack propagation.

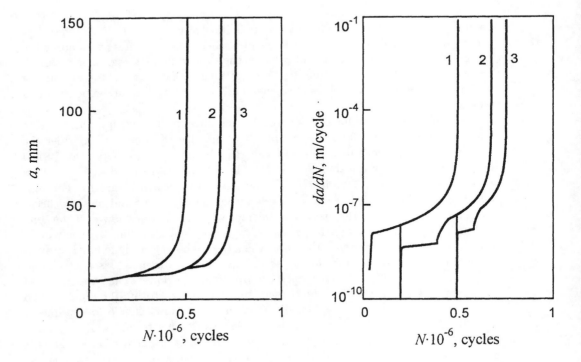

FIGURE 6.30
Crack size and growth rate histories under regular loading (1); single overloading (2); two overloadings (3).

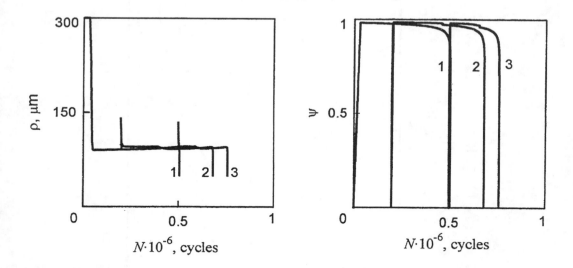

FIGURE 6.31
Crack tip radius and tip damage measure; notation as in Figure 6.30.

Some numerical results for a central crack in a plate under remote cyclic tension are presented in Figures 6.30-6.33. The loading process during the regular stage is taken with $\Delta\sigma_\infty = 50$ MPa, $R = 0.2$. At overloading we assume $\Delta\sigma_\infty = 300$ MPa. Material parameters in high-cycle fatigue are: $E = 200$ GPa, $\sigma_d = 5$ GPa, $\Delta\sigma_{th} = 250$ MPa, $m_\sigma = 4$, $\alpha = 1$, $\rho_s = 50\mu$m, $\rho_b = \lambda_\rho = 100\mu$m. During the elasto-plastic stage the material parameters are: $\sigma_0 = 500$ MPa, $\delta_f = 100\mu$m,

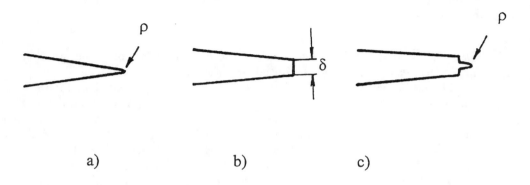

a)   b)   c)

FIGURE 6.32
Transition from regular crack tip (a) to plastically blunted tip (b), and vice versa (c).

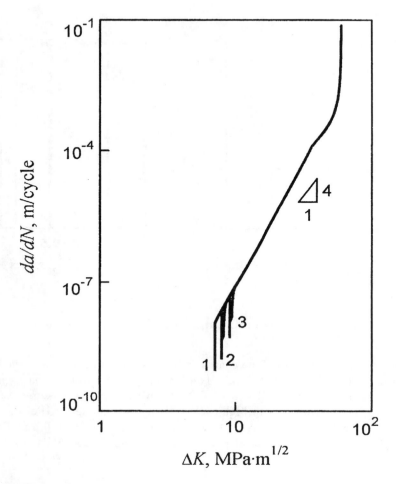

FIGURE 6.33
Crack growth rate diagram in presence of overloading; notation as on Figure 6.30.

$\delta_{th} = 0$, $\beta = 0.5$, $k = 0.5$. The width of the plate $w = 500$ mm, the initial crack half-length $a_0 = 10$ mm and the initial tip radius $\rho_0 = 300 \mu$m.

The crack size and crack growth rate histories are presented in Figure 6.30. Line 1 shows the crack growth under the regular loading ($\Delta\sigma_\infty = 50$ MPa). Line 2 corresponds to a single overloading after $N = 2 \cdot 10^5$ cycles with the subsequent return to the regular process. Line 3 is obtained for the case of two overloadings, at $N = 2 \cdot 10^5$ and $N = 5 \cdot 10^5$. The total fatigue life under the regular loading is equal to $5.05 \cdot 10^5$ cycles. This life is subject to increasing till $6.82 \cdot 10^5$ and $7.58 \cdot 10^5$ cycles under one and two overloadings, respectively.

The variation of the effective tip radius ρ and damage measure at the tip ψ is depicted in Figure 6.31. As the initial tip radius $\rho_0 > \rho_b$, the initial stage of fatigue damage is accompanied by the tip sharpening until the magnitudes are a little smaller than ρ_b. In the absence of overloadings (line 1), the process of crack growth proceeds at an approximately constant effective radius. Sharpening takes place only on the concluding stage. In the case of overloading with a distinct excursion into the elasto-plastic stage, a tip blunting occurs. The half of the crack tip opening displacement is shown in Figure 6.31a with vertical lines. After the return to the regular regime, we follow the high-cycle fatigue model. The latter, in general, is not compatible with the thin plastic zone model. By definition, the effective tip radius in elastic media does not exceed a certain magnitude, the "blunt" radius close to ρ_b. To match the two models, we assume that after overloading the crack begins to develop with the tip radius equal to $\delta/2$ (Figure 6.32). As to the tip damage, we assume that the plastic straining results in a new situation at the tip, at least when the cyclic damage is concerned (Figure 6.31b). In any case, both the effective tip radius and the tip damage measure begin to decrease rapidly when we are approaching the final failure.

The crack growth rate diagram in the presence of overloading is given in Figure 6.33. Line 1 corresponds to the regular regime, line 2 to a single overloading, line 3 to two overloadings. Each overloading results in the drop of the crack growth rate and temporary retardation. However, the duration of the retardation stage is short and practically nondistinguishable in the log scale.

6.10 Influence of Cyclic Softening on Fatigue Crack Growth

The effect of dispersed damage on fatigue crack growth is displayed in several ways. The most important factor is the influence of damage on material fracture toughness. Another effect, the change of material compliance, is less significant. The latter effect has been studied in Sections 4.10 and 4.11 with the application to elastic (more precisely, hypo-elastic) materials. It has been shown that the influence of increased compliance in the vicinity of the crack tips on the generalized driving forces may be treated as negligible. Generalized resistance forces vary more significantly. However, this effect is always present. The explanation of this phenomenon is obvious. Due to lower compliance, the stresses near the crack tips relax, and this results in retardation of the damage accumulation process.

One has to expect a similar phenomenon in low-cycle fatigue. Computational difficulties are enormous when the full-scale plasticity theory is concerned. However, the situation is much simpler for the thin plastic zone model. Neglecting the influence of damage on elastic properties outside the plastic zone, we take into consideration its influence on the tensile stresses within this zone. This means that we assume $\sigma = \sigma_0(\omega)$ at $a \leq |x| \leq a + \lambda$ where ω must be found as a solution of the damage accumulation equation (Figure 6.34a). It is essential that this type of influence differs from purely plastic softening and hardening which appear in monotonous loading.

The limit stress may depend on the damage measure ω within the process zone, for example, as follows:

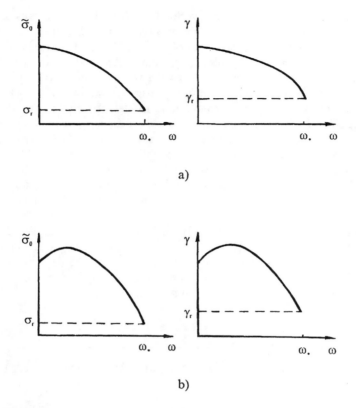

a)

b)

FIGURE 6.34
Limit stress and specific fracture work in the vicinity of the crack tip in case of softening (a) and temporary cyclic hardening (b).

$$\tilde{\sigma}_0 = \sigma_0 \left[1 - c \left(\frac{\omega}{\omega_*} \right)^{\beta} \right] \qquad (6.50)$$

Here σ_0 is the limit stress for nondamaged material; $\omega_* > 0$, $c > 0$ and $\beta > 0$ are material parameters. The choice of a right-hand side in (6.50) is rather arbitrary. In this case the equation describes softening. To include a cyclic hardening stage, we may assume that

$$\tilde{\sigma}_0 = \sigma_0 (1 - c_1 \omega^{\beta_1} + c_2 \omega^{\beta_2})$$

at $c_1 > 0$, $c_2 > 0$, $\beta_1 > 0$, $\beta_2 > 0$. Choosing appropriate numerical values, we could describe the increase of $\tilde{\sigma}_0$ at the first stage of damage accumulation, and the later decrease. A similar model can be used for the specific fracture work to include tip shielding because of microcracking (Figure 6.34b).

To develop an analytical technique for the case of limit stresses varying within the process zone, we must generalize the solution for the classical thin plastic zone model. In the framework of fracture mechanics, such a generalization has been considered by a number of authors [118, 153]. In fatigue the yield stress (or its equivalent) is not known beforehand. This stress is to be found as a result of a solution of a set of equations, including the equation for the length of the plastic zone. Hence the problem becomes more complicated than the corresponding problem of fracture mechanics. Here we follow the paper by Bolotin and Kovekh [35].

As the bulk of material is elastic, we can apply the superposition technique extending the solution of the problem for $\sigma_0 = \mathrm{const}$ upon the case when the limit stress $\tilde{\sigma}_0(x)$ varies within the plastic zone. Let $\tilde{\sigma}_0(x)$ be a differentiable function. The potential given in (6.15) is generalized as follows:

$$Z(z) = \frac{z}{(z^2 - b^2)^{1/2}} \left[\sigma_\infty - \frac{2\tilde{\sigma}_0(0)}{\pi} \cos^{-1}\left(\frac{a}{b}\right) - \frac{2}{\pi} \int_a^b \frac{d\tilde{\sigma}_0(\xi)}{d\xi} \cos^{-1}\left(\frac{\xi}{b}\right) d\xi \right] +$$

$$+ \frac{2\tilde{\sigma}_0(0)}{\pi} \tan^{-1}\left[\frac{a}{z}\left(\frac{z^2 - b^2}{b^2 - a^2}\right)^{1/2} \right] + \frac{2}{\pi} \int_a^b \frac{d\tilde{\sigma}_0(\xi)}{d\xi} \tan^{-1}\left[\frac{\xi}{z}\left(\frac{z^2 - b^2}{b^2 - \xi^2}\right)^{1/2} \right] d\xi$$

The condition that the stress singularity at $x = b = a + \lambda$, $y = 0$ is absent results in the equation with respect to λ:

$$\sigma(0) \cos^{-1}\left(\frac{a}{a+\lambda}\right) + \int_a^{a+\lambda} \frac{d\tilde{\sigma}_0(\xi)}{d\xi} \cos^{-1}\left(\frac{\xi}{a+\lambda}\right) d\xi = \frac{\pi\sigma_\infty}{2} \tag{6.51}$$

This equation is a generalization of the corresponding equation in the case when $\tilde{\sigma}_0 = \mathrm{const}$. Taking into account (6.51) we obtain

$$Z(z) = \frac{2\tilde{\sigma}_0(0)}{\pi} \tan^{-1}\left[\frac{a}{z}\left(\frac{z^2 - b^2}{b^2 - a^2}\right)^{1/2} \right] +$$

$$+ \frac{2}{\pi} \int_a^b \frac{d\tilde{\sigma}_0(\xi)}{d\xi} \tan^{-1}\left[\frac{\xi}{z}\left(\frac{z^2 - b^2}{b^2 - \xi^2}\right)^{1/2} \right] d\xi \tag{6.52}$$

A similar approach is applicable for the displacement $v(x)$ of the crack faces at $a \le x \le a + \lambda$. It is more convenient to use equation (6.19). Then

$$v(x, 0) = c\tilde{\sigma}_0(0)[(x - a)M(b, x, a) - (x + a)M(b, x, -a)] +$$

$$+ c \int_a^b \frac{d\tilde{\sigma}_0(\xi)}{d\xi}[(x - \xi)M(b, x, \xi) - (x + \xi)M(b, x, -\xi)]d\xi \tag{6.53}$$

where $c = 1/\pi E$ for the plane-stress state, and $c = (1 - \nu^2)/\pi E$ for the plane-strain state.

The next step is to determine the generalized forces. Equation (6.26) remains valid when $\sigma_0 = \tilde{\sigma}_0(x)$; but the result of substitution of equation (6.11), (6.52) and (6.53) in (6.26) is too cumbersome to present here. The direct numerical procedure is more appropriate in this situation. We use equation (6.38) for the generalized resistance force with $\psi(N) = \omega[a(N), N]$, and equation (6.48) for the damage measure $\omega(x, N)$.

The computational procedure includes the solution of equations (6.51)-(6.53), (6.26), (6.38), (6.48) and (6.50) in combination with the stability conditions $G < \Gamma$ for fixed cracks and $G = \Gamma, \partial G/\partial a < \partial \Gamma/\partial a$ for continuously propagating cracks. In general, the procedure is similar to that used in the former study, being complicated by involving equations (6.51)-(6.53).

The following numerical data are used: $E = 200$ GPa, $\sigma_0 = 500$ MPa, $\gamma_0 = 15$ kJ/m^2, $\sigma_d = 5$ GPa, $\Delta\sigma_{th} = 250$ MPa, $\omega_* = 1$, $m_\sigma = \beta = 4$, $c = 0.5$, $\alpha = 1$, $n = 0$. When no opposite is mentioned, it is assumed that $\Delta\sigma_\infty = 150$ MPa, $R = 0.5$, $a_0 = 1$ mm.

The distribution of the tensile stress range $\Delta\sigma$ during the initiation stage is shown in Figure 6.35a. The start of the crack occurs at $N = N_* = 3305$. The curves 1, 2, 3 are plotted for $N = 500, 1500$, and 2500. Curve 4 corresponds to $N = N_* = 3305$. The plastic zone consists of two sections. Within the first section the range $\Delta\sigma$ is uniformly distributed, decreasing with N due to microdamage. The remaining part is subjected to cyclic stresses that grow with $x - a_0$. At $x = a_0 + \lambda_p$ the stress range $\Delta\sigma$ attains the double "virgin" yield stress, $\Delta\sigma = 2\sigma_0$. Further,

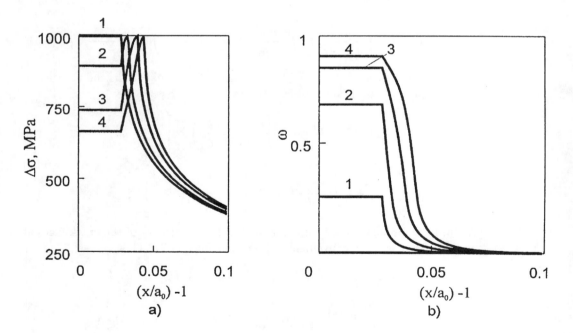

FIGURE 6.35
Tensile stress range (a) and damage measure (b) distribution at the initiation stage; lines 1, 2, 3, 4 are drawn for $N = 500, 1500, 2500$, and $N_* = 3305$, respectively.

the elastic distribution of reversible stresses takes place.

Additional information is presented in Figure 6.35b. There the distribution of the damage measure ω is plotted for the same cycle numbers as in Figure 6.35,a. The damage is constant within the cyclic process zone and then decreases rapidly. The far field damage is insignificant.

Figure 6.36a illustrates the stage of early crack growth. The first crack tip advancement at $N = N_*$ is jump-wise (section AB). The size of the first jump slightly exceeds the length of the process zone. The section BC corresponds to the nonstationary crack propagation when the crack growth rate is an oscillating function of the cycle number. An intermittent acceleration and retardation takes place. Some sections of the curve may be interpreted as jumps of crack growth smoothed because of a coarse mesh. However, the used mesh is sufficiently fine with steps equal to 1/50 of the length of the process zone. The nonstationary stage takes a comparatively short portion of the fatigue life. The regular growth begins in the vicinity of the point C in Figure 6.35a and proceeds until the final rupture.

The initial part of the crack growth rate diagram is given in Figure 6.36,b. The relationship between da/dN and ΔK contains several peaks that might be interpreted as corresponding to the jumps of the crack. Actually, at some points the rate da/dN changes in an order of magnitude. But it ought to be noted that this stage is a small part of the total diagram. In the discussed numerical example it is section $8.5 \leq \Delta K \leq 9.5 \ \mathrm{MPa \cdot m^{1/2}}$.

Compared, say, with Figure 6.15, Figure 6.36 exhibits a more nonsteady be-

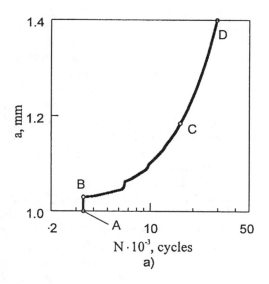
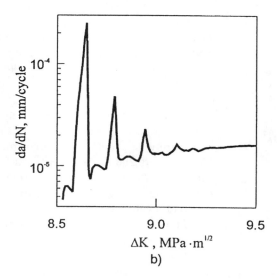

FIGURE 6.36
Early stage of crack growth: (a) crack history; (b) growth rate diagram.

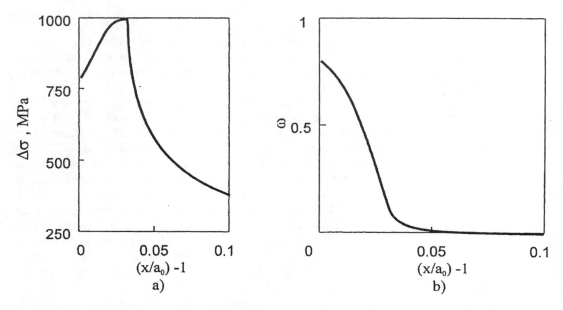

FIGURE 6.37
Tensile stress range (a) and damage measure (b) distribution at $N = 5 \cdot 10^4$ for the same data as in Figures 6.34 and 6.36.

havior of the crack tip at the beginning of crack growth. The cause is the choice of finer mesh and the refusal of any smoothing procedure. The general character of the process remains the same.

A typical distribution of $\Delta \sigma$ ahead of the crack tip during the regular growth stage is shown in Figure 6.37a. It is plotted at $N = 5 \cdot 10^4$. The stress range $\Delta \sigma$ varies within the process zone monotonously increasing from $\Delta \sigma = 800$ MPa at $x = a$ until $\Delta \sigma = 1000$ MPa at $x = a + \lambda_p$. The corresponding distribution of ω is shown in Figure 6.37b. This distribution is a result of the summation of all the damage accumulated in the past. The process of crack propagation is distinctly continuous.

Some results of parametrical analysis performed by Bolotin and Kovekh [35] are presented in Figures 6.38-6.40. The influence of the loading level is illustrated in Figure 6.38a plotted for three magnitudes of the applied stress range, $\Delta \sigma_\infty = 150, 175$ and 250 MPa (lines 1, 2, 3 respectively). In all cases $R = 0.5$. General character of curves changes insignificantly. There is only a quantitative difference, namely the increasing of the initiation stage as well as of the total fatigue life when the load level increases.

The corresponding crack growth rate diagrams are presented in Figure 6.38b. The solid curves show the averaged rate da/dN. The light curves show upper and lower envelopes of the results of direct numerical simulation. The details are not shown as they are the same as in Figure 6.37. The crack behavior is very similar to that of the so-called "small" cracks. In particular, one can observe the diminishing of the crack growth rate at the earlier stage when we discussed the process in terms of averaged magnitudes. Another peculiarity is a large scatter of

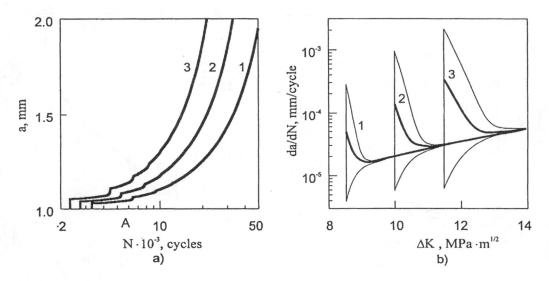

FIGURE 6.38
Influence of the load level on crack history (a) and growth rate diagram (b);
lines 1, 2, 3 are drawn for $\Delta\sigma_\infty = 150$, 200, 250 MPa at $R = 0.5$.

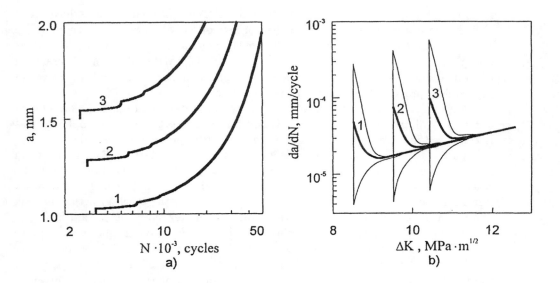

FIGURE 6.39
Influence of the initial crack size on crack history (a) and growth rate diagram
(b); lines 1, 2, 3 are drawn for $a_0 = 1$, 1.25, 1.5 mm and $\Delta\sigma_\infty = 150$ MPa.

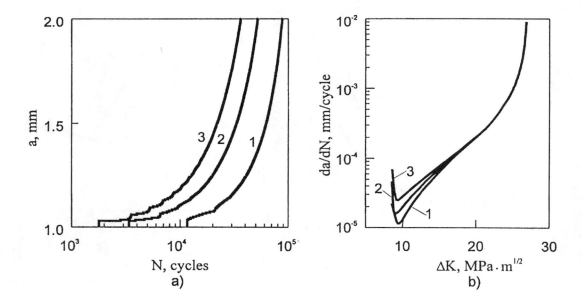

FIGURE 6.40
Influence of cyclic softening on crack history (a) and growth rate diagram (b);
lines 1, 2, 3 correspond to $c = 0$, 0.5, 0.8, respectively.

rates corresponding to the same magnitudes of ΔK. Similar features have been already discussed in Section 2.6 in the context of "small" cracks. It ought to be mentioned that the upper bound of the nonstationary behavior in Figure 6.38 does not exceed 12.5 MPa $\cdot$ m$^{1/2}$ while $K_c = (\gamma_0 E)^{1/2} \approx 55$ MPa $\cdot$ m$^{1/2}$ for the assumed numerical data. After the termination of the nonstationary stage, all the curves in Figure 6.38 practically coalesce in the same line.

The effect of the initial crack length is illustrated on Figure 6.39a. There curves 1, 2, 3 are drawn for $a_0 = 1, 1.25, 1.5$ mm, respectively. The applied stress is $\Delta\sigma_\infty = 150$ MPa at $R = 0.5$. The crack growth curves $a = a(N)$ deviate with a_0 significantly. However, the cycle numbers until the start and until the total rupture do not change to a large degree. The same conclusion is applicable to the crack growth rate diagrams (Figure 6.39b). The initial part of the diagrams exposes significant scatter. Depending on the initial crack size, the crack growth rate may differ in an order of magnitude for the same ΔK. Later this scatter vanishes.

It is of interest to estimate the effect of cyclic softening on fatigue crack growth. This effect has been introduced in equation (6.50) with the material constant c. At $c = 0$ the cyclic softening effect is absent. Both the specific fracture work γ and the yield stress $\tilde{\sigma}_0$ decrease when c increases. Some numerical results are given in Figure 6.40. They are obtained for $\Delta\sigma_\infty = 150$ MPa, $R = 0.5$, and $\beta = 4$. Three magnitudes of c are considered: $c = 0$ (curve 1), $c = 0.5$ (curve 2) and $c = 0.8$ (curve 3). The change of the initiation stage duration is drastic: the number N_* increases in an order of magnitude when we come from $c = 0$ to $c = 0.8$ (Figure 6.40a).

The change in the total fatigue life is not so significant. The greatest part of the difference originated at the earlier stages of the fatigue life. It is also seen when we look at the corresponding crack growth rate diagram (Figure 6.40b). The crack growth rate, even after being averaged with respect to local peaks, strongly depends on c in the initial part of the diagram. All the curves are merging when we go to the region of larger ΔK.

6.11 Summary

The thin plastic zone model has been used in this chapter to predict crack initiation and growth in elasto-plastic materials. Being comparatively simple, this model allows various generalizations that are necessary in fatigue analysis to include the influence of damage accumulation on material properties. In some aspects, the developed model may be interpreted as a model of low-cycle fatigue. Wide versatility of the model has been demonstrated.

The idea of treating the thin plastic zone model in a broader sense, covering both low-cycle and high-cycle fatigue seems attractive. However, the Paris exponent predicted by this model takes values between 2 and 3. Including plastic hardening (especially the isotropic one), we obtain higher magnitudes of the exponent. There is also the possibility of incorporating the threshold and related effects entering at lower load levels. On the other side, to describe fatigue under full-scale plastic straining, the application of plasticity theory in the full scope together with damage mechanics is necessary. It is the domain for further studies, and the difficulties of these problems cannot be overestimated.

Chapter 7

CRACK GROWTH IN
HEREDITARY MEDIA

7.1 Introductory Remarks

Crack nucleation and crack growth in polymers and other materials with hereditary behavior were studied by a number of authors [69, 155]. It is essential that cracks grow both under long-acting (constant or slowly varying) loads and under cyclic loading. In the first case one often speaks of static fatigue, in the second of cyclic fatigue. The latter includes classical high-cycle fatigue as well as low-cycle fatigue. Related phenomena occur in metals and metallic alloys at elevated temperatures loaded statically, cyclically, or both [89, 65, 119]. Structural failures due to crack growth in disks, blades and other components of power machinery are of essentially practical importance.

These types of fractures have much in common, and in the first aspect, because of the strong effect of the history of loading and deformation on the behavior of the system cracked body-loading. Constitutive stress-strain equations for polymers, rubber-like materials and composites with polymer matrices are, as a rule, linear, and for metallic materials at elevated temperatures nonlinear. However, this difference is not of primary importance. In principle, all phenomena of damage and fracture are strongly nonlinear. In the study of fracture of polymers, linear visco-elastic models are usually applied, and a number of results are based on the elasto-visco-elastic analogy and related approaches [53, 81, 130]. On the other hand, fracture of metals and alloys at elevated temperatures, as a rule, is preceded and accompanied by significant creep deformation (including creep damage, too). Moreover, many authors, beginning from Rabotnov and Kachanov, attributed macrocrack growth to the creep microdamage entirely. Namely, they assumed that a crack tip advances at a certain small distance when the microdamage creep measure at the crack tip attains a certain critical magnitude. In the simplest case, when a scalar measure ω

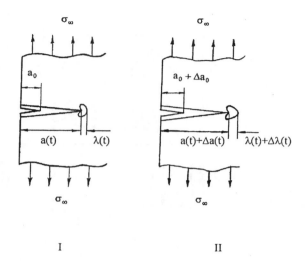

FIGURE 7.1
Isochronic variation of cracks in the presence of hereditary effects.

is used with values from the interval $[0,1]$, the condition of macrocrack propagation was assumed $\omega = 1$. Critical values smaller than unity, say, $\omega = 0.999$, are used to avoid singularities in some constitutive equations. Macrocrack growth is treated also as a result of coalescence of sets of micropores situated on the prolongation of the crack.

Even in the presence of creep and creep microdamage, such as diffusion and dislocation processes, micropore nucleation and coalescence, etc., the final fracture remains to be a macroscopic phenomenon relative to the global instability of the system cracked body-loading. It is typical for short-time loading, as well as for sustained ones. In the latter case, if the loading process is irregular, the final fracture can occur due to the instability born from a sudden overloading with a moderately low microdamage at the crack tip.

Evaluation of generalized driving forces for hereditary media meets some additional difficulties. The Griffith variations (G-variations) are by definition isochronic. It means that time is not subjected to variation in calculation of the virtual work. On the other hand, the history of loading is of essential significance for hereditary media. The only way to overcome this difficulty is to consider two identical and identically loaded bodies whose cracks are slightly different. The differences between the crack lengths play the part of initial G-variations and they carry on this part during the later development of cracks (Figure 7.1). Such an interpretation of G-variations is related to the two-specimen method in experimental fracture mechanics.

Because of hereditary properties, the relationship between characteristic times of various processes accompanying damage and fracture is of special importance. In addition to temporal parameters of loading, damage accumulation, and crack growth, characteristic times of hereditary behavior enter the picture. In the simplest case it is the only characteristic time such as the retardation time τ_r in the standard visco-elastic solid. The loading process, as usual, is characterized at least by two parameters, τ_s and τ_f, for sustained and cyclic load components, respectively. The characteristic damage time τ_d and the characteristic time of crack growth τ_a must also be included. The latter is assessed as $\tau_a = \lambda(da/dt)^{-1}$, where λ is a characteristic size of the process zone.

Thermal effects are essential in the fatigue of hereditary materials such as polymers. If the load frequency is high, the heat production must be taken into account with the resulting effect of temperature on material properties. To avoid additional complications, we will neglect thermal effects here; more precisely, we will consider isothermal processes assuming that material parameters are taken at the current temperature.

A typical situation in fatigue is as follows:

$$\tau_s \sim \tau_d \sim \tau_r, \quad \tau_f << \tau_r, \quad \tau_a << \tau_r \tag{7.1}$$

The first relationship means that the characteristic time of sustained (constant or slowly varying) loading is of the same order of magnitude as the characteristic times of damage accumulation and hereditary deformation. The second relationship requires that the frequency of the cyclic load component is high enough to treat material properties as "frozen" at least within several load cycles. The last relationship in (7.1) requires that crack growth must be sufficiently slow.

There are some special cases when hereditary effects are not so significant. It depends on the relationship between the above characteristic times and the duration T of the designed service life or test duration. When conditions (7.1) are satisfied, and $T << \tau_r$, material may be treated as elastic with instantaneous compliance parameters. In contrast, when $T >> \tau_r$, long-time compliance parameters may be used. The case $\tau_r \sim T$ is nontrivial even when the conditions (7.1) are satisfied.

Energy considerations are widely used in fracture mechanics because of their simplicity and apparent physical transparency. But they fail when we turn to hereditary media. In linear fracture mechanics, one distinguishes the stored (potential) energy which also includes the energy of external loads, and the energy irreversibly spent on crack propagation. In nonlinear fracture mechanics, an irreversible, dissipated part of energy is connected with plastic straining. However, the concept of the stored energy which could be spent on fractures, becomes vague when heredity is involved.

The situation is explained in an elementary example. A simple mechanical model is depicted in Figure 7.2. Two springs, 1 and 2, and the dashpot 3 form a system corresponding to a standard linear visco-elastic body. Additionally, a dry friction or cohesion element 4 enters the model. The symmetrical part of the system with elements 1′, 2′, and 3′ is included just to emphasize the fact that fracture takes place within the system. The start of sliding in the element 4 corresponds to the crack growth start. At this moment another part of the system begins to move that imitates crack propagation in the body.

The question is, which part of the stored energy can be spent in the form of the fracture work, i.e., to provide the start of sliding. Certainly, all energy accumulated in spring 1 is available. As to the energy stored in spring 2, its amount depends on history. When loading is very slow ($\tau_s >> \tau_r$), dissipation in the dashpot is small. In contrast, under rapid loading ($\tau_s << \tau_r$) the dissipation is significant,

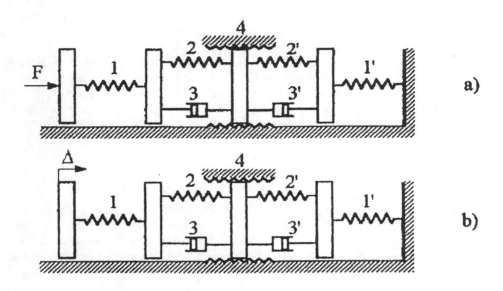

FIGURE 7.2
Elementary model of fracture in visco-elastic solid.

and the amount of energy in spring 2 is small. Moreover, the energy stored in this spring, in general, cannot be spent completely on the decohesion in element 4. The deformation and decohesion rates enter the picture there. This statement is valid both for force-controlled loading (Figure 7.2a) and displacement-controlled loading (Figure 7.2b).

Thus, we have seen that even in a simple model, such as given in Figure 7.2, the energy considerations are not elementary. To overcome difficulties, we must consider in detail all the processes of loading, deformation, and fracture. The principle of virtual work is certainly valid here; however, it is not so easy to apply. The reason has been mentioned already: virtual variations must be isochronic, i.e., they must relate to two neighboring states of the system at the same time moment, and this immediately involves all the issues of heredity.

7.2   Linear Visco-Elastic Media

The theory of linear visco-elasticity is frequently used to describe the response of polymer structural components to various load actions. There are some serious limitations when the linear theory is applicable to polymers. Because of high compliance, most polymers are subjected to significant deformation before the final fracture. This requires taking into account finite deformations and, perhaps, finite displacements. The theory becomes nonlinear in geometrical aspects. In addition, polymers exhibit nonlinear mechanical properties, especially in the domains of significant damage. As an example, the so-called "creasing" phenomenon in polymers should be mentioned. It is in fact the formation of an array of oriented microvoids and microcracks that produce an optical effect if a polymer is transparent. In some sense the crease zone in polymers is similar to the plastic zone in metallic alloys. It is a domain where mechanical properties are subjected to significant change. In particular, a strong anisotropy occurs in the crease zone. Some of these effects can be described in the framework of a model similar to the thin plastic zone model in elasto-plastic materials. However, the problem is not yet studied thoroughly even in the case of monotonous loading.

The situation in fatigue is a little different. Fatigue crack propagation in polymers usually takes place under comparatively low stress levels. Therefore, the resulting strains are not necessarily large. In addition, the final failure in fatigue may be rather close to quasi-brittle fracture even in the case when a material exposes high ductility under monotonous loading. In this chapter we discuss fatigue crack growth in visco-elastic media with the assumption that the stress-strain fields follow the linear theory of visco-elasticity.

Constitutive equations for linear visco-elastic media are frequently presented with the use of linear integral operators. If aging and other time deterioration processes are negligible, these operators are of Volterra's type. In particular, the relationships between stresses and strains in the case of uniaxial tension take the form

$$E_0\varepsilon(t) = \sigma(t) + \int_{-\infty}^{t} M(t-\tau)\sigma(\tau)d\tau$$

$$\frac{\sigma(t)}{E_0} = \varepsilon(t) - \int_{-\infty}^{t} N(t-\tau)\varepsilon(\tau)d\tau \qquad (7.2)$$

Here E_0 is the instantaneous Young's modulus; $M(t-\tau)$ and $N(t-\tau)$ are usually named creep (compliance) and relaxation kernels. In fact, the first of equations (7.2) describes the deformation under given stresses, and the second one describes the stress relaxation under given strains. The kernels $M(t)$ and $N(t)$ are normalized in (7.2).

Along with kernels, the corresponding compliance and relaxation functions are used. Let the state at $t < 0$ be natural, i.e., nonstressed and nonstrained. Integrating equations (7.2) by indent parts, we obtain

$$\varepsilon(t) = \int_0^t D(t - \tau) d\sigma(\tau)$$

$$\sigma(t) = \int_0^t R(t - \tau) d\varepsilon(\tau)$$

(7.3)

where standard notations for Stieltjes' integrals are used. The kernel in (7.3) satisfies conditions $D(0) \equiv 1/E_0$, $R(0) \equiv E_0$,

$$\frac{D(t)}{D(0)} = \int_0^t M(\theta) d\theta, \qquad \frac{R(t)}{R(0)} = \int_0^t N(\theta) d\theta$$

Evidently, $D(t - \tau)$ is equal to the strain at the instant t, produced with a step-wise unity stress increase at the instant $\tau < t$. Similarly, $R(t - \tau)$ is equal to the stress response to the step-wise unity strain change. It is expedient to separate the initial step of loading, and to treat the further loading as a differentiable process. Then equations (7.2) may be re-written as

$$\varepsilon(t) = D(0)\sigma(t) + \int_{0+}^t D(t - \tau) \frac{d\sigma(\tau)}{d\tau} d\tau$$

$$\sigma(t) = R(0)\varepsilon(t) + \int_{0+}^t R(t - \tau) \frac{d\varepsilon(\tau)}{d\tau} d\tau$$

(7.4)

A number of analytical presentations are used in the theory of visco-elasticity. Among them are functions based on mechanical models composed of linear elastic springs and linear viscous dashpots, such as Kelvin's, Maxwell's and Voight's models. The behavior of these models is described by linear differential equations with constant coefficients. Then compliance and relaxation functions are sums of exponential time functions. For example, the behavior of the standard visco-elastic solid is described with the differential equation

$$E_\infty \varepsilon + E_0 \tau_0 \frac{d\varepsilon}{dt} = \sigma + \tau_0 \frac{d\sigma}{dt}$$

(7.5)

where E_∞ is the long-time (equilibrium) Young's modulus. The time constant τ_0 is a characteristic relaxation time. The compliance function and compliance kernels take the form

$$D(t) = \frac{1}{E_\infty} + \frac{E_0 - E_\infty}{E_0 E_\infty} \exp\left(-\frac{t}{\tau_r}\right)$$

$$M(t) = \frac{E_0 - E_\infty}{E_0 \tau_r} \exp\left(-\frac{t}{\tau_r}\right)$$

(7.6)

with the retardation time $\tau_r = \tau_0 E_0 / E_\infty$. The corresponding resolvent functions are

$$R(t) = E_\infty + (E_0 - E_\infty)\exp(-t/\tau_0)$$

$$N(t) = \frac{E_0 - E_\infty}{E_0 \tau_0}\exp\left(-\frac{t}{\tau_0}\right) \tag{7.7}$$

Equations (7.5)-(7.7) can be generalized including higher derivatives in differential equations and obtaining, correspondingly, additional exponential terms in visco-elastic functions. Another way is to generalize the kernels multiplying them by power functions of time. As an example, consider the compliance kernel

$$M(t) = \frac{(E_0 - E_\infty)t^{\alpha-1}}{E_0 \theta^\alpha \Gamma(\alpha)}\exp\left(-\frac{t}{\tau_r}\right) \tag{7.8}$$

where $\alpha > 0$, and $\Gamma(\alpha)$ is the gamma function. Evidently, it is like the kernel $M(t)$ from equation (7.6) and transfers to the latter at $\alpha = 1$. The compliance function corresponding to equation (7.8) is as follows:

$$D(t) = \frac{1}{E_\infty} + \frac{E_0 - E_\infty}{E_0 E_\infty}\frac{\gamma(\alpha, t/\sigma_r)}{\Gamma(\alpha)} \tag{7.9}$$

Here notation $\gamma(\alpha, n)$ is used for the non-complete gamma function, i.e.,

$$\gamma(\alpha, n) = \int\limits_0^\alpha u^{n-1}\exp(-u)du$$

To expand these models upon three-dimensional stress-strain states, it is sufficient to replace scalar functions in (7.2)-(7.4) with their tensor equivalents. For example, the first equation (7.3) is generalized as follows

$$\varepsilon_{\alpha\beta}(t) = \int\limits_0^t D_{\alpha\beta\gamma\delta}(t - \tau)d\sigma_{\gamma\delta}(\tau). \tag{7.10}$$

A fourth-rank compliance tensor function $D_{\alpha\beta\gamma\delta}(t)$ enters equation (7.10). In an isotropic case this tensor function is expressed with the use of two scalar functions. Usually, strains and stresses are divided into spherical and deviatorical components: $\varepsilon_{\alpha\beta} = \varepsilon_0\delta_{\alpha\beta} + e_{\alpha\beta}$, $\sigma_{\alpha\beta} = \sigma_0\delta_{\alpha\beta} + s_{\alpha\beta}$, $\varepsilon_0 = \varepsilon_{\gamma\gamma}/3$, $\sigma_0 = \sigma_{\gamma\gamma}/3$. Then visco-elastic functions are introduced corresponding to volume and shear elastic moduli. Assuming incompressibility, only the shear compliance function remains in (7.10) and related equations. Another assumption, frequently used in engineering analysis, is to put Poisson's ratio $\nu = \text{const}$. Then only a single compliance function remains corresponding to Young's modulus.

If aging and related phenomena are significant, visco-elastic functions $M(t-\tau)$, $N(t-\tau)$, $D(t-\tau)$ and $R(t-\tau)$ of the time difference $t - \tau$ must be replaced with functions of two time variables: $M(t,\tau)$, $N(t,\tau)$, etc. One of the ways to specify such functions is to assume that at least a part of the coefficients entering equations (7.6)-(7.9) depends on time explicitly or implicitly. For example, they might depend on a certain damage function that is governed with a damage accumulation equation and, therefore, they are varying in time.

For the evaluation of fatigue and related damage, we need a passable model of microdamage accumulation. As the microdamage level grows both with time t and cycle number N, we introduce two scalar measures ω_s and ω_f with magnitudes from the segment $[0, 1]$. The damage accumulation in a hereditary medium is a hereditary process, too. The simplest equations with allowance for heredity are

$$\omega_s(t) = \frac{1}{t_c} \int\limits_0^t B_s(t - \tau) F_s[\sigma(\tau)] d\tau$$

$$\omega_f(N) = \int\limits_0^N B_f(N - \nu) F_f[\sigma(\nu)] d\nu \tag{7.11}$$

where $F_s[\cdot]$ and $F_f[\cdot]$ are nonlinear functions of the stress $\sigma(\tau)$ or, in general, of the stress tensor. The kernels $B_s(t - \tau)$ and $B_f(N - \nu)$ take into account the history of loading and crack growth processes. The time constant t_c enters the first of equations (7.11) according to dimensional considerations.

In special cases equations (7.11) may be replaced with proper differential equations. Later on we use the threshold-power laws of microdamage accumulation:

$$\frac{\partial \omega_s}{\partial t} = \frac{1}{t_c} \left(\frac{\sigma_r - \sigma_{th}}{\sigma_s} \right)^{m_s}, \quad \frac{\partial \omega_f}{\partial N} = \left(\frac{\Delta\sigma - \Delta\sigma_{th}}{\sigma_f} \right)^{m_f} \tag{7.12}$$

Here σ_r is a representative stress responsible for static fatigue, $\Delta\sigma$ is a representative (effective) stress range responsible for cyclic fatigue, σ_s and σ_f are corresponding resistance stresses, and σ_{th} and $\Delta\sigma_{th}$ are threshold resistance stresses. All these magnitudes, as well as the exponents m_s and m_f are supposed to be material parameters, generally, dependent on the currently attained level of microdamage and temperature. Equations (7.12) are valid at $\sigma_r > \sigma_{th}$, $\Delta\sigma > \Delta\sigma_{th}$. If any of these inequalities are violated, the right-hand side of the corresponding equation must be put to zero.

Let the measures ω_s and ω_f be normalized in such a way that they may be treated as additives. Their sum $\omega = \omega_s + \omega_f$ is interpreted as a total microdamage measure with values from the interval $[0, 1]$. Then equations (7.12) result in

$$\frac{\partial \omega}{\partial t} = \frac{1}{t_c} \left(\frac{\sigma_r - \sigma_{th}}{\sigma_s} \right)^{m_s} + f \left(\frac{\Delta\sigma - \Delta\sigma_{th}}{\sigma_f} \right)^{m_f} \tag{7.13}$$

where $f(t)$ is the cycle number per time unit (load frequency). Note that the interaction between the two kinds of microdamage is taken into account in (7.13) implicitly. In fact, σ_s and σ_{th} may depend on ω_f. Vice versa, σ_f and $\Delta\sigma_{th}$ may depend on ω_s.

7.3   Generalized Driving Forces for Linear Visco-Elasticity

Linear visco-elastic medium is the simplest kind of hereditary media, and it is natural that a number of publications are dedicated to this topic. The widely known success achieved in linear visco-elasticity with the use of the correspondence (Volterra's) principle has encouraged the authors to search for the solution using the analogy between linear elastic and linear visco-elastic problems. Some authors, using this philosophy, went rather far into nonlinear visco-elasticity. Not trying to present a survey of previous publications in fracture mechanics of linear visco-elastic bodies, the author prefers to send the readers to the survey paper [81] and a more recent survey in the book [155].

The first problem in the study of crack growth in visco-elastic media is the evaluation of generalized driving forces in the presence of growing cracks. For simplification, consider a single one-parameter crack with the size a, say, a mode I crack with the length $2a$ in an unbounded body under the applied stress σ_∞.

Several approaches were suggested to determine the strain energy release rate using heuristic considerations relating to Volterra's principle. In [81] it was supposed that

$$G(t) = D(\lambda/c)K^2(t) \tag{7.14}$$

where $D(t)$ is the visco-elastic compliance for the considered crack mode and stress-strain state, and $K(t)$ is the transient stress intensity factor. The compliance is taken at the characteristic time $\tau_c = \lambda/c$, where λ is the scale parameter assumed as a material constant, and $c \equiv da/dt$ is the crack growth rate. The question arises, what are the numerical values of λ for real polymers. It was estimated that for a polyurethane rubber λ is of the order 10^{-10} m, i.e., very small, making the continuum interpretation doubtful. In addition, this size is small compared with the size of the observed process zone.

An alternative heuristic equation was proposed in the paper [130]:

$$G(t) = \int_{[0,t]} D(t - \tau)d[K^2(\tau)] \tag{7.15}$$

Another equation has the form

$$G(t) = K(t) \int_{[0,t]} D(t - \tau)dK(\tau) \tag{7.16}$$

In contrast to equation (7.15), equation (7.16) allows that the stress field around a moving crack is elastic, and the strain field is equal to the hereditary transform with the kernel $D(t - \tau)$. But equation (7.16) is also an heuristic guess. In the two extreme cases, i.e., for very rapidly and very slowly growing cracks, equations (7.15) and (7.16) provide the same reasonable result. For example, for the mode I crack in the plane-strain state

$$\frac{K^2(t)(1 - \nu_0^2)}{E_0} \leq G(t) \leq \frac{K^2(t)(1 - \nu_\infty^2)}{E_\infty}$$

where E_0 and ν_0 are instant values of Young's modulus and Poisson's ratio, and E_∞ and ν_∞ are corresponding long- time (equilibrium) values. Hence equations (7.15) and (7.16), as results of interpolation between the two extremal cases, may be used as a suitable approximation at least for certain material models, loading processes and fracture modes. Moreover, under some conditions these equations provide rather close numerical results.

To compare equations (7.15) and (7.16), the following artificially stated problem is used. Let a body be loaded at the instant $t = 0$, and the loading remain stationary at $t > 0$. A crack begins to grow at the same instant $t = 0$ with a given rate c that is assumed to be constant. Present the constitutive equations of linear visco-elastic media in the integral form. Namely, assuming Poisson's ratio $\nu = $ const, let strains $\varepsilon_{\alpha\beta}(t)$ and displacements $u_\alpha(t)$ be equal to the integral transforms of strains $\varepsilon_{\alpha\beta}^0(t)$ and displacements $u_\alpha^0(t)$ in a corresponding elastic body with the uniaxial compliance $D(0) = E_0^{-1}$. Then

$$D(0)\varepsilon_{\alpha\beta}(t) = \int_0^t D(t - \tau)d\varepsilon_{\alpha\beta}^0(\tau)$$

$$D(0)u_\alpha(t) = \int_0^t D(t - \tau)du_\alpha^0(\tau)$$

(7.17)

and, after integration by parts,

$$\varepsilon_{\alpha\beta}(t) = \varepsilon_{\alpha\beta}^0(t) + \int_0^t M(t - \tau)\varepsilon_{\alpha\beta}^0(\tau)d\tau$$

$$u_\alpha(t) = u_\alpha^0(t) + \int_0^t M(t - \tau)u_\alpha^0(\tau)d\tau$$

(7.18)

where $M(t) = D'(t)/D(0)$ is the normalized compliance kernel. Computations were performed in [22] for mode I cracks in an unbounded plate of linear visco-elastic material. "At infinity" either the uniformly distributed tensile stress σ_∞, or the uniformly distributed tensile strain ε_∞, is applied at $t = 0$ and kept constant at $t > 0$. For stress distribution the standard Inglis-Williams formulas were used with regular terms not omitted.

The normalized compliance kernel was taken according to equation (7.8). At $0 < \alpha < l$ the kernel $M(t)$ in (7.8) has a weak singularity. At $\alpha = 1$ we obtain the standard visco-elastic material with the uniaxial constitutive equation (7.5). The compliance function entering equations (7.18) is given in equation (7.9).

For further computations, re-write equations (7.15) and (7.16) for the case of initial step loading:

$$G(t) = D(t)K^2(0) + \int\limits_{0+}^{t} D(t-\tau)\frac{d[K^2(\tau)]}{d\tau}d\tau \tag{7.19}$$

$$G(t) = D(t)K^2(0) + K(t)\int\limits_{0+}^{t} D(t-\tau)\frac{dK(\tau)}{d\tau}d\tau \tag{7.20}$$

The comparison of numerical results obtained with the use of equations (7.19) and (7.20) is presented in Figures 7.3-7.4. It is assumed that the crack has an elliptical shape with semi-axes a and $b \ll a$. The crack tip advances with the constant rate $c \equiv da/dt$ and with the constant tip radius $\rho = b^2/a$. The magnitudes of G are normalized with respect to $G_0 = \sigma_\infty^2 \pi a_0(1-\nu^2)/E_0$. The ratio $\eta = c\tau_r/a_0$ of the crack growth rate to the characteristic rate of hereditary deformation (a_0 is the initial semi-length of the crack) is used as a parameter. The ratio G/G_0 is plotted against the normalized crack extension $(a-a_0)/a_0$.

Figure 7.3 is drawn for $\alpha = 1/2$ and Figure 7.4 for $\alpha = 1$. Parts (a) and (b) of these figures correspond to the ratios of moduli $E_0/E_\infty = 2$ and 10, respectively. The dotted lines are drawn for cracks in elastic media with the instant modulus E_0 (the lower line) and the equilibrium modulus E_∞ (the upper line). Comparison is done for ratios η equal to 10 (a rapid crack growth), 1 (the crack growth rate comparable with the rate of visco-elastic deformation) and 0.1 (a slow crack growth). For each case two lines are presented in Figures 7.3 and 7.4, and those twains are labelled 1, 2, and 3, respectively. As a rule, the upper lines refer to equation (7.19) and the lower ones to equation (7.20). It is rather unexpected that both analytical presentations for the driving force G yield such close numerical results. The points in Figures 7.3 and 7.4 notated with special symbols will be discussed a little later.

As has been mentioned, equations (7.19) and (7.20) are purely heuristic and, rigorously speaking, incorrect. The attempts to substantiate these and similar equations do not seem convincing. Both the Volterra principle and energy considerations fail when generalized driving forces in their rigorous sense are concerned. However, equations (7.19) and (7.20) are attractive because of their simplicity. These equations give rather close numerical results, at least for the considered problems and assumed numerical data. Equation (7.19) is a little easier to handle, both in analytical and numerical calculations. In any case, it is expedient to compare the results provided by these equations with the results of direct computation of generalized driving forces as has been defined in equation (3.8).

Consider a Griffith crack in a plate of visco-elastic material. In the plane-stress case the stress field coincides with that in a similar elastic plate. To find the strain and displacement fields, we must apply the integral transform. Then the work of external and internal forces must be calculated for two identical and identically loaded plates with slightly different crack sizes at the same time moment (Figure 7.1). All the history of loading, straining, and crack growth must be taken into account. This analysis requires numerical integration both in time and in space being rather laborious. Some results obtained by Bolotin and Minokin [22] are presented hereafter.

The generalized force $G(t)$ in the two-dimensional case and the absence of volume forces is defined as follows:

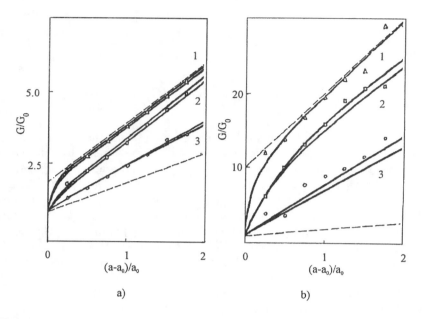

FIGURE 7.3
Generalized driving forces for normalized crack growth rate $\eta = 10$, 1, and 0.1 (lines 1,2,3): (a) $\alpha = 1/2$, $E_\infty/E_0 = 2$; (b) $\alpha = 1/2$, $E_\infty/E_0 = 10$.

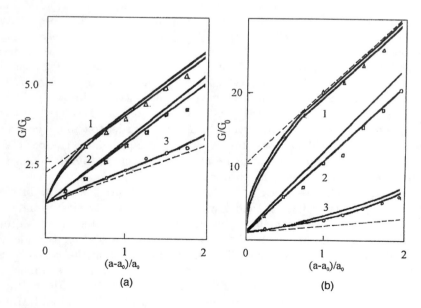

FIGURE 7.4
The same as in Figure 7.3 at $\alpha = 1$.

$$G(t) = \int\limits_C p_\alpha(t) \frac{\partial u_\alpha(t)}{\partial a} ds - \int\limits_\Omega \frac{\partial w(t)}{\partial a} d\Omega \qquad (7.21)$$

Here $u_\alpha(t)$ is the displacement field, and w is the density of the total work produced by internal forces at the moment t, i.e.,

$$w(t) = \int\limits_0^t \sigma_{\alpha\beta}(\tau) d\varepsilon_{\alpha\beta}(\tau) \qquad (7.22)$$

In equation (7.21) Ω is the integration domain, and C is its boundary. Equation (7.21) is similar to (3.22). But because of the presence of heredity and dissipation, the boundary C must be moved as far as possible from the crack domain to diminish the error.

The finite-difference approximation of equation (7.21) takes the form

$$G(t)\Delta a = \int\limits_C p_\alpha(t) \left[u_\alpha^{II} - u_\alpha^I \right] ds - \int\limits_\Omega \left[w^{II}(t) - w^I(t) \right] d\Omega \qquad (7.23)$$

where G-variation is replaced with the initial difference Δa of crack sizes of bodies I and II (compare with Figure 7.1). This difference is to be taken as small as possible. The right-hand side of equation (7.23) contains the differences of large, but very close, quantities. Therefore, a reasonable compromise must be used in the choice of Ω and Δa.

Consider, as above, the crack growth under stresses σ_∞ subjected at $t = 0$ and sustained later as constant. The crack growth rate $c = da/dt$ is assumed constant. In the further computations the restructuring of the coordinate mesh was used with the minimal mesh size domain moving with the crack tip. This minimal size was used as the difference Δa in (7.23). It means that the initial half-lengths of the cracks were assumed a_0 and $a_0 + \Delta a$.

Some numerical results are presented in Figures 7.3-7.4 where the points notated with special symbols were obtained with the use of (7.23). As was shown earlier, the two competing equations for the generalized force, i.e., (7.19) and (7.20), give rather close numerical results. It appears that they differ in the same order of magnitude as the a priori error of the computational realization of equation (7.23). Moreover, the computed points are scattered, due to the rather coarse mesh, in the same order of magnitude.

At $\eta = 0.1$, i.e., when the crack grows relatively slowly compared with the hereditary processes in the body, the values of $G(t)$ approach with time rather closely the line $G(t) = K^2(t)/E_\infty$. At $\eta = 10$ the crack is propagating much faster than the hereditary deformation is developing. The difference between $G(t)$ and its instant modulus magnitude is small at the beginning but increasing during crack growth.

Similar numerical results were obtained for other types of loading, including nonstationary step-wise ones. Figures 7.5 and 7.6 are drawn for the case of strain-controlled loading. A plate containing an elliptical crack with initial semi-axes ratio $a_0/b_0 = 10$ is subjected at $t = 0$ to the given nominal strain ε_∞. Later on, this strain is kept constant. As formerly, it is assumed that the crack begins to grow

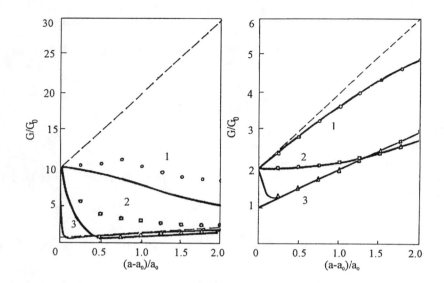

FIGURE 7.5
Generalized driving forces in the case of strain-controlled loading; notation as in Figure 7.3.

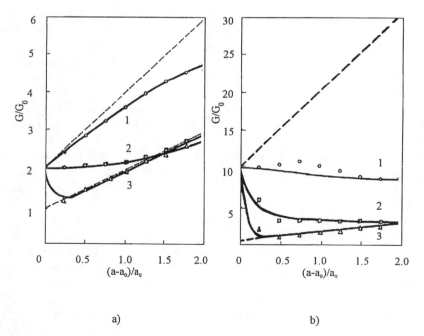

a)

b)

FIGURE 7.6
The same as in Figure 7.5 at $\alpha = 1$.

at $t = 0$ with the constant rate $c \equiv da/dt$. The semi-axis b varies in time in such a way that the radius of curvature on the crack tip $\rho = b^2/a$ remains constant. To apply equation (7.25), the Kolosov-Muskhelishvili formulas are used together with the integral time transform. The results are plotted in Figures 7.5 and 7.6 with special symbols. The solid lines are drawn according to equation (7.19) with $K(t)$ calculated as for a mathematical slit. The force G is normalized to the characteristic value $G_0 = E_0 \varepsilon_\infty^2 \pi a_0$. Note that Figure 7.5 corresponds to $\alpha = 1/2$, Figure 7.6 to $\alpha = 1$ for two moduli ratios, $E_0/E_\infty = 2$ and $E_0/E_\infty = 10$.

When the crack growth rate is not very high, we observe the decrease of $G(t)$ during the initial stage. The cause is stress relaxation. Later on, because of the increase of $G(t)$ crack growth begins. Deviation of results computed with equation (7.23) seems more significant than in the former numerical examples. It might be a result of a rather coarse approximation of the crack shape.

7.4   Crack Growth under Sustained Loading

Three types of crack propagation in visco-elastic solids are to be distinguished: crack growth under sustained, in particular, constant loading; fatigue crack growth under cyclic loading; crack growth under combined loading, i.e., the combination of static and cyclic fatigue. The first type of crack growth, named static fatigue, was discussed in papers [22, 53, 132]. In this section we treat the process of crack growth in linear visco-elastic bodies in the framework of the general theory developed in the previous chapters. Both direct numerical approach and the quasistationary approximation are applied. In the latter case we reduce the problem to integro-differential equations with respect to crack size.

Consider a one-parameter mode I crack with the size $a(t)$. Assume that the stress distribution in the body coincides with that in an elastic body under similar loading and the same crack size $a(t)$. The crack (non)propagation is governed, as usual, by the relationship

$$G(t) \lessgtr \Gamma(t) \tag{7.24}$$

Let the generalized driving force be evaluated as in equation (7.19), and the generalized resistance force as

$$\Gamma = \Gamma_0(1 - \psi^\alpha) \tag{7.25}$$

Here Γ_0 is the generalized resistance force for the nondamaged material, $\psi(t) \equiv \omega[a(t), t]$ is the microdamage measure at the tip, and $\alpha > 0$.

The equation of damage accumulation under sustained loading has the form of the first equations in (7.11), namely

$$\omega_s(t) = \frac{1}{t_c} \int_0^t B_s(t - \tau) F_s[\sigma(\tau)] d\tau \tag{7.26}$$

In particular, when $B_s(t) \equiv 1$, $F_s(\sigma) = [(\sigma - \sigma_{th})/\sigma_s]^m$, we come to the equation

$$\frac{\partial \omega_s}{\partial t} = \frac{1}{t_c}\left(\frac{\sigma - \sigma_{th}}{\sigma_s}\right)^{m_s} \tag{7.27}$$

If the "self-healing" effect of damage is significant, the simplest form of the kernel in (7.26) is

$$B(t) = \exp\left(-\frac{t}{\tau_h}\right) \tag{7.28}$$

with the characteristic time constant τ_h. The characteristic tip radius, which in fact is a certain measure of stress concentration near the damaged crack tip, is governed by an equation similar to (4.17):

$$\frac{d\rho}{dt} = \frac{\rho_s - \rho}{\lambda_\rho}\frac{da}{dt} + (\rho_b - \rho)\frac{d\psi}{dt} \tag{7.29}$$

Material parameters ρ_s, ρ_b, and λ_ρ have the same interpretation as in the equations mentioned above.

The problem stated in equations (7.24)-(7.29) is similar to that for linear elastic solids. However, it is more complex because of the strong dependence on history entering, in particular, in the form of integrals upon the time intervals $(-\infty, t]$ or $[t_0, t]$. The computational algorithm contains several iteration procedures with repetitive integration upon the crack growth and damage accumulation histories.

Some numerical examples are presented in Figures 7.7-7.9. The model of standard visco-elastic solid is used with the parameters $E_0 = 2E_\infty = 1$ GPa, $\tau_r = 10^2$ s,

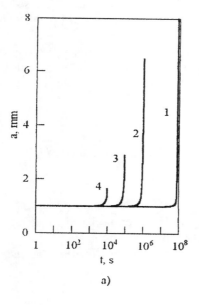

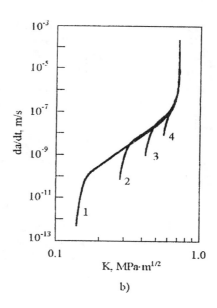

a)   b)

FIGURE 7.7
Crack growth histories (a) and growth rate diagrams (b) at $\sigma_\infty = 2.5, 5, 7.5,$ and 10 MPa (lines 1,2,3, and 4).

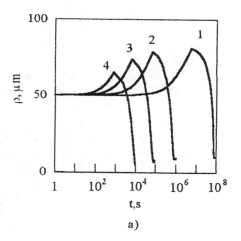

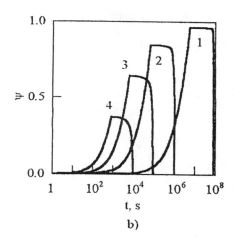

a)   b)

FIGURE 7.8
Effective tip radius (a) and tip damage measure (b) for the same data as in Figure 7.7.

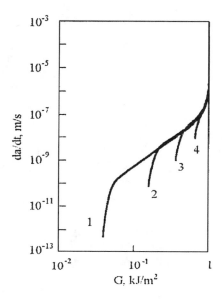

FIGURE 7.9
Crack growth rate in the function of a generalized driving force for the same data as in Figure 7.7.

$\gamma_0 = 1$ kJ/m^2. These magnitudes are of an order that is typical for most industrial polymers at room temperature. There are no sufficient experimental data concerning other parameters. To obtain the numerical results for growth rates which are of the same order as empirical ones, we assume that $\sigma_s = 100$ MPa, $\sigma_{th} = 10$ MPa, $t_c = 10^3$ s, $m_s = 4$, $\alpha = 1$. Then, for the damage modelled by equation (7.27) and typical stresses, say $\sigma_\infty = 10$ MPa, $a/\rho = 10^2$, the time until damage at the crack tip is of the order 10^3 s. When a/ρ increases, the life-time rapidly decreases. Characteristic lengths in equation (7.29) are chosen as in the elastic case, i.e., $\rho_s = 10$ μm, $\rho_b = \lambda_\rho = 100$ μm. These magnitudes are of the order of crease zone sizes which also seems natural to assume.

Figures 7.7-7.9 are plotted for a central crack in a plate under remote tension. The plate width and crack half-length are $w = 100$ mm and $a_0 = 1$ mm; the initial tip radius is $\rho_0 = 50$ μm. Figure 7.7a shows the crack growth history at $\sigma_\infty = 2.5$, 5, 7.5, and 10 MPa (lines 1, 2, 3, and 4, respectively). The same notations are used in Figure 7.7b, where the rate da/dt in meters per second (m/s) is plotted against the stress intensity factor K (subscript indicating the fracture mode is omitted). The time histories of the effective tip radius ρ and the tip damage measure ψ are presented in Figure 7.8. The general tendency is the same as in the elastic case (compare, e.g., with Figure 4.5).

Figure 7.7b is analogous to common fatigue crack growth rate diagrams, having a similar sigmoidal shape. In this connection, a question of practical importance arises, which load parameter is preferable to draw growth rate diagrams for static fatigue in polymer materials. The lines in Figure 7.7b corresponding to different applied loads exhibit divergence. There is a reason to search another correlating load parameter, and the best candidate is the generalized driving force G. An attempt to draw such diagrams is presented in Figure 7.9 for the same data as in Figure 7.7. The "current" driving force determined by equation (7.15) is chosen as the correlating parameter in Figure 7.9. The divergence of rate curves is of the same order as in Figure 7.7b. The cause is, that, in the considered numerical example, the characteristic heredity time τ_r is small compared with the times at final fracture. Thus, the driving force $G(t)$ does not deviate from $G_\infty = K^2/E_\infty$ during almost all of the fatigue life. This means that both K and G_∞ seem to be suitable correlating load parameters. There is a peculiarity in Figures 7.7-7.9 typical for crack growth rate diagrams in the presence of hereditary deformation. The slope in Figure 7.7b differs from the exponent m_s in equation (7.27) except, maybe, a short section before the final failure. This may be attributed to the change of material compliance during the process of crack growth. The rate da/dt refers to visco-elastic material while the stress intensity K is a load parameter for linear elastic material.

The latter observations may be partially explained using the coarse approximation similar to that discussed in Sections 3.9 and 4.8. Actually, equation (7.27) gives the following estimate for the tip damage measure:

$$\psi(t) \sim \omega_{ff}(t) + \frac{\lambda_p}{t_c}\left(\frac{da}{dt}\right)^{-1}\left(\frac{\sigma_t - \sigma_{th}}{\sigma_s}\right)^{m_s} \qquad (7.30)$$

Here $\omega_{ff}(t)$ is the far field damage, λ_p is the process zone size assumed constant, and σ_t is a characteristic stress within this zone. Substituting (7.30) in (7.25) and putting $\Gamma(t) = \Gamma_0(1 - \psi^\alpha)$ equal to $G(t)$, we obtain

$$\frac{da}{dt} = \frac{\lambda_p}{t_c} \left(\frac{\sigma_t - \sigma_{th}}{\sigma_s} \right)^{m_s} \left[\left(1 - \frac{1}{\Gamma_0} \int_0^t D(t - \tau) d \left[K^2(\tau) \right] \right)^{1/\alpha} - \omega_{ff}(t) \right]^{-1} \quad (7.31)$$

Equation (7.31) is similar to (4.15). However, it is an integro-differential equation: the crack size $a(t)$ enters its right-hand side not only with $\sigma_t(t)$, but also with $K(t)$, i.e., under the integration sign. The stress intensity factor $K(t)$ enters equation (7.31) in a more complex way than in the purely elastic case. The exponent m_s is not present so distinctly in crack growth rate diagrams when we deal with visco-elastic media. Some deviations are perhaps a result of time evolution of the process zone size λ_p.

A question of interest is the role of "self-healing" effects that are expected, for example, in thermoplastics, especially at elevated temperatures. The simplest analytical model of this effect is given in equation (7.28). Combining (7.26) and (7.28), we come to the equation of damage accumulation

$$\omega_s(t) = \frac{1}{t_c} \int_0^t \exp \left(-\frac{t - \tau}{\tau_h} \right) \left[\frac{\sigma(\tau) - \sigma_{th}}{\sigma_s} \right]^{m_s} d\tau$$

The influence of τ_h is illustrated in Figures 7.10-7.11 plotted for the same data as Figures 7.7-7.9. The difference is that the retardation time $\tau_r = 10^3$ s. The applied stress is fixed $\sigma_\infty = 7.5$ MPa while τ_h is subject to variation: $\tau_h = 10^5, 1.8 \cdot 10^5$ s; and $\tau_\infty \to \infty$ (in the latter case we return to the former model). Crack growth histories and growth rate diagrams are presented in Figure 7.10 where lines 1, 2, and 3 correspond to the listed above magnitudes of τ_h. As one might expect, when τ_h decreases, the total fatigue life increases. In particular, no crack propagation occurs at $\tau_h = 10^5$ s (see line 1 in Figure 7.10a). The origin of this effect seems to be evident: due to "self-healing", the tip damage is less at lower τ_h, and the resistance force is higher. As a result, we have $G < \Gamma$ any $t > 0$.

However, the situation is not so simple because the damage rate $d\psi/dt$ enters equation (7.29) for the effective tip radius ρ. This radius controls the tip stress σ_t which, in turn, controls damage accumulation. Some details of this complex mechanism can be observed in Figure 7.11 where time histories are drawn both for ρ and ψ. Note that in case 1 the tip blunting and damage accumulation stop attaining a certain level; it is the case where, due to "self-healing", the crack does not propagate at all.

7.5   Combination of Sustained and Cyclic Loading

Turning to crack growth under cyclic loading, we must stress that the damage in polymers produced by cyclic loading is, as a rule, accompanied by the damage due to sustained (constant or slowly varying) components of loading. Moreover,

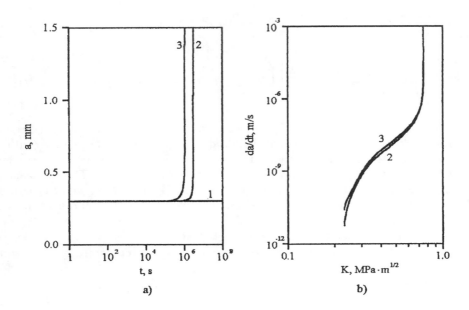

FIGURE 7.10
Influence of "self-healing" on crack growth histories (a) and growth rate diagrams (b) at $\tau_h = 10^5, 1.8 \cdot 10^5$ s, and $\tau_\omega \to \infty$ (lines 1, 2, and 3).

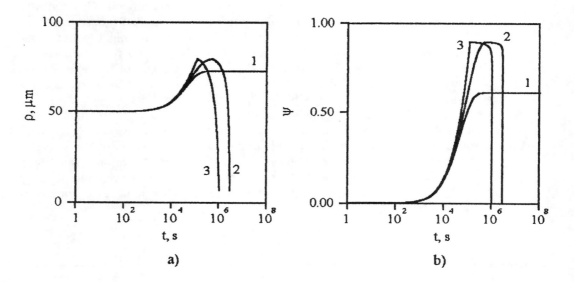

FIGURE 7.11
Influence of "self-healing" on effective tip radius (a) and tip damage measure (b) for the same data as in Figure 7.10.

the phenomenon of fatigue in polymer materials (as well as in fatigue of metals at elevated temperatures) combines the features of both static and cyclic fatigue.

Equation (7.21) is valid both for slowly varying and regular cyclic loading. Its application requires cumbersome computations even in the case of static fatigue. In the case of cyclic loading, when the number of cycles is counted in thousands and more, the direct use of equation (7.21) is hardly possible. Therefore, we have to turn to some additional assumptions which are expected to give a rather good approximation for generalized driving forces.

Even the simplified equations, say, equation (7.15), require laborious computations when the cycle number is high. Some averaging procedures are of use to reduce the scope of routine computational work. The actual loading process consists, as a rule, of two components. One of them is constant or slowly varying within the time segments short enough to treat other variables as "frozen". The second component is cyclic, and the corresponding stresses vary in time rapidly enough to treat the stress-strain field generated by this component as an elastic one with the compliance corresponding to the load frequency. As a result, we split the stress-strain field into two parts, that corresponding to slowly varying loading, and that corresponding to cyclic loading. It is obvious that such splitting requires a certain relationship between characteristic times relating to the processes of loading, hereditary deformation, damage accumulation, and crack growth.

Let the applied stresses be given as

$$\sigma_\infty = \sigma_{\infty,m} + \sigma_{\infty,a} \sin \theta t \tag{7.32}$$

The mean stress $\sigma_{\infty,m}$, the amplitude $\sigma_{\infty,a}$, and the frequency θ, in general, are slowly varying functions of time. Respectively, the stress intensity factor varies as follows:

$$K = K_m + K_a \sin \theta t \tag{7.33}$$

Substituting (7.33) in (7.15), we find the instantaneous magnitudes of the driving force $G(t)$. Its maximal (within a cycle) magnitudes enter the conditions (7.24) of crack behavior.

We are primarily interested in maximal magnitudes. We describe them by a slowly varying upper envelope denoted here as $G_+(t)$ The objective is to obtain the simplest way to assess $G_+(t)$ without going into details of the time variation of $G(t)$ within each cycle whose number is usually very high.

An averaging procedure is useful here. Actually, when the characteristic duration of cycles τ_f, say the period $2\pi/\theta$ of functions in (7.32) and (7.33), is small compared with other characteristic temporial parameters, we may split $G_+(t)$ into two parts. The first part corresponds to the slowly varying process, the second one to the change of the fields within each cycle.

Consider the formula for the driving force given in equation (7.15). Then

$$G(t) = \int\limits_0^t D(t - \tau) d\left[K^2(\tau)\right] \tag{7.34}$$

where $K(t)$ is defined in (7.33). The envelope of $G(t)$ may be presented as

$$G_+(t) = \int\limits_0^t D(t-\tau)d\left[K_s^2(\tau)\right] + D(\tau_f)K_f^2(t) \qquad (7.35)$$

where $K_s(t)$ is the slowly varying component of $K(t)$ and $K_f(t)$ is the cyclic component. The second term in the right-hand side of (7.35) contains the compliance $D(\tau_f)$ corresponding to the time $\tau_f \ll \tau_r$. This term, evidently, is just the energy release rate for the cyclic component.

The splitting of $K^2(t)$ into two parts is not unique; however these parts are connected as $K_f^2 = K_{\max}^2 - K_s^2$. One of the approaches is to assume that

$$K_s^2 = K_m^2, \qquad K_f^2 = 2K_m K_a + K_a^2 \qquad (7.36)$$

Another way is to equalize K_s^2 to the mean square of the current $K^2(t)$. In the case of equation (7.33) we obtain

$$K_s^2 = K_m^2 + \frac{1}{2}K_a^2, \quad K_f^2 = 2K_m K_a + K_a^2 \qquad (7.37)$$

Intuitively, both approaches are acceptable. This can be checked by numerical experimentation. The comparison of results is given in Figures 7.12-7.14 when the generalized forces and their envelopes are drawn for the standard linear visco-elastic body at $E_0 = 2E_\infty$ (case a) and $E_0 = 10E_\infty$ (case b). It is assumed that $E_0 = 1$ GPa, $\tau_r = 10^3$ s, $a_0 = 0.1$ mm, $f = 10^{-2}$ Hz, $\sigma_{\infty,m} = 100$ MPa, $\sigma_{\infty,a} = 100/3$ MPa, i.e., $R = 0.5$. For comparison, equations (7.35)-(7.37) are taken.

Figure 7.12 is obtained for the case of fixed crack tips. Lines labelled with numeral 1 are drawn according to equation (7.35) with the use of equations (7.36) and (7.37). The numerical results are very close; sometimes the lines practically coincide. Line 2 corresponds to the "static" component of the generalized driving force

$$G_s(t) = \int\limits_0^t D(t-\tau)d\left[K_s^2(\tau)\right] \qquad (7.38)$$

with $K_s(t)$ taken as in (7.36). One can observe a satisfactory agreement between the result computed by direct integration in (7.24) and by the two approximate approaches to the splitting of load processes.

The case of a propagating crack is illustrated in Figure 7.13 drawn for the same data as in Figure 7.12. The only difference is that it is assumed $a(t) = a_0 + ct$ where $a_0 = 0.1$ mm, $c = 10^{-8}$ m/s. The same notation is used as in Figure 7.13.

At last, Figure 7.14 presents the generalized driving forces for the case of displacement controlled loading. It is assumed that the applied strain is $\varepsilon_\infty = \varepsilon_{\infty,m} + \varepsilon_{\infty,a}\sin\theta t$ at $\varepsilon_{\infty,m} = 10^{-2}$ and $\varepsilon_{\infty,a} = 0.5 \cdot 10^{-2}$, and the crack is propagating as in the case of Figure 7.13. Line 1 is drawn according to (7.35) and (7.36), line 2 according to (7.38). Stress relaxation dominates in the beginning of crack growth. Later on, the generalized driving force tends to increase due to the crack propagation. It is distinctly seen in Figure 7.14b.

The agreement between numerical results is rather satisfactory, in any case, for the used data. In some sense, these results may be interpreted as evidence of

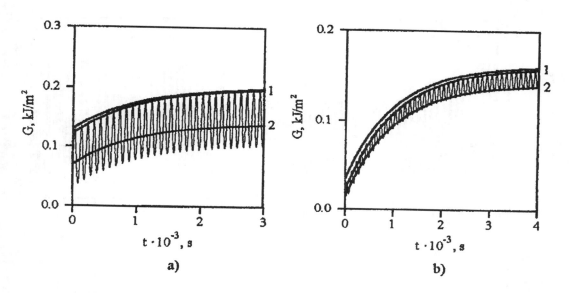

FIGURE 7.12
Current magnitudes of the generalized driving force, its approximate envelope
(1) and "static" component (2) for fixed cracks at $E_0 = 2E_\infty$ (a) and $E_0 = 10E_\infty$
(b).

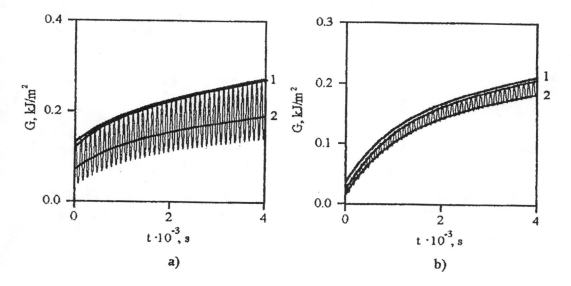

FIGURE 7.13
The same as in Figure 7.12 for moving crack tip.

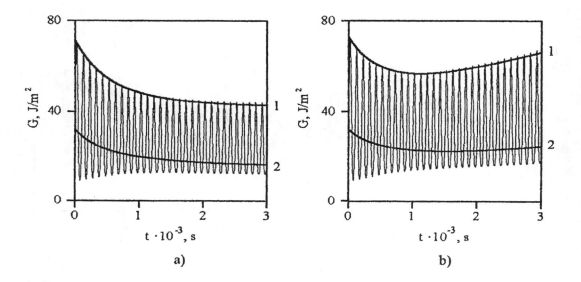

FIGURE 7.14
The same as in Figure 7.12 for displacement-controlled loading.

the validity of equation (7.15) which, as a matter of fact, is purely heuristic. The validity of this equation for constant or slowly varying loading has already been demonstrated by direct comparison with the results of laborious direct computations. On the other hand, the applicability of Irwin's equation for cyclic loading at $\tau_f \ll \tau_r$ is also beyond criticism. Summing up, we may conclude that all mentioned approaches to the evaluation of generalized driving force may be applied, at least at the current stage of study. When the cycle number in consideration is not very high, equation (7.35) is applicable without any further simplifications. When the cycle numbers are counted in millions, equation (7.35) economizes the computation time.

7.6   Fatigue Crack Growth in Polymer Materials

All the preparatory material for the prediction of fatigue crack growth in linear visco-elastic bodies has been presented in Sections 7.2-7.5. The set of governing relationships contains the principal relationship (7.24) which, in terms of the envelope $G_+(t)$, takes the form

$$G_+(t) \lessgtr \Gamma(t) \tag{7.39}$$

The left-hand side of (7.39) is given in (7.34), and the right-hand side in (7.25) with the additional assumption that the damage measures ω_s and ω_f are additive. Then

$$\Gamma = \Gamma_0 \left[1 - (\psi_s + \psi_f)^\alpha \right] \tag{7.40}$$

Neglecting the "healing" effects, we use equation (7.13) for damage accumulation. The evolution of conditions at the tips during crack growth and damage accumulation are described in equation (7.29).

Most numerical data that have been used in the above numerical study are used in the following computations, too. Let us list them again, supplementing with the data for parameters relating to cyclic fatigue: $E_0 = 2E_\infty = 1$ GPa, $\tau_r = 10^2$ s, $\gamma_0 = 1$ kJ/m^2, $\sigma_s = \sigma_f = 100$ MPa, $\sigma_{th} = \Delta\sigma_{th} = 10$ MPa, $t_c = 10^3$ s, $m_s = m_f = 4$, $\alpha = 1$, $\rho_s = 10$ μm, $\rho_b = \lambda_\rho = 100$ μm. The initial conditions are kept the same in all examples, namely $a_0 = 0.5$ mm, $\rho_0 = 50$ μm and zero initial damage in all the body, including the tip zones. The plate is subjected to uniform cyclic loading, and the crack propagates in mode I. Not to involve additional complications that are not necessary here, we assume that the plate width is large compared with the crack length $2a$. It means that we consider the Griffith problem in fatigue for visco-elastic solids (compare with Figure 4.1).

The most interesting point, perhaps, is the influence of the load ratio on the fatigue crack propagation, precisely the comparative contribution of sustained and cyclic components of loading in fatigue life. Because of entering of two temporal variables, time t and cycle number N, three series of computations are performed. In the first series, the applied stress component $\sigma_{\infty,m}$ is kept constant whereas the other component, $\sigma_{\infty,a}$ is subjected to variation. To the contrary, the component $\sigma_{\infty,m}$ is changed in the second series at the constant amplitude $\sigma_{\infty,a}$. Since the comparative contribution of those components strongly depends on the load frequency f, the latter parameter is also subjected to variation on a wide scale. This requires the third series of numerical experimentation.

Figures 7.15 and 7.16 are obtained at $\sigma_{\infty,m} = 5$ MPa and $\sigma_{\infty,a} = 1, 3, 5$ MPa (lines 1, 2, and 3). The load frequency is $f = 0.1$ Hz in all the cases. The crack growth histories in the function of the cycle number are presented in Figure 7.15a, and the growth rate diagrams in Figure 7.15b. In the latter case, the growth rate is interpreted as da/dN, and the range ΔK is used as a correlating load parameter. The evolution of the effective tip radius ρ and the tip damage measure ψ is shown in Figure 7.16.

The crack growth diagrams look rather agreeable when we plot da/dN versus the range ΔG of the driving force (Figure 7.17a). However, the slope of curves, which is almost the same along the major part of the diagrams, differs from $m_s/2$ or $m_f/2$. The attempt to plot the range da/dt in the function of G_+ (Figure 7.17b) fails: the divergence of curves becomes even larger than in Figure 7.15b.

The influence of the mean applied stress $\sigma_{\infty,m}$ is illustrated in Figures 7.18 and 7.19. The above numerical data are used: the difference is that $\tau_r = 10^3$ s, $\sigma_{\infty,a} = 5$ MPa, $\sigma_{\infty,m} = 5, 7.5, 10$, and 12.5 MPa (lines 1, 2, 3, and 4, respectively). Crack growth histories and growth diagrams (the latter in terms of da/dt and $K_{\max} = K_m + K_a$) are given in Figure 7.18. The evolution of the effective tip radius ρ and the tip damage measure ψ can be observed in Figure 7.19.

It is of interest to compare the contribution of the two components of damage, ψ_s and ψ_f, in summed damage ψ whose behavior, as one can see in Figure 7.19b, is similar to that in the absence of static fatigue. Two cases are demonstrated in

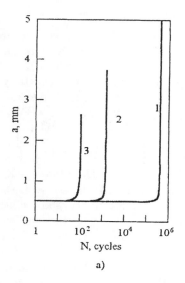
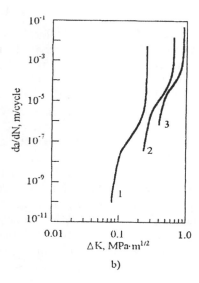

FIGURE 7.15
Crack growth histories under cyclic loading (a) and growth rate diagrams (b) at $\sigma_{\infty,m} = 5$ MPa, $\sigma_{\infty,a} = 1$, 3, and 5 MPa, $f = 0.1$ Hz (lines 1, 2, and 3, respectively).

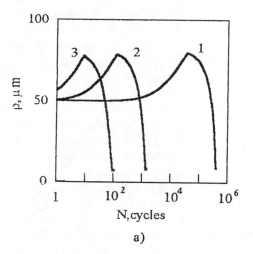
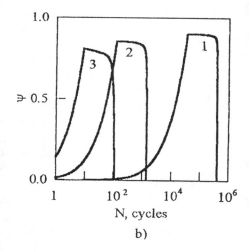

FIGURE 7.16
Effective tip radius (a) and tip damage measure (b) for the same data as Figure 7.15.

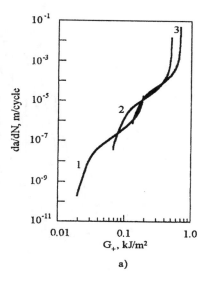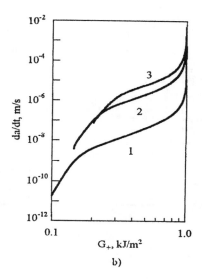

FIGURE 7.17
Crack growth rates plotted versus the range of the generalized force (a) and its
maximal magnitude (b) for the same data as in Figure 7.15.

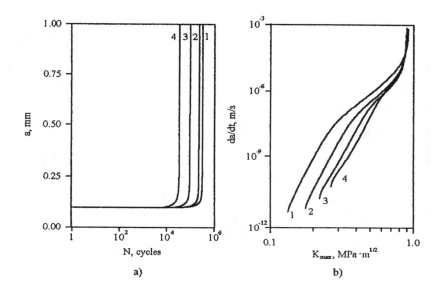

FIGURE 7.18
Crack growth histories under cyclic loading (a) and growth rate diagrams at
$\sigma_{\infty,a} = 5$ MPa, $\sigma_{\infty,m} = 5$, 7.5, 10, and 12,5 MPa (lines 1, 2, 3 and 4).

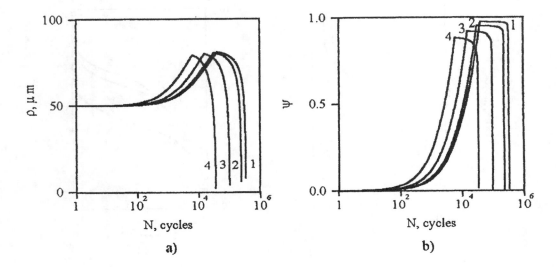

FIGURE 7.19
Effective tip radius (a) and tip damage measure (b) for the same data as in Figure 7.18.

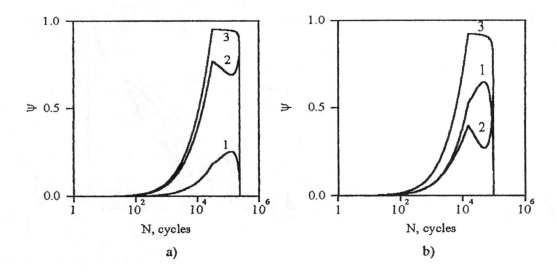

FIGURE 7.20
Component of damage measure at $\sigma_{\infty,m} = 10$ MPa, $\sigma_{\infty,a} = 5$ MPa (a) and $\sigma_{\infty,m} = 12.5$ MPa, $\sigma_{\infty,a} = 5$ MPa (b): (1) static; (2) cyclic; (3) summed.

Figure 7.20. Figure 7.20a is depicted for $\sigma_{\infty,a} = 2.5$ MPa, $\sigma_{\infty,m} = 7.5$ MPa. Line 1 corresponds to static damage ψ_s, line 2 to cyclic damage ψ_f, and line 3 to total damage, $\psi = \psi_s + \psi_f$. During the entire process, the cyclic component dominates upon the static one. Figure 7.20b is drawn for the case when $\sigma_{\infty,a} = 2.5$ MPa at $\sigma_{\infty,m} = 10$ MPa. In this case we observe the domination of static damage (line 1). The summed damage measure history is very similar to that in Figure 7.20a. Note that Figure 7.20a and b correspond to cases 2 and 3 in Figure 7.18, respectively. The cases 1 and 4 are extremal in this series of computations. In case 1 almost all the damage is produced by the cyclic component of loading, in case 4 by the static one.

The effect of load frequency is illustrated in Figures 7.21 and 7.22. In this case, both $\sigma_{\infty,m}$ and $\sigma_{\infty,a}$ are kept constant, namely, $\sigma_{\infty,m} = 10$ MPa, $\sigma_{\infty,a} = 2.5$ MPa. This means the applied load ratio $R = \sigma_\infty^{\min}/\sigma_\infty^{\max} = 0.6$. The only parameter subjected to change is the load frequency. Lines 1, 2, and 3 correspond to frequencies $f = 10^{-2}$, 10^{-1}, and 1 Hz. Compared with the assumed retardation time ($\tau_r = 10^3$ s), even the lowest frequency satisfies condition (7.1).

The cycle number is chosen as a temporial variable in Figure 7.21a where the rate da/dN is plotted versus the stress intensity range ΔK. In contrast, Figure 7.21b presents the same process in terms of time t. The rate da/dt is plotted versus the maximal stress intensity $K_{\max}$. For the chosen numerical data, the diagrams in Figure 7.21a are more compact. On the contrary, the frequency effect is significant when we consider da/dt as a function of $K_{\max}$. It is, of course, an almost obvious conclusion.

The envelope G_+ of the generalized driving force and the range ΔG of this force within a cycle also might be among adequate load parameters for growth rate diagrams. It is illustrated in Figure 7.22 obtained for the same numerical data as Figure 7.21. There is no considerable difference in the scattering of results presented in Figures 7.21 and 7.22.

We do not enter here into further verbal discussion of numerical results gathered in Figures 7.15-7.22. Appropriate conclusions could be made by comparison of growth rate diagrams presented in these figures.

7.7   Creep Deformation and Damage Accumulation in Metals

Delayed damage and fracture (creep) of metals and metallic alloys are of primary significance in many domains of engineering such as power production and have been thoroughly studied during the last decades both experimentally and theoretically. A number of mechanisms are involved in creep. Among them are diffusion and dislocation phenomena, formation of microvoids and micropores, their coalescence with the initiation of macroscopic cracks, and crack propagation in the already damaged material.

Many authors follow the concept that creep damage and fracture can be described in the framework of continuum damage mechanics. Beginning from the pioneering works by Kachanov (1958) and Rabotnov (1963), a number of papers

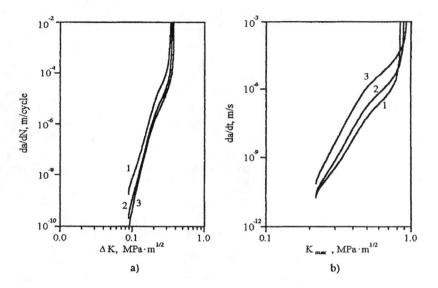

a)   b)

FIGURE 7.21
Growth rate diagrams at $\sigma_{\infty,m} = 5$ MPa, $\sigma_{\infty,a} = 2.5$ MPa, and load frequency $f = 10^{-2}, 10^{-1}$, and 1 Hz (lines 1, 2 and 3).

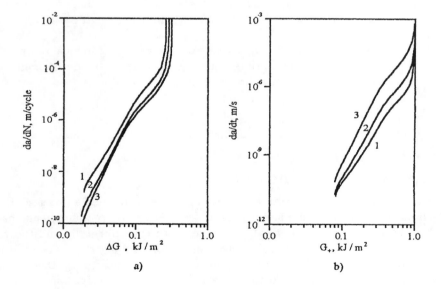

a)   b)

FIGURE 7.22
Growth rates in the function of the range (a) and upper envelope (b) of the generalized driving force for the same data as in Figure 7.21.

were published in which authors attribute microcrack propagation to a certain damage measure. It was assumed that a crack tip advances when the damage measure at the tip (or at a certain point ahead of the tip) attains a certain critical level. In the simplest case, when the scalar measure ω is used with values from $[0,1]$, the condition of crack growth is taken as $\omega_* = 1$. To avoid singularities, another critical level, say $\omega_* = 0.99$, has been used in some publications.

Another approach to creep damage is based on the concepts of micromechanics. This approach includes the analysis of birth, growth and coalescence of microdefects and their further interaction. For example, macrocrack propagation may be considered as a result of coalescence of an array of microscopic defects situated along the crack path.

The aim of the analysis presented below is to demonstrate the applicability of the general theory to the case of delayed fracture associated with creep. As an introduction, a very brief survey is given of the constitutive equations most frequently used in creep analysis [126].

The equations of creep deformation are usually taken in the form

$$\dot{\varepsilon}_{jk} = \frac{1+\nu}{E}\dot{s}_{jk} + \frac{1-2\nu}{E}\dot{\sigma}_0\delta_{jk} + \frac{3}{2}B\sigma_e^{\beta-1}s_{jk} \qquad (7.41)$$

Here $\sigma_0 = 1/3\sigma_{jj}$ is the mean (hydrostatic) stress; $s_{jk} = \sigma_{jk} - 1/3\sigma_0\delta_{jk}$ is the deviatioric stress tensor; $\sigma_e = 3/2(s_{jk}s_{jk})^{1/2}$ is the stress intensity. Differentiation with respect to time is denoted by dots. The first two terms in the right-hand side of (7.41) describe elastic straining with Young's modulus E and Poisson's ratio ν. The last term characterizes the straining due to creep. Here $B > 0$ and $\beta \geq 1$ are material parameters. At $\beta = 1$ the material is linear visco-elastic. In the one-dimensional case, equation (7.41) yields Norton's law for stationary creep:

$$\dot{\varepsilon} = \frac{\dot{\sigma}}{E} + B\sigma^\beta \qquad (7.42)$$

Material parameters E, ν, B and β depend on temperature. This dependence is of primary importance in creep. In fatigue, both static and cyclic, these parameters also depend on the current damage field that expands upon a significant domain around and ahead of the crack.

Let us assume that the damage is described by a single scalar measure ω similar to that used in the former sections of this book. The equation of damage accumulation

$$\frac{\partial\omega}{\partial t} = \frac{D\sigma^m}{(1+m)(1-\omega)^n} \qquad (7.43)$$

is similar to equations (4.3). The only difference is the entrance of the parameter D which is a characteristic rate of damage accumulation. This is not only a contribution to tradition. Presenting the damage equation as (7.43), we can use experimental data from literature [126, 119]. As a rule, the exponent n in (7.43) is taken as positive to describe the damage attributed to tertiary creep. Usually it is assumed that $n = m$. In general, all material parameters in (7.43) depend on temperature. The representative stress σ in equation (7.43) must be chosen taking into account material behavior in creep. A rather wide scope of materials can be covered if we insert in (7.43) the characteristic stress

$$\sigma = \chi\sigma_0 + (1 - \chi)\sigma_e \tag{7.44}$$

When $\chi = 1$, it means that the material is sensitive to the mean stress σ_0, and the damage is mostly accompanied by the micropore nucleation. If shear mechanisms dominate, it is sensible to assume that $\sigma \approx \sigma_e$, i.e., $\chi \approx 0$. In typical situations $0 < \chi \leq 1$.

Material compliance certainly depends on the local damage. The simplest analytical model is

$$E = E_0(1 - \omega)^{n_E}, B = B_0(1 - \omega)^{n_B} \tag{7.45}$$

where E_0 and B_0 correspond to the nondamaged material; $n_E \geq 0$, $n_B \geq 0$ are material parameters. By the way, fixing the magnitudes of n_E and n_B, e.g., putting $n_E = n_B$, we could assess the damage measure ω by measurement of elastic and inelastic components of material compliance in the process of creep straining. Then the measure ω has a certain physical meaning.

7.8   Generalized Driving Forces in Creep

The definition of the generalized driving forces given in Section 3.4 and used throughout this book remains valid in creep damage problems. In two-dimensional problems equation (3.22) is valid. Similar to the case of linear visco-elasticity, equation (7.23) is applicable for the numerical estimation of driving forces for one-parameter cracks in the presence of creep [25].

In this section, using rather simple model examples, we will study the influence of creep and damage on the generalized driving force. In particular, the comparison with the J-integral approach generalized upon non-elastic phenomena will be presented along with analyzing the influence of integration domains on the final numerical results. The easiest way to perform such an analysis is to consider stationary (fixed) cracks as well as cracks propagating with constant rates. The process of real crack growth associated with creep will be discussed in the next sections.

Consider a rectangular plate with length $2l$, width $2b$ and the central crack whose initial length is $2a_0$ (Figure 7.23). The plate is subjected to the remote stress whose is applied at $t = 0$ and then is kept constant. Temperature is assumed constant also. Equation (7.23) for numerical assessment of the generalized driving force is

$$G(t)\Delta a(t) = \int_0^b \sigma_\infty(t)[v_{II}(x, t) - v_I(x, t)]dx - \int_0^l \int_0^b [w_{II}(x, y, t) - w_I(x, y, t)]dxdy \tag{7.46}$$

Here $v(x, t)$ is the displacement in y-direction at the plate side $y = l$; $w(x, y, t)$ is the work density of internal forces attained at the moment t.

Application of equation (7.46) is very laborious. It includes the parallel solution of creep problems for two bodies, I and II, with cracks of slightly different dimen-

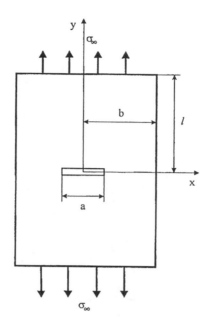

FIGURE 7.23
Plate with central crack under remote sustained tension.

sions (Figure 7.1). In contrast to the case of linear visco-elasticity, the full-scale application of computational mechanics is required here [2, 104]. As a competitor of $G(t)$, the J-integral generalized upon nonlinear problems is frequently used. In the two-dimensional case we have

$$J(t) = \int\limits_C \left(wn_j - p_\alpha \frac{\partial u_\alpha}{\partial x_j} \right) ds \tag{7.47}$$

where the same notation is used as in (1.24) or (3.25). An essential difference from the classical case is that in this case $w(x, y, t)$ is the density of work performed by internal forces during the entire history of loading, damage accumulation, and crack growth. It means that the numerical procedure is required to evaluate the integral in (7.47) similar to that in the case of $G(t)$. In addition, the question is open of how to choose the integration contour C. Inelastic strains and damages are spread upon the whole plate. This means that we must integrate over all the domain $x \le b$, $y \le l$. Four integration contours near the crack tip are shown in Figure 7.24.

It is widely known that the J-integral is hardly the best load parameter correlating with the crack growth rate in the presence of creep. Heuristic modifications of the J-integral are usually applied, such as the C-integral. Instead of energy and work, the corresponding time derivatives enter the C-integral. As a matter of fact, compared with the J-integral which may be interpreted as an energy flux, the C-integral and its relatives are a kind of a power flux. One cannot make a comparison

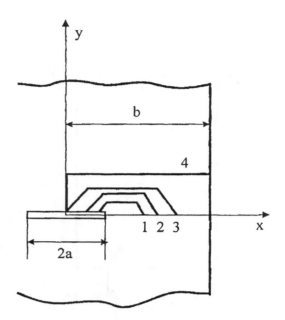

FIGURE 7.24
Four integration contours around the crack tip.

with generalized forces in the sense of analytical mechanics. However, it seems that these load parameters exhibit a rather fair correlation with experimental data [126].

Some numerical results for $E_0 = 150$ GPa, $\nu = 0.3$, $B_0 = 1.22 \cdot 10^{-21}$ [(MPa)^5 $\cdot$ s]$^{-1}$, $\beta = 5$ are presented in Figures 7.25 and 7.26. They are plotted for $a_0 = 10$ mm, $b = 60$ mm, $l = 80$ mm, $\sigma_\infty = 75$ MPa. At this stage of study, the influence of damage is neglected.

The magnitudes of the J-integral computed with the use of integration contours 1, 2, 3, and 4 are shown in Figure 7.25 by lines 1-4. Line 5 corresponds to the driving force estimated by equation (7.46). Two items can be observed in Figure 7.25. First, it is a significant difference between $G(t)$ and $J(t)$. Second, there is a strong dependence of the J-integral on the choice of the integration contour. However, the lines for the contour 4 and the line 5 practically coincide. This means that the domain of significant plastic straining is practically surrounded by contour 4.

The driving generalized forces for cracks uniformly growing in time at $\sigma_\infty = 150$ MPa are shown in Figure 7.26. Three crack growth rates are considered given by the nondimensional parameter $\eta = ct_c/a_0$. Here $c \equiv da/dt$, and t_c is the characteristic time of creep straining estimated as $t_c = 100$ hours. Lines 1, 2, 3 are drawn for $\eta = 0.1, 1$, and 1 mm $\cdot$ hour $^{-1}$, respectively. They correspond to the crack growth rates $c = 0.01, 0.1$, and 1 mm $\cdot$ hour $^{-1}$.

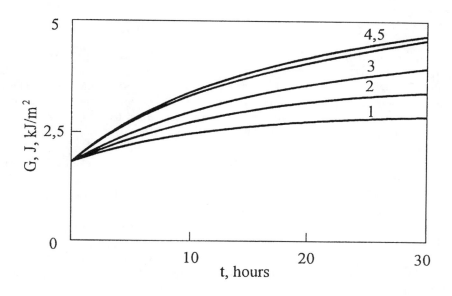

FIGURE 7.25
Time histories of the *J*-integral for four integration contours (lines 1,2,3,4) and the generalized driving force (line 5) for a fixed crack.

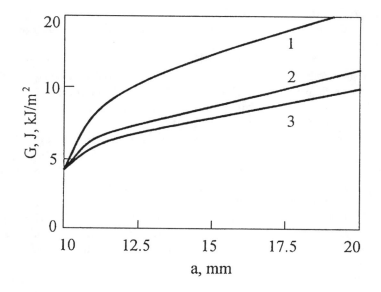

FIGURE 7.26
Time histories of the generalized driving force (1,2,3) and the *J*-integral (1',2',3') for crack growth rates 0.01, 0.1 and 1 mm·hour^{-1}.

7.9   Crack Growth Associated with Creep

Modeling of crack growth is based, as usual, on equation (7.24). The driving force $G(t)$ and the damage measure are determined by direct numerical calculation. The further procedure follows in the same way as in the case of linear visco-elasticity.

The numerical results for a central crack in a plate are presented in Figures 7.27-7.31. They are obtained for the data corresponding to a stainless steel at the averaged temperature $T = 810$ K: $E_0 = 150$ GPa, $\nu = 0.3$, $B_0 = 1.3 \cdot 10^{-24}$ [(MPa) $^\beta \cdot$ s] $^{-1}$, $\beta = 7.3$, $\chi = 0.5$, $n_E = n_B = 0$. It is easy to estimate the characteristic stress and characteristic time for creep deformation. Presenting the last term in (7.42) as $(\sigma/\sigma_c)^\beta t_c^{-1}$, we obtain that $\sigma_c = 324$ MPa at $t_c = 100$ hours. The damage accumulation equation is taken, following [126], in the form (7.43) at $D_0 = 7.5 \cdot 10^{-21}$ (MPa) $^{-m}$ s^{-1}, $m = n = 6.2$. The generalized resistance force is taken as $\Gamma = \gamma_0(1 - \psi)$ at $\gamma_0 = 15$ kJ/m^2. The applied stress is $\sigma_\infty = 75$ MPa, and the initial crack size $a_0 = 10$ mm.

The relationship between the generalized forces is shown in Figure 7.27. Line 1 depicts the driving force, and line 2 the resistance force. The crack growth initiation takes place at $t_* = 17$ hours. Later on, the equality $G(t) = \Gamma(t)$ holds until the final rupture at $t_{**} = 64$ hours.

The damage measures are also depicted in Figure 7.27. Line 3 shows the damage measure at the crack tip, line 4 the damage in the far field. At the moment of crack growth start, the tip damage attains the level $\psi \approx 0.7$. The far field damage is increasing monotonously during the crack growth. It does not exceed the level $\omega = 0.12$ at the final failure, whereas the tip damage drops until $\psi \approx 0.20$. It is also seen in Figure 7.28 where the distribution of the damage measure ω along the crack path during all the failure life is shown. The dotted line corresponds to damage at the crack tips.

The tensile stress distribution ahead of the crack tip is shown in Figure 7.29. In the initial moment (line 1) the material is deformed elastically. At the moment of crack growth initiation (line 2), the lessening of maximal stress is observed as well as the shift of the maximum ahead of the tip. Lines 3 and 4 in Figure 7.29 correspond to later stages of crack growth.

Some complementary information is given in Figures 7.30 and 7.31. Figure 7.30 illustrates the effect of the plate width on fatigue life and tip damage history. The applied stress is $\sigma_\infty = 150$ MPa, and the fracture toughness $\gamma_0 = 30$ kJ/m^2. Lines 1,2,3, and 4 are drawn for $w = 40, 60, 100$ mm, and $w \to \infty$.

Figure 7.31 shows the influence of applied stresses in terms of driving forces (Figure 7.31a) and tip damage measures (Figure 7.31b). Lines 1,2,3 are drawn for $\sigma_\infty = 100, 150$, and 200 MPa at $w = 60$ mm. Evidently, the applied stresses' level strongly affects fatigue life, both their initiation stage and that of crack growth.

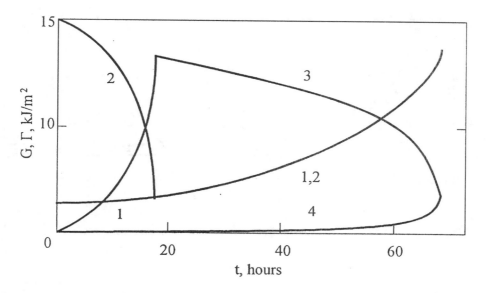

FIGURE 7.27
Time histories of generalized force (1), generalized resistance force (2), tip damage measure (3), and far field measure (4) for a crack growing under a sustained load.

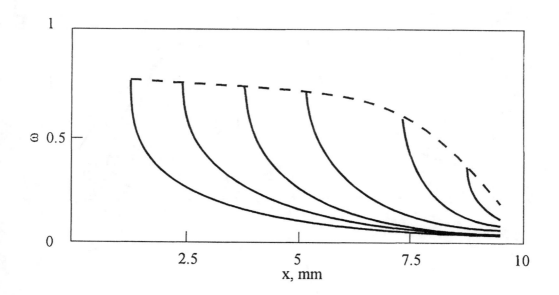

FIGURE 7.28
Evolution of damage distribution ahead of crack during fatigue life.

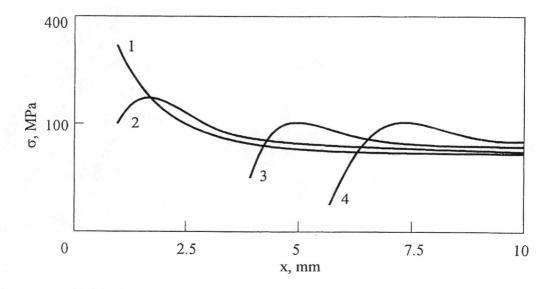

FIGURE 7.29
Evolution of stress distribution ahead of crack: (1) initial; (2) at crack growth
start; (3) and (4) during following crack growth.

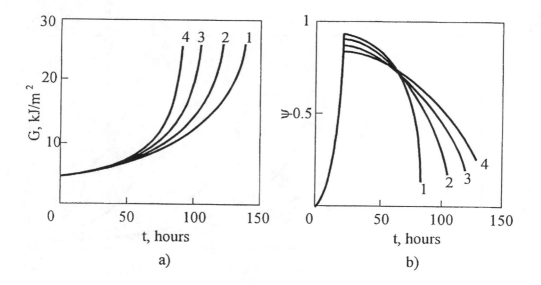

FIGURE 7.30
Effect of plate width on generalized driving force (a) and tip damage measure
(b); lines 1,2,3, and 4 correspond to $b = 40, 60, 80$, and $w \to \infty$.

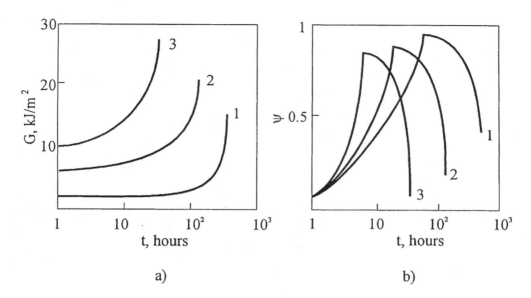

a)   b)

FIGURE 7.31
Effect of applied stress level on generalized driving force (a) and tip damage measure (b); lines 1,2,3 correspond to $\sigma_\infty = 100,\ 150,\ 200$ MPa.

7.10 Creep Crack Growth in Turbine Disks

The application of the theory to turbine disks subjected to loading by centrifugal forces and thermal fields varying along the radius has been presented in the paper by Minakov [103]. The constitutive equations (7.41), (7.45) and damage equations (7.43), (7.44) were used for disks of material similar to those considered in the above numerical examples. The principal difference in numerical data is that damage exponent $m = 6$ in equation (7.43) and the specific fracture work is taken $\gamma_0 = 80$ kJ/m^2. The latter magnitude corresponds to comparatively high fracture toughness. In particular, the critical stress intensity factor at $E_0 = 150$ GPa is $K_c = (\gamma_0 E_0)^{1/2} \approx 109$ MPa $\cdot$ m$^{1/2}$.

The disk is schematically shown in Figure 7.32. The internal disk radius is $R_1 = 50$ mm, the external $R_2 = 300$ mm. The rotation speed is 50 revolutions per second, the temperature at the disk periphery is about 873 K and at the internal surface is about 573 K. The external surface is loaded by given stresses $\sigma_\infty = 50$ MPa; the internal surface is free of loading. The influence of temperature on material properties, in particular on magnitudes of B and D, is taken into account according to experimental data supplemented by some ad hoc assumptions.

Two cases are considered: a crack with length a_1 growing from the internal surface, and an external crack with length a_2 (Figure 7.32). The case of an internal crack is illustrated in Figures 7.32-7.37. At $a_0 = 10$ mm the initial magnitude of the driving force is $G = 10.4$ kJ/m^2. Thus, the initial state of the system is subequilibrium. The crack growth initiation takes place at $t_* = 14.5$ hours. Then

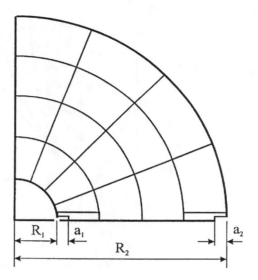

FIGURE 7.32
Schematization of a turbine disk with radial cracks.

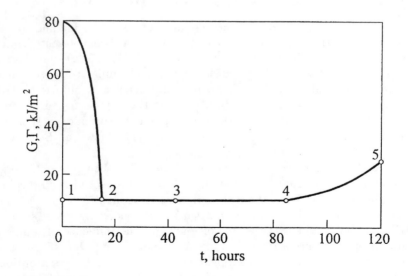

FIGURE 7.33
Relationship between driving force (1) and resistance force (2) for an internal radial crack in the disk.

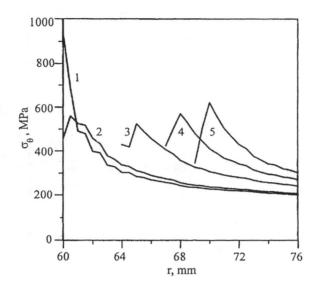

FIGURE 7.34
Distribution of circumferential stresses ahead of a propagating internal crack;
lines 1,2,3,4,5 correspond to times numbered in Figure 7.13.

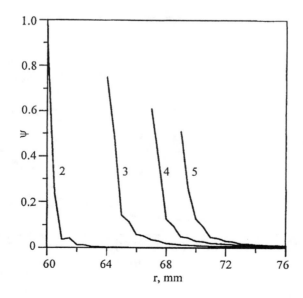

FIGURE 7.35
Distribution of damage ahead of internal crack; notation as in Figure 7.34.

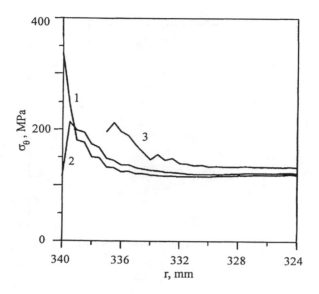

FIGURE 7.36
Relationship between generalized driving and resistance forces for an external radial crack in the disk.

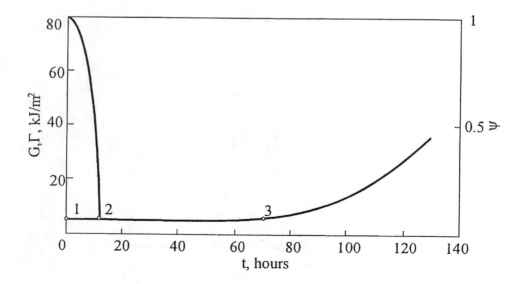

FIGURE 7.37
Distribution of circumferential stresses ahead of external crack; lines 1,2,3 correspond to times numbered in Figure 7.36.

the continuous crack propagation takes place under the condition $G(t) = \Gamma(t)$ (Figure 7.33). The final fracture occurs at $t_* \approx 140$ hours.

The evolution of the circumferential stress at the crack tip is shown in Figure 7.34. Line 1 corresponds to the initial state when the disk is deformed elastically. Line 2 is drawn for the crack growth start moment. The latter evolution is presented by lines 3, 4, and 5. The distribution of the damage measure is shown in Figure 7.35 where the same numeration of lines is used as in Figure 7.34.

The early crack growth is mostly controlled by damage at the crack tip. The contribution of creep straining is comparatively small. Approximately at $t = 100$ hours the influence of creep becomes significant, and this results in the increase of the generalized driving force. In addition, the contribution of the far field damage becomes more significant. The combination of these factors results in the acceleration of crack growth.

The case of external cracks is a little different. At the initial length $a_0 = 20$ mm with the magnitude of the driving force $G = 1.34$ kJ/m^2. The time histories of circumferential stresses (Figure 7.34) are similar to those for internal cracks (Figure 7.34). Line 1 corresponds to the initial loading, line 2 to the crack growth initiation, and line 3 is drawn for $t = 75$ hours. The stress level, in general, is lower than that for internal cracks. However, when the crack enters a more damaged and compliant domain, the crack growth rate significantly increases. In addition, the temperature at the periphery is higher; it results in higher damage and deformation rates. As a result, the total fatigue life in the case of an internal crack $a_0 = 10$ mm and that in the case of an external crack $a_0 = 20$ mm are of the same order of magnitude. It is illustrated in Figure 7.37 where crack growth histories are shown for internal and external cracks (lines 1 and 2, respectively).

7.11 Summary

The application of the theory of fatigue crack growth to bodies with material exhibiting hereditary mechanical properties was presented in this chapter. Among them are polymers under sustained and/or cyclic loading as well as metallic alloys subjected to creep. Some ways were demonstrated to overcome difficulties arising from the necessity of dealing with isochronic variations in the presence of hereditary deformations.

Crack growth in linear visco-elastic materials under sustained and cyclic loading was considered to predict the life until the fracture of structural components. The influence of various material and loading parameters on crack propagation in polymers was studied. In particular, the interaction between slowly varying and cyclic components of the loading process was investigated. Crack growth in metals associated with creep was studied also, and creep fatigue in turbine disks was discussed as an illustrative example. The final results were discussed in terms of predicted fatigue lives and crack growth diagrams.

Many issues of fatigue in hereditary media are beyond the given discussion. Among them is the problem of accounting for finite deformations, especially for thermoplastics with their nonconventional straining near the crack tips. Another problem is accounting for thermal effects such as the heat production in polymers under high-frequency loading and its influence on mechanical properties. These and related problems are not at all passed unnoticed. In principle, they might be incorporated in the scheme of the theory without changing the general methodology.

Chapter 8

ENVIRONMENTALLY AFFECTED FATIGUE AND RELATED PHENOMENA

8.1 Interaction of Mechanical and Environmental Factors in Fatigue

Environmental factors, most of which are of a non-mechanical nature, have an influence on crack initiation and propagation both under cyclic and sustained mechanical loading. The typical case is the corrosion fatigue when the interaction of cyclic loading and chemical, electro-chemical or bio-chemical actions takes place. Moist air, natural water, and other active media, as a rule, intensify the process of fatigue crack growth. Concentration of the active agent, temperature and electro-chemical conditions are among the most important factors in corrosion fatigue along with the parameters of cyclic mechanical loading. A similar situation takes place when the loading is stationary or varying slowly. In these cases, one talks on stress corrosion cracking.

A number of other phenomena are to be listed which are related to corrosion fatigue and stress corrosion cracking. Many gaseous media, both natural and used in industry, produce a significant influence on a material's resistance to mechanical damage. In particular, they affect the specific fracture work and other parameters characterizing fracture and fatigue toughness. A typical phenomenon is the hydrogen embrittlement in metals and metallic alloys which changes their mechanical properties due to diffusion of hydrogen from the surface. This results in the acceleration of fatigue crack growth and decrease of resistance to final fracture. A similar effect is produced by intensive irradiation.

Related phenomena are observed in nonmetallic materials such as polymers and polymer-based composites, and ceramic materials (both natural and artificial). In the absence of strong mechanical actions, one says aging or deterioration. When these processes are accompanied by significant cyclic or sustained loading, we deal with the phenomena similar to corrosion fatigue and stress corrosion cracking.

Corrosion, aging, embrittlement and other deterioration effects are the subject of electro-chemistry as well as of material science. These phenomena are controlled by a number of parameters whose influence is sometimes controversial and not easy to predict. There are a lot of publications concerning corrosion, corrosion fatigue and stress corrosion cracking. Most of them deal with experiments and their interpretation in terms of material science, chemistry or electro-chemistry. A survey of publications in this topic can be found in [93]. The character of these publications is frequently, from the viewpoint of solid mechanics, too esoteric. As for analytical approaches to the prediction of damage due to corrosion fatigue and stress corrosion cracking, they are based, as a rule, on a hypothetical principle of additivity or its slight modifications.

The objective of this chapter is to show that the theory of corrosion fatigue and related environmentally affected phenomena enter the framework of the theory of fatigue crack propagation developed in this book. Actually, we treat crack propagation as a result of interaction of the global balance of forces and energy in the system cracked body-loading with the process of dispersed damaged accumulation. The latter is considered irrespective of the source and the special feature of damage. Staying within a pure phenomenological approach to damage similar to that of continuum damage mechanics, it is easy to widen the range of damages in consideration. One could include chemical, electro-chemical, radiational, biological, etc. types of damage. For each type we have to introduce special measures and equations governing the evolution of these measures in time.

Concentrations, fluxes, temperature, etc. enter here as control parameters. Special equations are also needed to describe the influence of damage on mechanical properties. Some information can be extracted from the available experimental data. However, it is impossible to abandon various ad hoc assumptions which have to be checked in the future. In any case, it must be stressed that we do not intend to go into the topics which are rather far from mechanics of solids and structures leaving the intrinsic phenomena of material damage to the specialists in material science [52, 58, 93, 96, 128].

Even in the framework of the mechanical statement of the problem, there are very specific complications. Consider, in particular, corrosion fatigue. It is obvious that the main chemical factor controlling damage accumulation near the crack tip is the content of the active agent near the crack tip. Since the "applied" content is given at the surface, near the entrance into the crack, the problem arises of agent transport through the crack to its tip. The tip is propagating in time, and its propagation is accompanied by blunting and sharpening. The crack opening displacements are subjected to cyclic variation including crack closure, etc. It means that the agent transport within a crack is an essentially nonstationary process, and a number of parameters entering the model of this process have not yet been studied. The above statements should clarify our position when we need additional assumptions to close the set of governing equations.

8.2   Transport of Environmental Agents within Cracks

Later on, we will characterize the content of an active environmental agent in the vicinity of the crack tip and, generally, in the material of a body with a scalar variable. This variable depends on the cycle number (or on the physical time) as well as on the position of the material particle under consideration. For brevity, we name this variable just content and denote it $c(x, t)$ or $c(x, N)$. In the simplest cases, such as when a solution of a single chemically active substance is concerned, the content $c(x, t)$ is just the concentration of the solution. In the case of humid air this variable is referred to as the water vapor content. However, ambient air or, say, sea water contains a number of aggressive components which cannot be characterized by a single variable. In addition, several mechanisms are, as a rule, present in the corrosion-assisted damage such as anodic dissolution, oxidation, repetitive ruptures of passive films, hydrogen embrittlement, etc. Not only chemical content, but temperature, pressure and electro-chemical interaction between the active medium and the material of a body are important factors. Not to go into excessive details, we will deal in this chapter with a single variable for environmental conditions. The modeling of hydrogen affected degradation that plays a significant role in most environmentally assisted damage will be discussed in Section 8.11.

The content at the crack tip $c_t(t) \equiv c[a(t), t]$, in general, differs from that at the crack mouth denoted here $c_\infty(t)$. There is the transport of the agent through the crack which must be described by a corresponding equation. This equation is assumed one-dimensional with respect to the content $c(x, t)$ where x is the coordinate measured along the crack (Figure 8.1). Crack growth and meandering of the crack trajectory must be taken into account as well as the boundary conditions at the moving crack tip. The latter vary in time due to damage accumulation, crack blunting and sharpening, passive film formation and its rupture, etc. In corrosion fatigue, when a crack is "breathing" under cyclic loading, the pulsating fluid or gas flow is to be taken into account along with the diffusion process. Although chemical and electro-chemical processes near the crack tips were recently studied in detail [93, 128], there are not enough experimental data to use an adequate hydrodynamical model for the penetration of active agents into fatigue cracks. Hence, we turn to heuristic models trying to include most relevant factors in consideration.

In stress corrosion cracking, when the transport of the agent proceeds monotonously, quasistationary models seem to be acceptable. One of the simplest models is given by equation

$$\frac{dc_t}{dt} = \frac{c_a - c_t}{\tau_D} + \frac{c_b - c_t}{\lambda_D}\frac{da}{dt}. \tag{8.}$$

Several governing parameters enter this equation. Among them there are the characteristic length λ_D and the characteristic time τ_D (the subscript D means diffusion). The third parameter c_a characterizes the stationary concentration attained for a fixed crack at $t \gg \tau_D$. This parameter depends on the entrance content c_∞ and the current crack length a, for example, as follows:

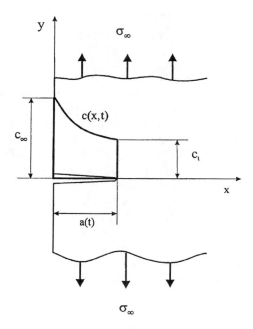

FIGURE 8.1
Distribution of active agent within a crack.

$$c_a = c_\infty \left(1 + \frac{a}{a_\infty} \right)^{-n_a}$$

(8.2)

Here $a_\infty > 0$ and $n_a \geq 0$ are some constants depending on material and environment properties. The fourth parameter in (8.1), c_b, is a certain characteristic content for rapidly growing cracks. It is natural to suppose that $0 \leq c_b \leq c_a$.

Equations (8.1) and (8.2) present a sensible scheme of transport within a growing crack under the condition that λ_D and τ_D depend on temperature, agent parameters, electro-chemical potential in the system agent-metal, etc. When the crack tip is fixed, and the initial condition is $c_t(0) = c_\infty$, equation (8.1) describes the evolution of the tip concentration as

$$c_t(t) = c_a + (c_\infty - c_a) \exp \left(-\frac{t}{\tau_D} \right)$$

(8.3)

At $c_t(0) = 0$, i.e. when the crack tip is initially "dry", we obtain that

$$c_t(t) = c_a \left[1 - \exp \left(-\frac{t}{\tau_D} \right) \right]$$

(8.4)

In any case, the solution of equation (8.1) is elementary.

Three typical examples are schematically presented in Figure 8.2. Figure 8.2a shows the evolution of the content when the tip is fixed. Line 1 corresponds to the initial condition $c_t(0) = c_\infty$, i.e., to equation (8.3), line 2 to the initial condition

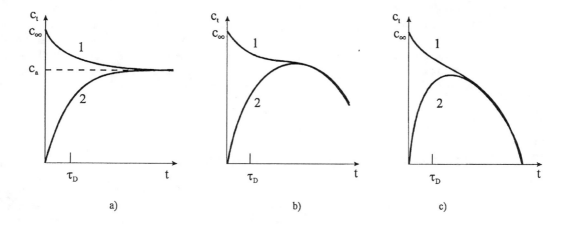

FIGURE 8.2
Active agent content at the tip of cracks: (a) fixed, (b) slowly growing, (c) rapidly growing; lines 1 and 2 correspond to different initial conditions.

$c_t(0) = 0$, i.e., to equation (8.4). The dashed line presents the characteristic content given in (8.2). Figure 8.2b illustrates the case of a slowly propagating crack. When the crack growth rate is high, the transport of the agent to the tip becomes difficult. It is illustrated in Figure 8.2c for the case $c_b = 0$. The same notation is used in all three cases.

When loading is cyclic, we have to make some changes in equations (8.1) and (8.2). The hydrodynamics of the agent flow in one-dimensional channels with the pulsating cross section is elementary. But to apply a fluid dynamics model, we have to know in detail the change of the crack profile and the crack effective cross section within each loading cycle, not to mention the electro-chemical conditions at the tip.

To avoid complications that do not seem necessary at the presented stage of study, we will use equation (8.1) for the cyclic process interpreting $c_t(t)$ as an average content at the tip within a cycle. Then the frequency of loading θ becomes an additional control parameter. We might assume that all the parameters, $\lambda_D, \tau_D, a_\infty$ and n_a entering equations (8.1) and (8.2) depend on θ. If we prefer to eliminate the frequency effect from all the listed parameters but one, the characteristic length parameter a_∞ is perhaps the best candidate. In fact, the most important effect of cyclic loading is crack breathing. Thus we include this effect in equation (8.2).

In further numerical examples we assume that

$$a_\infty(\theta) = a_\infty(0) \left(1 + \frac{\theta}{\theta_\infty}\right)^{n_\theta} \tag{8.5}$$

where θ_∞ is a characteristic frequency, and $n_\theta \geq 0$. At $\theta \ll \theta_\infty$ we come to equation (8.2) used for quasistationary transportation. When the loading frequency increases, the intensification of mixing takes place within the crack hollow. This effect results in increasing of the characteristic penetration length $a_\infty(\theta)$. At $\theta \gg \theta_\infty$ the magnitude of $a_\infty(\theta)$ becomes large compared with the running crack length $a(t)$. It means that for high frequencies, slowly growing cracks, and sufficiently long loading times, $c_t(t)$ approaches the applied content c_∞. However, all these assumptions, as those given in equations (8.1) and (8.2), are purely heuristic and could be replaced with other appropriate assumptions.

8.3 Corrosion Damage Accumulation

In corrosion fatigue and related phenomena, the current material state depend on two and more damage measures. Along with the pure mechanical damage, damage as a result of environmental agents must be introduced. Stress corrosion cracking and corrosion fatigue will be considered in the next sections by assuming that the environmental damage may be described by single scalar measure ω_c similar to the mechanical measure ω. To avoid misunderstandings, we supply the mechanical damage measures produced by static and cyclic load components with special subscripts, denoting them ω_s and ω_f, respectively.

In contrast to the mechanical damage which is spread upon all the future crack trajectories, i.e., both in the near and far fields, corrosion damage is localized in the vicinity of the crack tip. The thickness of the localization layer is usually very small, being of the order of the oxide film thickness. Beyond this layer only mechanical damage is present. The situation is illustrated in Figure 8.3a. There the distribution of damage ahead of an edge crack is shown under quasistatic loading. The measure ω_s attains the maximal level within the process zone $a \leq x \leq a + \lambda_p$ while the measure ω_c is localized at $0 \leq x \leq \lambda_c$. Except for very short cracks, we may assume $\lambda_c < \lambda_p$. In any case, the linear approximation of the corrosion damage distribution is acceptable such as

$$\omega_c(x,t) = \psi_c(t)\left(1 - \frac{x-a}{\lambda_c}\right) \tag{8.6}$$

At $x > a + \lambda_c$ we put $\omega_c \equiv 0$. Then the corrosion damage will be described by the tip measure $\psi_c \equiv \omega_c[a(t), t]$. For damage due to static and cyclic fatigue we introduce measures $\psi_s(N) \equiv \omega_s[a(N), N]$ and $\psi_f(t) \equiv \omega_f[a(t), t]$.

From the above considerations it follows that the time evolution of the corrosion uamage measure may be described by equation

$$\frac{d\psi_c}{dt} = F(c_t, \psi_s, \psi_f, \psi_c) \tag{8.7}$$

with the right-hand side depending on the agent content and the damage measures at the crack tip. In particular, the power-threshold law is convenient

$$\frac{d\psi_c}{dt} = \frac{1}{t_c}\left(\frac{c_t - c_{th}}{c_d}\right)^{m_c} \tag{8.8}$$

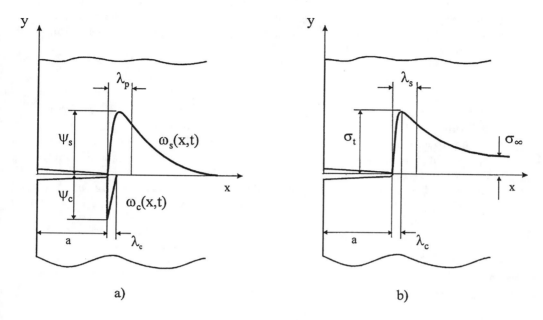

FIGURE 8.3
Distribution of damage (a) and tensile stresses (b) in environmentally assisted fatigue.

with material parameters c_d, c_{th}, m_c, and characteristic time t_c.

Dependence of the right-hand side in (8.7) on ψ_s and ψ_c can be included considering, say, c_d and c_{th} as functions of these variables. Equation

$$\frac{d\psi_c}{dt} = \frac{1}{t_c} \left(\frac{c_t - c_{th}}{c_d} \right)^{m_c} (1 - \psi_s - \psi_f)^{-n_c} \qquad (8.9)$$

at $n_c > 0$ is one of the possible ways to take into account the vulnerability of mechanically damaged materials to corrosion. On the other hand, corrosion damage enhances the process of mechanical damage accumulation. For example, the equation similar to (8.9)

$$\frac{\partial \omega_s}{\partial t} = \frac{1}{t_c} \left(\frac{\sigma - \sigma_{th}}{\sigma_d} \right)^{m_s} (1 - \omega_c)^{-n_s} \qquad (8.10)$$

at $n_s > 0$ takes into account this effect.

Obviously, equations (8.8)-(8.10) are purely phenomenological and can be replaced with other relationships. In any case, the equations of damage accumulation must contain a minimal number of material parameters. These parameters must distinctly enter the final results, even numerical ones, to assess the numerical magnitudes of parameters from macroscale experimental data.

It ought to be stressed that the above equations as well as those for the agent transport within a crack are valid for fixed temperature and pressure conditions.

The impact of the latter in environmentally assisted damage is considerable. A number of models have been suggested to include these factors using the notions of physical kinetics. For example, one could present the resistance stress σ_d in (8.10) as follows:

$$\sigma_d = \sigma_d^0 \exp\left(-\frac{A}{kT}\right)$$

Here σ_d^0 is a material constant, A is the activation energy, T is the absolute temperature, k is the Boltzmann constant. The temperature dependence of pressure p_t at the tip and the tensile stress σ_s also can be introduced similarly by the choice of appropriate equations for activation energy. A systematic discussion of damage and fracture in terms of kinetics can be found in the book by Krausz and Krausz [88]. In the studies by Mura [105] mechanical aspects are involved along with the solid physics approach. All these matters are important but are beyond the margins of this book.

8.4   Stress Corrosion Cracking

We begin to demonstrate the application of the theory from one of the simplest cases when crack growth takes place under constant or slowly varying, sustained loading. Apart from the possibility of using, in contrast to corrosion fatigue, the only temporal variable (physical or operational time), this problem is simpler when we consider the active agent behavior within the crack hollow. However, the problem is not elementary, in particular, because of the interaction between mechanical and nonmechanical (chemical or electro-chemical) damages.

Consider the mode I edge crack in a plane-strain state (Figure 8.3). Denote the crack length a, and the applied stress σ_∞. The latter is supposed to be a slow time function, $\sigma_\infty(t)$. The crack is one-parametrical with G-coordinate $a(t)$. The general condition of its (non)propagation is, as usual,

$$G(t) \lessgtr \Gamma(t) \tag{8.11}$$

Here $G(t)$ is the generalized driving force, and $\Gamma(t)$ is the corresponding resistance force.

The choice of the mechanical model for material deformation is of crucial significance. A domain of nonelastic straining is certainly present at the tip of each crack. In corrosion, additional complications occur because of the change of mechanical properties within the corrosion process zone. These effects can be described with the thin plastic zone model under the condition that the influence of nonmechanical factors on mechanical properties is included. At the same time, one cannot miss the advantages given by the model of linear elastic material. Introducing an effective tip radii, we practically include the stress redistribution due to plasticity, and include all types of damage, mechanical and nonmechanical ones.

Equations (4.17), (7.29), etc. show the possibility of modeling the interaction between different effects on stress distribution ahead of the tip. The natural generalization upon stress corrosion cracking is as follows:

$$\frac{d\rho}{dt} = \frac{\rho_s - \rho}{\lambda_\rho}\frac{da}{dt} + (\rho_b - \rho)\frac{d\psi_s}{dt} + (\rho_c - \rho)\frac{d\psi_c}{dt} \qquad (8.12)$$

As in (4.17), the first term in the right-hand side describes tip sharpening when the crack growth rate increases. The second term describes the tip blunting due to mechanical damage. The third term is new. It corresponds to the contribution of corrosion in tip blunting. As earlier, ρ_s is the "sharp" magnitude of the effective tip radius; ρ_b is the "blunt" magnitude. To diminish the number of material parameters, one might assume that $\rho_c = \rho_b$; however this issue is rather vague. At least, setting $\rho_c = \rho_b$, we replace (8.12) by equation

$$\frac{d\rho}{dt} = \frac{\rho_s - \rho}{\lambda_\rho}\frac{da}{dt} + (\rho_b - \rho)\frac{d}{dt}(\psi_s + \psi_c) \qquad (8.13)$$

The additivity of environmentally affected damage measures occurs in (8.13).

When the effective tip radius is known, and the material is considered as homogeneously linear elastic, equations (4.17) may be used to calculate the tensile stresses ahead of the crack. The literal interpretation of the stresses fails here because of the strong nonhomogeneity within the corrosion process zone. If the length λ_c of this zone is small compared with the length λ_p of the mechanical process zone, we may refer the maximal stress σ_t to the remote border of the corrosion process zone. This interpretation is illustrated in Figure 8.3b. Evidently, it is valid when $\lambda_c \ll \lambda_p$. In any case, going into a discussion of internal contradictions when a coarse phenomenological model is concerned seems excessive.

Returning to the principal relationship (8.11), let us neglect the contribution of damage into the generalized driving force. Then in the plane-strain state

$$G = \frac{K^2(1 - \nu^2)}{E} \qquad (8.14)$$

where the subscript corresponding to the mode I is omitted.

The generalized resistance force is, as usual, connected with the specific fracture work. In the considered problem we promptly set $\Gamma = \gamma$. Similar to equations (4.6), (5.31), etc., the specific fracture work may be given as

$$\gamma = \gamma_0 \left[1 - \chi\left(\omega_s^{\alpha_s} + \omega_c^{\alpha_c}\right)\right] \qquad (8.15)$$

at $\alpha_s > 0$, $\alpha_c > 0$. Parameter $0 < \chi \le 1$ characterizes the residual fracture toughness for completely damaged material when $\omega_s^{\alpha_s} + \omega_c^{\alpha_c} = 1$. Then equation (8.15) gives

$$\gamma_{res} = \gamma_0(1 - \chi) \qquad (8.16)$$

In the case when additivity of damage is expected, we set

$$\gamma = \gamma_0 \left[1 - \chi\left(\omega_s + \omega_c\right)^\alpha\right] \qquad (8.17)$$

at $\alpha > 0$. In the latter case the generalized resistance force is

$$\Gamma = \Gamma_0 \left[1 - \chi\left(\psi_s + \psi_c\right)^\alpha\right] \qquad (8.18)$$

where $\Gamma_0 \equiv \gamma_0$. Later on, in numerical simulation, we use equations (8.17) and (8.18).

The set of governing relationships for stress corrosion cracking is compiled of equations (8.1) and (8.2) for active agent transport, equations (8.8)-(8.10) for damage accumulation, equation (8.13) for effective tip radius evolution, and equations (8.11), (8.14), and (8.18) which describe the general energy (forces) balance in the system. The tensile stresses acting perpendicular to the crack trajectory are evaluated by equation (5.5) with the correction factor Y for an edge planar crack.

8.5   Numerical Simulation

In the calculation presented below, we mostly follow numerical magnitudes of material parameters used in previous chapters. In particular, we assume $E = 200$ GPa, $\nu = 0.3$, $\gamma_0 = 20$ kJ/m^2, $\alpha = \chi = 1$. In equation (8.10) we put $\sigma_d = 10$ GPa, $\sigma_{th} = 250$ MPa, $m_s = 4$, $n_c = 0$, $t_c = 10^3$ s. Because of the lack of direct experimental data, the choice of parameters related to corrosion is rather arbitrary. We assume that in equations (8.1), and (8.2) describing the transport within a crack, $\tau_D = 10^3$ s, $\lambda_D = 10$ mm, $a_\infty = 100$ mm, $n_a = 1$. In equation (8.6) and (8.8) for corrosion damage we put $c_{th} = 0$, $m_c = 4$, $\lambda_c = 100$ μm, and in equation (8.13) for tip conditions $\rho_s = 10$ μm, $\lambda_\rho = \rho_b = 100$ μm. We characterize the content of an aggressive agent at the crack mouth by a normalized variable, c_∞/c_d. When no contrary is mentioned, the initial conditions are as follows:$a_0 = 1$ mm, $\rho_0 = 50$ μm, $c(x) = c_\infty$ at all $0 \le x \le a_0$, and no initial damage is present. We suppose that the threshold effects are taken into account by the choice of σ_{th}. Some aspects of near-the-threshold cracking will be discussed separately.

Time history of the normalized content at the tip, c_t/c_d, is presented in Figure 8.4a. The applied stress is constant, $\sigma_\infty = 100$ MPa. Lines 1, 2, 3, 4 are drawn for $c_\infty/c_d = 0.25, 0.5, 0.75, 1.0$. When a crack becomes deeper, the transport of the agent to the tip meets difficulties, and the tip content decreases in time.

The evolution of the effective tip radius ρ is presented in Figure 8.4b. In general, the picture is similar to that in the case of "pure" fatigue, say, to that given in Figure 4.4b. The initiation stage, when the crack tip is fixed, as well as the growth stage, is distinctly seen following the radius evolution.

The effect of corrosion in the considered case is significant. In particular, the duration of the initiation stage t_* varies as $t_* = 0.26 \cdot 10^6, 0.16 \cdot 10^5, 3.2 \cdot 10^3$, and $1.0 \cdot 10^3$ when the normalized content c_∞/c_d varies from 0.25 to 1.0. This is distinctly seen in Figure 8.4b where the changes in curve conduct correspond to the moments of crack growth start. The total fatigue life varies, respectively, as $t_{**} = 3.1 \cdot 10^7, 6.6 \cdot 10^6, 2.4 \cdot 10^6$, and $1.1 \cdot 10^6$. Compared with purely mechanical fatigue, we observe the early beginning of crack propagation. The stage of crack growth takes the major part of the fatigue life.

Damage accumulation at the tip is illustrated in Figure 8.5. The summed tip damage $\psi = \psi_s + \psi_c$ is plotted against time in Figure 8.5a for the same numerical data (and the same notation) as in Figure 8.4. The evolution of the summed damage measure is also similar to that in "pure" fatigue. However, the distribution

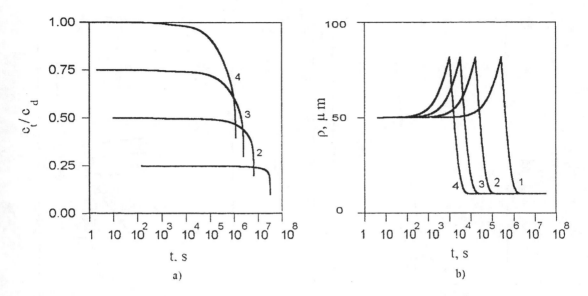

FIGURE 8.4
Evolution of tip content (a) and effective tip radius (b) at $c_\infty/c_d = 0.25$, 0.5, 0.75, and 1.0, (lines 1, 2, 3, and 4, respectively).

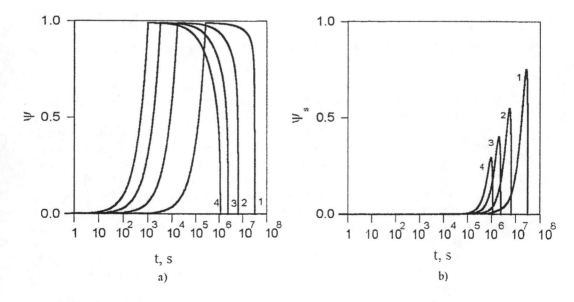

FIGURE 8.5
Evolution of the summed tip damage (a) and its mechanical component (b); notation as in Figure 8.4.

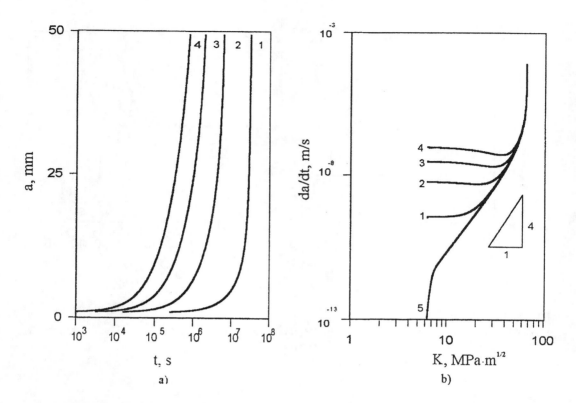

FIGURE 8.6
Crack growth histories (a) and growth rate diagrams (b); notation as in Figure 8.4.

of damage between the components is of some interest. This distribution can be observed comparing Figures 8.5a and 8.5b. The latter presents the evolution of the component ψ_s. The contribution of this component is small and decreases with the content c_∞. The corrosion component ψ_c is easy to assess as a difference between ψ and ψ_s that is plotted in Figure 8.5a and 8.5b. As a whole, the stress corrosion cracking remains to be controlled by purely mechanical conditions (8.11).

Crack growth histories at various c_∞/c_d are depicted in Figure 8.6a. The growth rate is enhanced when the content of the aggressive agent increases. The diagram similar to those widely used in the treatment of laboratory tests and in fatigue life prediction is shown in Figure 8.6b. The crack growth rate da/dt measured in meters per second is plotted there versus the stress intensity factor K. The results for the natural environment ($c_\infty/c_d = 0$ in our notation) are given in Figure 8.6b with the line labelled 5.

Some parametrical analysis has been performed to assess the role of load level and initial conditions. The final results are presented in Figure 8.7 in the form of crack growth rate diagrams. Figure 8.7a is obtained for $c_\infty/c_d = 0.5$, $a_0 = 1$ mm, and $\sigma_\infty = 75, 100, 125$, and 150 MPa. Figure 8.7b is obtained for $c_\infty/c_d = 0.5$,

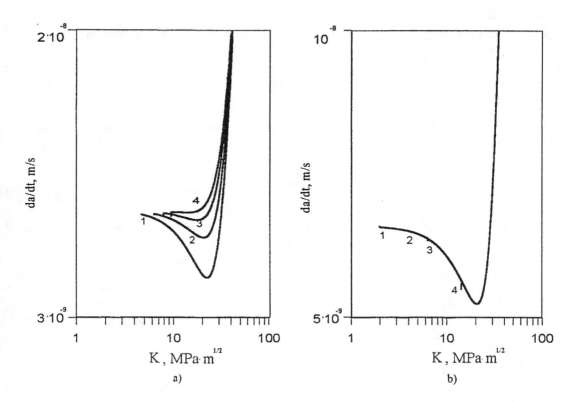

FIGURE 8.7
Growth rate diagrams for various load levels (a), and various initial conditions
(b); notation as in Figure 8.4.

$\sigma_\infty = 100$ MPa and $a_0 = 0.1, 0.5, 1$, and 5 mm (lines 1, 2, 3, and 4, respectively).

As one would expect, under the same environmental and loading conditions, the change of the initial crack length produces a significant effect only on the first section of diagrams. The most striking observation is a distinct drop of growth rates that occurs in the middle part of the diagrams, especially when the load level is low, say, $\sigma_\infty = 25$ MPa. Here the growth rate is controlled mostly by corrosion. In fact, the initial sections of the diagram are almost independent of the load level. Later on, the transport of the agent becomes hindered, and the content at the crack tip becomes less. At the last stage all curves in Figures 8.6 and 8.7 are merging, at least when one looks on them in the log − log scale.

8.6   Threshold Problem in Stress Corrosion Cracking

As in other studies of fatigue, the initial stage of stress corrosion cracking is of practical interest. This stage relates to comparatively low stress levels and/or small initial crack lengths.

There are several approaches to the problem staying in the framework of solid mechanics. The threshold stress intensity factor in "pure" fatigue is within the interval

$$(Y/2)\sigma_{th}(\pi\rho_s)^{1/2} \le K_{th} \le (Y/2)\sigma_{th}(\pi\rho_b)^{1/2} \tag{8.19}$$

For the above used numerical data, this gives $0.8\,\text{MPa}\cdot\text{m}^{1/2} \le K_{th} \le 5\,\text{MPa}\cdot\text{m}^{1/2}$. The threshold values of stress intensity factor K_{scc} in stress corrosion cracking is of the same order of magnitude. These values usually refer to the residual fracture toughness of the material subjected to corrosion damage.

Cracking under complete corrosion begins when the equality $G = \Gamma_{res}$ is primarily attained. Here Γ_{res} is the residual magnitude of the resistance force. In our case $\Gamma_{res} = \gamma_{res}$ where γ_{res} is given in equation (8.16). As a result, we obtain that the relationship between the stress corrosion cracking threshold K_{scc} and the common fracture toughness K_c takes the form

$$K_{scc} \sim K_c(1 - \chi)^{1/2} \tag{8.20}$$

As the threshold is usually rather low, the magnitudes of χ are close to unity. For example, to obtain $K_{scc} = 0.05K_c$ we must assume $\chi = 1 - 25 \cdot 10^{-4}$. The interpretation of threshold effects given in equation (8.20) is perhaps necessary when we try to include explicitly the residual fracture toughness of thin corroded films. However, it seems more expedient to put, as in the above computations, $\omega_{res} = 0$ and to study the initial sections of crack growth diagrams. Some numerical results for early crack growth are shown in Figure 8.8 and 8.9.

In terms of stress intensity factors, threshold effects are expected in the domain between 1 and $5\,\text{MPa}\cdot\text{m}^{1/2}$ when common metal alloys are concerned. In this aspect Figure 8.7b already contains some features of the threshold effect, in particular, at $a_0 = 0.1$ mm.

Figure 8.8a is drawn for the initial crack depths $a_0 = 0.1, 0.3$ and 0.5 mm (lines 1, 2, and 3, respectively) at $\sigma_\infty = 50$ MPa, $c_\infty/c_d = 1.0$. The smallest initial depth corresponds to the stress intensity factor $K = Y\sigma_\infty(\pi a_0)^{1/2}$ which is a little less than the lower estimate given in (8.19). At $a_0 = 0.1$ mm we obtain (in the context of the discussed model) that $K_{th} \approx K_{scc}$. Figure 8.8b shows the crack growth rate diagrams at $a_0 = 0.1$ mm and $\sigma_\infty = 50, 75, 100$ MPa (lines 1,2,3, respectively).

The crack growth diagrams presented in Figures 8.7 and 8.8 exhibit nonsteady behavior. Such peculiarities are usually observed in short fatigue cracks [90, 102]. In our case, the bottom of the "pit" is in the domain where the damage due to corrosion begins to decrease while the mechanical component of damage is yet comparatively small. There is evidence that such features of crack growth diagrams are met in laboratory experiments. The time histories of damage measures ψ_s and ψ_c are

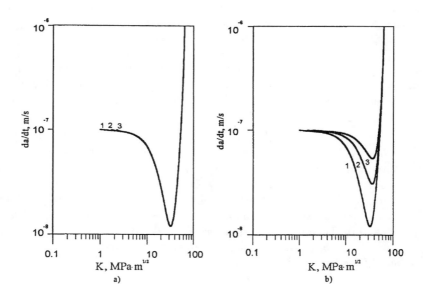

FIGURE 8.8
Growth rate diagrams: (a) for $a_0 = 0.1$, 0.3, 0.5 mm; (b) for $\sigma_\infty = 50$, 75, 100 MPa (lines 1, 2, 3, respectively).

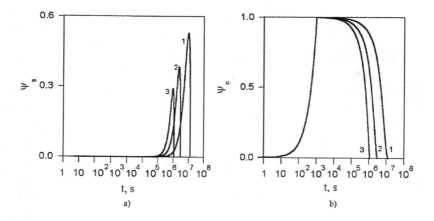

FIGURE 8.9
Damage measures at the moving tip in the case of small initial crack depth; notation as in Figure 8.8a.

shown in Figure 8.9. The numerical data and notation of curves are the same as in Figure 8.8a. Similar to Figure 8.5, corrosion damage dominates everywhere except the final stage when the cracks become sufficiently deep.

There is an alternative approach to the threshold problem in stress corrosion cracking. For example, in the book by Bolotin [19] the threshold parameters such as c_{th} in (8.8) and (8.9), as well as σ_{th} in (8.10), have been considered as functions of current conditions at the tip. For instance, equation

$$c_{th} = c_{th}^0 (1 - \psi_c)^{-p_c} (1 - \psi_s)^{p_s} \qquad (8.21)$$

with $p_c > 0$, $p_s > 0$ takes into account the influence of damage on the corrosion threshold which in the absence of damage is equal to c_{th}^0. Two competitive mechanisms, formation of passive films and formation of juvenile surfaces, are covered by equation (8.21). Change of the tip radius also affects the thresholds, both with respect to corrosion and to mechanical damage. Then the right-hand side in equation (8.21) and in the similar equation for σ_{th} will contain the ratio ρ/ρ_c of the effective tip radius ρ to a certain material parameter ρ_c. Some details can be found in the paper [17] and in the book [19].

8.7   Corrosion Fatigue

From the methodological point of view, there is no significant difference in modeling of crack growth under sustained or cyclic loading. Some complications are involved because the corrosion fatigue is frequency dependent. The load frequency influences both the transport of the agent within the "breathing" crack and the process of damage accumulation. The presence of two temporal variables, the physical (calendar or operating) time and the cycle number, makes the picture more complicated.

In addition, some complications are involved from the effects produced by average loads. These effects are important also in "pure" mechanical fatigue. But they enter there in the form of crack closure and related phenomena which may be taken into account by an appropriate choice of threshold resistance stresses and/or effective stress ranges. In corrosion fatigue, we meet the combination of effects inherent to stress corrosion cracking under sustained loading and to the damage due to cyclic loading. The situation is similar to that in cracking of polymers and in creep fatigue (Chapter 7).

The set of governing relationships, as usual, consists of the equations of agent transport, the equations for damage accumulation, crack tip evolution, stress distribution ahead of the tip, and the relationships between the generalized forces. Namely, equations (8.1), (8.2) and (8.5) describe the transport within a crack "breathing" with the frequency $f = \theta/2\pi$. Corrosion damage is described by equation (8.8) or its generalization, equation (8.9). For the damage measure ω_s referred to the sustained component of loading, equation (8.10) is used.

To include the cyclic damage ω_f, rewrite equation (8.10) as follows:

$$\frac{\partial \omega_s}{\partial t} = \frac{1}{t_c} \left(\frac{\sigma - \sigma_{th}}{\sigma_d} \right)^{m_s} (1 - \omega_s - \omega_f - \omega_c)^{-n_s} \qquad (8.22)$$

As to the cyclic damage, we assume that it is governed by equation

$$\frac{\partial \omega_f}{\partial t} = \left(\frac{\Delta\sigma - \Delta\sigma_{th}}{\sigma_f} \right)^{m_s} (1 - \omega_s - \omega_f - \omega_c)^{-n_f} \tag{8.23}$$

with the stress range $\Delta\sigma$ and material parameters σ_f, $\Delta\sigma_{th}$, m_f, and n_f. A similar generalization is natural to introduce with respect to the corrosion damage measure using instead of (8.9) the following equation:

$$\frac{d\psi_c}{dt} = \frac{1}{t_c} \left(\frac{c_t - c_{th}}{c_d} \right)^{m_c} (1 - \psi_s - \psi_f - \psi_c)^{-n_c} \tag{8.24}$$

Let the effective tip radius ρ satisfy the equation similar to (8.12)

$$\frac{d\rho}{dt} = \frac{\rho_s - \rho}{\lambda_\rho} \frac{da}{dt} + (\rho_b - \rho)\frac{d\psi_s}{dt} + (\rho_f - \rho)\frac{d\psi_f}{dt} + (\rho_c - \rho)\frac{d\psi_c}{dt}$$

or, in the case of common "blunt" radius ρ_b

$$\frac{d\rho}{dt} = \frac{\rho_s - \rho}{\lambda_\rho} \frac{da}{dt} + (\rho_b - \rho)\frac{d}{dt}(\psi_s + \psi_f + \psi_c) \tag{8.25}$$

In all the cases, $dN = f dt$ where f is the frequency counted in cycles per time unit. Since the cyclic damage is growing with the cycle number, as is presented in (8.23), the frequency implicitly enters all the above equations (8.22)-(8.25).

Considering the simplest model of an edge planar crack, we employ equation (8.14) for the generalized driving force. Equation (8.18) for the generalized resistance force must be supplemented with a term describing the contribution of cyclic damage. We use the equation

$$\Gamma = \Gamma_0[1 - \chi(\psi_s + \psi_f + \psi_c)^\alpha] \tag{8.26}$$

supposing that damage measures satisfy the additivity law. The closing relationship, as earlier, is given in (8.11).

A detailed numerical study of corrosion fatigue crack growth has been presented recently in the paper by Bolotin and Shipkov [39]. Hereafter we follow, in general, this paper with a minor change of some numerical parameters.

Let us, for simplicity, omit the contribution of the sustained component of damage. In fact, this component is small when the mean and cyclic components of loading are of the same order of magnitude. Material parameters are: $E = 200$ GPa, $\nu = 0.3$, $\gamma_0 = 20$ kJ/m^2, $\chi = \alpha = 1$, $\sigma_f = 10$ GPa, $\Delta\sigma_{th}^0 = 250$ MPa, $m_f = m_c = 4$, $n_f = n_c = 0$, $\rho_s = 10$ μm, $\lambda_\rho = \rho_b = \rho_f = \rho_c = 100$ μm, $t_c = 10^3$ s. It is assumed that $\Delta\sigma_{th} = \Delta\sigma_{th}^0(1-R)$, $c_{th} = 0$. In equation (8.1) we put $\lambda_D = 10$ mm, $\tau_D = 10^3$ s, $c_b = 0.5\, c_a$, in equation (8.2) $a_\infty = 100$ mm, $n_a = 1$, and in equation (8.5) $n_\theta = 1$, $f_\infty = 1$ Hz (hereafter all the frequencies are counted in cycles per second).

Most of the following computations are performed at $\Delta\sigma_\infty = 100$ MPa, $R = 0.2$, and the load frequency $f = 1$ Hz. The initial data are: $a_0 = 1$ mm, $\rho_0 = 50$ μm, $c(a_0, 0) = c_\infty$, $c(x, 0) = 0$, and $x > a_0$. Four applied content ratios are considered: $c_\infty/c_d = 0.25$, 0.5, 0.75 and 1.0. The corresponding curves are labelled in Figures 8.10 and 8.12 as 1, 2, 3, and 4.

The evolution of the tip content c_t in the function of N is shown in Figure 8.10a, and that of the effective tip radius ρ in Figure 8.10b. Figure 8.10 is similar to

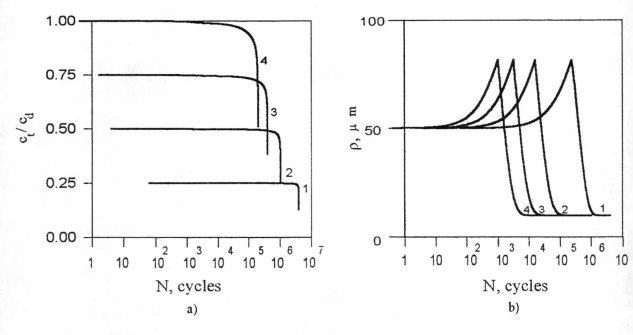

FIGURE 8.10
Evolution of components of damage and of the total damage measure at c_∞/c_d = 0.25, 0.5, 0.75, and 1.0 (lines 1, 2, 3, and 4, respectively).

Figure 8.4 obtained for stress corrosion cracking. The cycle numbers at the crack growth start are $N_* = 0.24 \cdot 10^6$, $0.16 \cdot 10^5$, $0.32 \cdot 10^4$, and $0.99 \cdot 10^3$. These numbers are small compared with the total fatigue life estimated as $N_{**} = 0.39 \cdot 10^6$, $1.01 \cdot 10^6$, $0.4 \cdot 10^6$, and $0.2 \cdot 10^6$, respectively.

Time histories of damage measures ψ_f, ψ_c and $\psi = \psi_c + \psi_f$ are shown in Figure 8.11. Cases a, b, c, and d are plotted for $c_\infty/c_d = 0.25$, 0.5, 0.75, and 1.0. In the beginning, corrosion damage dominates in all four cases. The measure ψ_c is slowly decreasing while the measure ψ_f begins to grow. Sometimes peculiarities are observed such as the nonmonotonous behavior which is seen in Figure 8.11a. It is the result of interaction between several mechanisms and may refer to the formation of juvenile surfaces. Near the final failure the mechanical component of damage dominates even when the agent content is high. In all the cases the summed tip damage drops rapidly when we are approaching the final failure.

The crack growth histories $a(N)$ and the crack growth diagrams (this time as the rate da/dN in the function of the stress intensity factor range ΔK) are presented in Figure 8.12. The initial parts of the diagrams in Figure 8.12b, except that for $c_\infty/c_d = 0.25$, contain the "plateaus" and even the "pits". Then the curves tend to merge. The Paris exponent at the last sections of the diagrams is close to $m_f = 4$. This indicates again that the contribution of corrosion in total damage becomes smaller for longer cracks. It is not only a result of higher stress concentration for deeper cracks, but also of weaker transportation of the active agent to the remote crack tip.

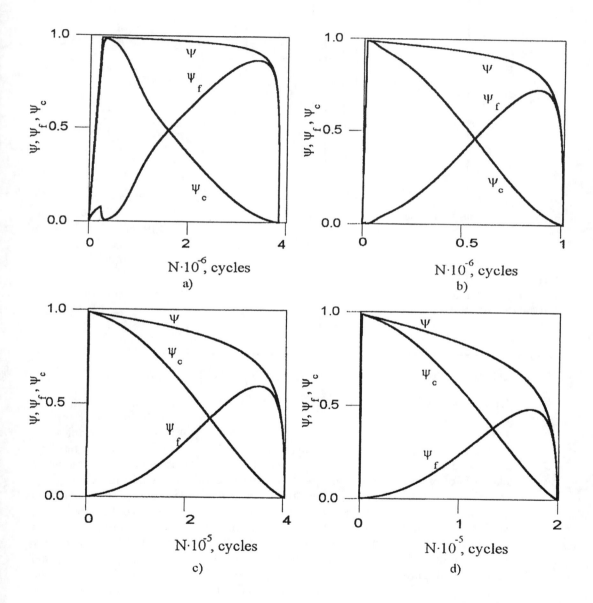

FIGURE 8.11
Evolution of damage measures at $c_\infty/c_d = 0.25$, 0.5, 0.75, and 1.0 (cases a, b, c, and d).

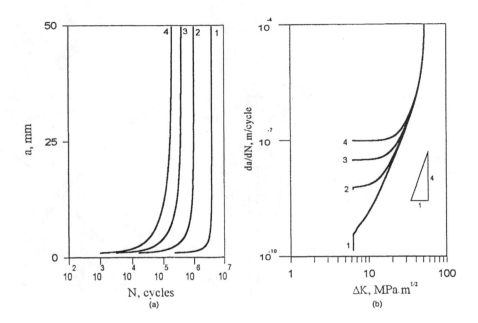

FIGURE 8.12
Crack growth histories (a) and growth rate diagrams (b) in corrosion fatigue (notation as in Figure 8.10).

8.8   Frequency Effects in Corrosion Fatigue

One of the complications in studies of corrosion fatigue has been mentioned in the beginning of Section 8.7. It is certainly the corollary of entering two temporal variables. When the load frequency is slowly varying in time, both variables can be used to describe corrosion fatigue. Some mechanisms are developing in physical time, and others are strongly cycle-dependent. Hence accounting for frequency effects is of practical significance both in experimental studies and in design.

Since the load frequency enters the equations of agent transport, in equation (8.5) in particular, the frequency effect must be discussed in terms of the tip content c_t. Figure 8.13 illustrates the influence of frequency f on the ratio c_t/c_d at $c_\infty/c_d = 1$. Lines 1, 2, 3, 4 are drawn for frequencies $f = 0.1$, 1, 10 and 100 Hz, respectively for the same numerical data as in Section 8.7. Time is chosen as the temporal variable in Figure 8.13a, and the cycle number in Figure 8.13b. The higher the frequency, the shorter is the fatigue life counted in time units, and the longer when it is counted in cycle numbers. This conclusion is quite obvious from the physical viewpoint. However, it is rather unexpected that the lines in Figure 8.13a corresponding to various frequencies intersect.

The crack growth rate diagrams are presented in Figure 8.14 in two different forms, using da/dt and da/dN as the rate measures. In both cases the rate is

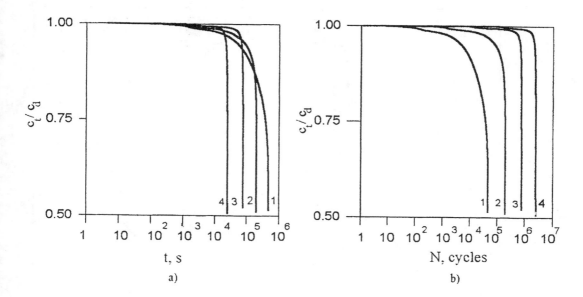

FIGURE 8.13
Tip content in function of time (a) and cycle number (b) at load frequencies f = 0.1, 1, 10, and 100 Hz (lines 1, 2, 3, and 4, respectively).

considered as a function of the stress intensity factor range ΔK. In Figure 8.14,a all the curves initiate from a close vicinity, then the divergence of curves becomes drastic. They only converge again near the final failure. On the contrary, in Figure 8.14b the divergence is significant on the early stage when the contribution of corrosion is significant. The Paris exponent is nearly $m_f = 4$ at this stage.

The above conclusion is not at all of a universal nature. In particular, the behavior of curves depends on the chosen numerical data. In our case, the characteristic length a_∞ in equation (8.5) is assumed too high while the closure effect is not included. The relationship between two components of damage also depends on the numerical values of the resistance parameters in equations (8.22)-(8.24).

As a rule, the contribution of average, slowly varying load components of stresses cannot be neglected in corrosion fatigue. This issue is taken into account in equations (8.22)-(8.26). There three damage measures are presented, among them one that originates from sustained loading. It means that corrosion fatigue crack growth includes, in general, the stress corrosion cracking component. The contribution of this component strongly depends on the frequency. Comparing the right-hand sides in (8.22) and (8.23), we see that the damage measures ω_s and ω_f will be of the same order of magnitude at

$$\left(\frac{\sigma}{\sigma_d} \right)^{m_s} \sim f t_c \left(\frac{\Delta \sigma}{\sigma_f} \right)^{m_f}$$

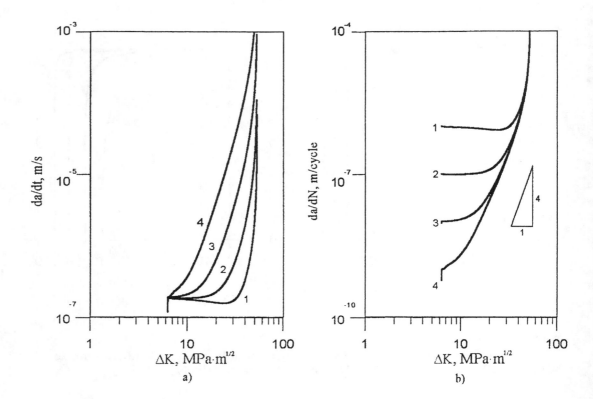

FIGURE 8.14
Crack growth rate diagrams in corrosion fatigue referred to time (a) and cycle
number (b); notation as in Figure 8.13.

This relationship shows that the final results depend on a number of various parameters. When $\sigma \sim \Delta\sigma$, $\sigma_d \sim \sigma_f$ and $m_s \sim m_f$ this equation results in $ft_c \sim 1$. In the former numerical results it has been assumed $t_c = 10^3$ s at the lowest load frequency $f = 0.1$ Hz. Thus $ft_c = 10^2$. Evidently, under these circumstances we might neglect the contribution of stress corrosion cracking compared with the contribution of cyclic fatigue. When load frequency is of the order $f = 10^{-3}$ Hz and less, both components are equally important. We do not discuss here the influence of average stress or minimal stress of the cycle which has a completely different origin. The tacit agreement has been used here that the stress ratio $R = \sigma_\infty^{\min}/\sigma_\infty^{\max}$ enters the threshold resistance stress or, maybe, other material characteristics. In the case when crack closure and related effects are necessary to be taken into account, the models developed in Sections 4.6 and 4.7 may be used.

8.9 Quasistationary Approximation

To predict the crack growth associated with enviromental effects, differential equations are often used which may be considered as generalizations of the Paris-Erdogan equation. Sometimes such equations are developed with the use of a superposition principle. Namely, the crack growth rate is determined as a sum of rates corresponding to two separated processes, the cracking under mechanical loading and under environmental effects. Because of a strong interaction between these processes such an approach is too coarse and frequently leads to discrepancies with experimental data. A number of attempts have been made to avoid the direct superposition and to take into account interaction effects. A survey of literature can be found in [52, 58, 88, 93, 128]. Most of the referred literature deals with corrosion fatigue.

In quasistationary approximation, we estimate the crack growth rate da/dt within time intervals Δt sufficiently short to treat other slowly varying variables as "frozen." In other words, these variables are treated as constants within such intervals. To make a decision, we must compare characteristic times of change for the most important variables.

Let λ be a characteristic size of the process zone, say,

$$\lambda = \max\{\lambda_s, \lambda_f, \lambda_c\}$$

Here λ_s and λ_f are the sizes of the process zones associated with sustained and cyclic loading, respectively, and λ_c is associated with environmental effects. Then the characteristic time for the crack growth is

$$\Delta t \sim \lambda (da/dt)^{-1} \tag{8.27}$$

The time parameter given in (8.27) must be compared with other characteristic times. We may state that the quasistationary approximation is applicable if

$$\Delta t \ll \min\{\tau_D, \tau_s, \tau_c, \tau_e, \tau_\rho\}, \tag{8.28}$$

where τ_D is the characteristic time for the agent transport; τ_s, τ_f, and τ_c are the characteristic times for damage accumulation processes of a various nature; τ_ρ is the characteristic time for the change of the effective tip radius. These times are evaluated as follows:

$$\tau_s \sim \left|\frac{d\psi_s}{dt}\right|^{-1}, \quad \tau_f \sim \left|\frac{d\psi_f}{dt}\right|^{-1}, \quad \tau_c \sim \left|\frac{d\psi_c}{dt}\right|^{-1}, \quad \tau_\rho \sim \frac{1}{\lambda}\left|\frac{d\rho}{dt}\right|^{-1} \tag{8.29}$$

Usually we may assume that τ_ρ is of the same order as one of the time parameters τ_s, τ_f, or τ_c. Consequently, the condition given in equation (8.28) can be simplified. When equations (8.27) and (8.29) are used, we come to conditions:

$$\frac{da}{dt} \ll \min \left\{ \frac{\lambda}{\tau_D}, \frac{\lambda_s}{t_c} \left(\frac{\sigma_t - \sigma_{th}}{\sigma_s} \right)^{m_s}, \frac{\lambda_f}{f} \left(\frac{\Delta\sigma_t - \Delta\sigma_{th}}{\sigma_f} \right)^m, \right.$$

$$\left. \frac{\lambda_c}{t_c} \left(\frac{c_t - c_{th}}{c_d} \right)^{m_c} \right\}$$

Further calculations are similar to those given in Section 4.8, 6.4, and 7.4. Let us demonstrate them for the case when equations (8.8) and (8.23) are used, at $n_s = n_f = 0$, and the specific fracture work is given in equation (8.26) at $\alpha = \chi = 1$. Then equations (8.8), (8.22) and (8.23) take the form

$$\frac{\partial\omega_s}{\partial t} = \frac{1}{t_c} \left(\frac{\sigma - \sigma_{th}}{\sigma_s} \right)^{m_s}, \quad \frac{\partial\omega_f}{\partial N} = \left(\frac{\Delta\sigma - \Delta\sigma_t}{\sigma_f} \right)^{m_f},$$

$$\frac{\partial\omega_c}{\partial t} = \frac{1}{t_c} \left(\frac{c - c_{th}}{c_d} \right)^{m_c} \tag{8.30}$$

By temporarily "freezing" all the variables in the right-hand sides of (8.30), we obtain

$$\psi_s \approx \frac{\lambda_s}{t_c} \left(\frac{da}{dt} \right)^{-1} \left(\frac{\sigma_t - \sigma_{th}}{\sigma_s} \right)^{m_s},$$

$$\psi_f \approx \lambda_f \left(\frac{da}{dN} \right)^{-1} \left(\frac{\Delta\sigma_t - \Delta\sigma_{th}}{\sigma_f} \right)^{m_f},$$

$$\psi_c \approx \frac{\lambda_c}{t_c} \left(\frac{da}{dt} \right)^{-1} \left(\frac{c_t - c_{th}}{c_d} \right)^{m_c}$$

where the representative values of σ_t, $\Delta\sigma_t$ and c_t are taken at the crack tip. For simplicity, the far field damage is neglected. On the other hand, the equilibrium condition $G = \Gamma$ gives

$$\psi_s + \psi_f + \psi_c = \left(1 - \frac{G}{\Gamma_0} \right)^{1/\alpha}$$

A combination of the above equations results in the final equation

$$\frac{da}{dt} = \frac{1}{t_c} \frac{\lambda_s \left(\dfrac{\sigma_t - \sigma_{th}}{\sigma_s} \right)^{m_s} + f t_c \lambda_f \left(\dfrac{\Delta\sigma_t - \Delta\sigma_{th}}{\sigma_f} \right)^{m_f} + \lambda_c \left(\dfrac{c_t - c_{th}}{c_d} \right)^{m_c}}{\left(1 - \dfrac{G}{\Gamma_0} \right)^{1/\alpha}} \tag{8.31}$$

Equation (8.31) is in some aspects a generalization of the equation that has been derived primarily in the paper by Bolotin [17]. From first sight, equation

(8.31) looks like a modification of the superposition approach that is improved with the addition of a denominator in its right-hand side. When $G \ll \Gamma_0$, the equation really looks like a result of the superposition approach. But, in fact, it is not so. The agent content at the tip c_t, the tip opening stress σ_t, its range $\Delta\sigma_{th}$, and, maybe, the characteristic lengths λ_s, λ_f and λ_c depend on the current magnitudes of the crack length $a(t)$ and the effective tip radius $\rho(t)$. It means that equation (8.31) is to be considered together with equation (8.1) for the transport process equation, and equation (8.25) for the tip radius evolution during crack growth. As a result, though the governing equation becomes a little simpler than the initial set of equations, the gain is rather questionable. The only advantage is transparency; equation (8.31) exhibits explicitly the contribution of each type of damage as well as the tendency to instability with increasing the driving force. We have situation similar to that discussed in previous chapters for different types of deterioration, and different material properties.

8.10   Application of the Thin Plastic Zone Model

Until now, we considered corrosion fatigue and stress corrosion cracking, treating the material as linear elastic but subject to damage. However, the choice of material properties is very wide. Elasto-plastic and visco-elastic behavior may be easily included in the analysis. The main difference is in the evaluation of generalized driving forces and stresses along the crack trajectories.

Consider, for example, the case when material properties are described in the framework of the thin plastic zone model. The problems discussed in Chapter 6 are related mostly to central mode I cracks for which the corresponding solution of the fracture mechanics problem is available in the analytical form. Turning to environmentally assisted cracking, we deal, as a rule, with cracks originating from the surface, i.e., with edge cracks. There are several solutions of fracture mechanics problems for edge cracks with the use of the thin plastic zone model, though these solutions are numerical. The survey of studies in this topic can be found in the book by Pluvinage [118].

In this study, to evaluate the generalized driving force, we use equation (6.27) for a central crack with the correction multiplier Y^2. Hence Y is the correction factor in formula $K = Y\sigma_\infty(\pi a)^{1/2}$. Thus, we assume that

$$G = \frac{8Y^2\eta\sigma_0^2 a}{\pi E}\left[\log\cos\left(\frac{\pi\sigma_\infty}{2\sigma_0}\right) + \frac{\pi\sigma_\infty}{2\sigma_0}\tan\left(\frac{\pi\sigma_\infty}{2\sigma_0}\right)\right] \qquad (8.32)$$

where the notation of Chapter 6 is used. To avoid additional complications, let the plate width be large compared with the crack depth. Then $Y \approx 1.12$.

For the stress range $\Delta\sigma$ within a cycle at $x \geq a$, $y = 0$ (Figure 6.4), we use equation (6.39)

$$\Delta\sigma = \begin{cases} 2\sigma_0, & a \leq x \leq a + \lambda_p \\ \dfrac{4\sigma_0}{\pi} \cot^{-1}\left[\dfrac{a}{x}\left(\dfrac{x^2 - (a+\lambda_p)^2}{(a+\lambda_p)^2 - a^2}\right)^{1/2}\right], & x > a + \lambda_p \end{cases} \tag{8.33}$$

with the length of the process zone

$$\lambda_p = a\left[\sec\left(\frac{\pi\Delta\sigma_\infty}{4\sigma_0}\right) - 1\right] \tag{8.34}$$

Equations (8.33) and (8.34) coincide with equations (6.39) and (6.4), respectively. The latter are valid for central cracks. When an edge crack is long enough but its tip is far from the opposite body border, equations (8.33) and (8.34) may be considered as the first approximations without going into laborious numerical computations on all fatigue life.

The following analysis is based on equations (8.32)-(8.34) as well as on the equations used in the previous sections. Among them are the equation of transport within a crack, the equations of damage accumulation, and the general relationship between generalized forces controlling the process of crack (non)propagation. Equations similar to (8.12) or (8.25) do not enter the statement of the problem because the tip stresses are controlled by plastic straining.

Some numerical results are presented in Figures 8.15 and 8.16. The material is assumed linear elastic with Young's modulus $E = 200$ GPa and Poisson's ratio $\nu = 0.3$ elsewhere except forr the thin plastic zone with the yield limit $\sigma_0 = 500$ MPa. The specific fracture work for nondamaged material is $\gamma_0 = 20$ kJ/m^2. The static component of damage is neglected. As to the cyclic component, the following material parameters are assumed: $\sigma_f = 10$ GPa, $\Delta\sigma_{th} = 250$ MPa, $m_f = 4$. In equations (8.1), (8.2), and (8.5) we put $\tau_D = 10^3$ s, $a_\infty = 1$ mm, $f_\infty = 1$ Hz, $c_b = 0.5\,c_a$. The parameter characterizing the agent content at the crack entrance is the ratio c_∞/c_d where c_d has the same meaning as in equation (8.8). We assume in (8.8) that $c_{th} = 0$, $m_c = 4$, $t_c = 10^3$ s. Crack growth histories and growth rate diagrams are presented in Figure 8.15 for $\Delta\sigma_\infty = 100$ MPa, $R = 0.2$, $f = 1$ Hz and the initial conditions $a_0 = 1$ mm, $c_t(x,0) = 0$ at $0 \leq x \leq a_0$. Lines 1, 2, 3 and 4 are drawn for $c_\infty/c_d = 0.25, 0.5, 0.75$, and 1.0. The only striking distinction from, say, Figure 8.6 is the initial jump-wise propagation as a result of the rupture of essentially damaged material in the process zone. Therefore, all the curves in Figures 8.15b are of a "tail-up" type (compare with Figures 6.15, 6.16, etc.).

The evolution of various kinds of damage at the crack tip is shown in Figure 8.16a for $c_\infty/c_d = 0.25$, and in Figure 8.16b for $c_\infty/c_d = 1$. When the agent content is low, we observe the domination of cyclic damage during the entire fatigue life. At high content we observe a different pattern except the vicinity of the final failure when the cyclic damage begins to grow. In general, Figure 8.16 has much in common with Figure 8.11 obtained under the assumption that loading is sustained and material is deformed elastically.

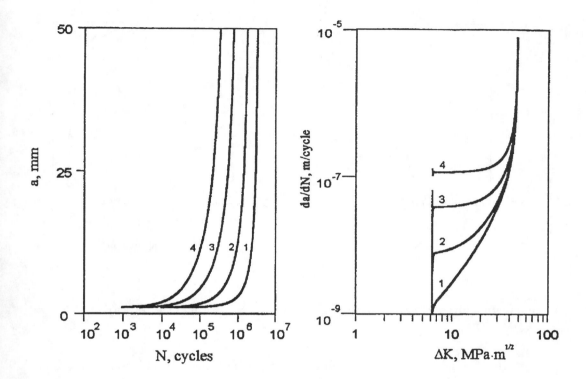

FIGURE 8.15
Crack growth histories (a) and growth rate diagrams in the presence of a thin plastic zone at $c_\infty/c_d = 0.25$, 0.5, 0.75, and 1.0 (lines 1, 2, 3 and 4).

8.11 Hydrogen Embrittlement and Related Phenomena

One of the most frequent environmentally assisted damage is caused by hydrogen interaction with metals and metallic alloys. Both free and chemically bounded hydrogen can be the source of degradation of material properties and, as a result, of structural failure related to stress corrosion cracking and corrosion fatigue. Hydrogen effect is an important component of the fatigue phenomena accompanied by electrolysis, say, in the presence of water or hydrogen sulfide. Even the biological environment generates hydrogen and produces effects which are often referred to as hydrogen embrittlement.

The hydrogen problems in technology, such as in aircraft engineering and energy production, have been rather widely discussed in the literature. Most authors consider the problems from the viewpoint of material science with the impact of electro-chemical and metallurgical aspects. The survey given by Hirth and Johnson

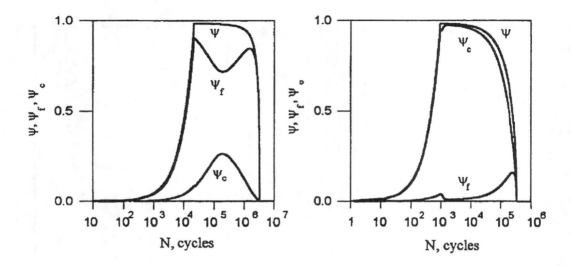

FIGURE 8.16
Damage measures at the moving crack tip in corrosion fatigue: (a) at $c_\infty/c_d = 0.25$; (b) at $c_\infty/c_d = 1.0$.

[70], being more than 20 years old, seems comprehensive and useful even now. Mechanical aspects were discussed by Cherepanov [50]. In this section, a model of hydrogen assisted damage and cracking will be discussed in the framework of the general theory of fatigue.

The term hydrogen embrittlement is rather conditional. It covers a number of different mechanisms associated with the diffusion of hydrogen ions into metals. Decrease of fracture toughness and yield stress near the contact surface certainly can be attributed to embrittlement as they signify the loss of ductility. But there are additional degradation phenomena. Among them are void nucleation and blister formation as the result of hydrogen diffusion. Evidently, the latter process may be attributed to microdamage. From the phenomenological viewpoint, they may be described similar to the damage produced by purely mechanical actions. However, it is difficult, staying in the framework of an analytical model, to draw the border between hydrogen-induced and chemical components of corrosion phenomena. In some aspects, hydrogen embrittlement allows even simpler modeling than, say, the corrosion with the formation of oxide film. The reason is that the diffusion of hydrogen into metals proceeds rapidly compared with common crack growth rates. The thickness of the damaged layer in metals is of the order of 10 μm and the characteristic time of saturation is usually measured in minutes. There is evidence that material properties such as yield limit and Young's modulus attain their regular magnitudes at approximately 10 μm ahead of the crack. This means that we may consider the tip damage measure as a function of the current hydrogen content at the tip. In typical situations, the carrier of the internal hydrogen is gaseous or liquid. In hydrogen technology, when a metal structure stays in long contact with the free hydrogen, other damage mechanisms occur such as intensive decarburization when carbon steels are concerned. We do not discuss such situations here.

The set of governing equations consists of relationships between generalized forces, damage accumulation equations, crack tip conditions and equations describing the effect of damage on generalized forces on the specific fracture work in particular.

Both the linear elastic model and the thin plastic zone model may be used. In the first case we describe the tip conditions by the effective tip radius ρ that depends on the crack growth and damage accumulation rates. As in other sections of this book, we treat the stress σ_t corresponding to elastic stress concentration as a representative stress. In the case of hydrogen embrittlement, it can be interpreted as the tensile stress at the distance λ_h ahead of the actual tip. Here λ_h is the thickness of the embrittlemented layer. In the case of the thin plastic zone, thickness λ_h takes a small portion of the plastic zone length. A linear approximation for the stress distribution at $a \leq x \leq a + \lambda_h$ may be assumed for simplicity in both cases.

Let us discuss briefly the (quasi)linear elastic model in terms of single-parameter cracks. The condition of crack (non)propagation is, as usual, taken in the form (8.11). The damage accumulation under sustained loading, in the simplest case, is described as follows:

$$\frac{\partial \omega_s}{\partial t} = \frac{1}{t_c} \left(\frac{\sigma - \sigma_{th}}{\sigma_s} \right)^{m_s}, \quad \frac{\partial \omega_c}{\partial t} = \frac{1}{t_c} \left(\frac{c - c_{th}}{c_d} \right)^{m_c} \tag{8.35}$$

These equations are similar to (8.9) and (8.10). The entering material parameters have a similar meaning. To take into account hydrogen embrittlement, we have to use non-zero initial conditions for damage. If the hydrogenization process is fast enough compared with further crack growth, we refer to embrittlement as an additive component of the damage measure which depends only on the hydrogen content at the tip. The formula

$$\omega_h(x, t) = \psi_h(t)(1 - \xi/\lambda_h) \tag{8.36}$$

similar to (8.6) presents the simplest example. Here $\xi = x - a$, $0 \leq \xi \leq \lambda_h$ and ψ_h is the tip damage due to hydrogen embrittlement. At $\xi > \lambda_h$ we put $\omega_h \equiv 0$. The current hydrogen content at the tip is governed by equations similar to (8.1) and (8.2) where c_∞ is the hydrogen content at the crack mouth.

The stress distribution ahead of the crack is estimated as for a shallow slit with an effective tip curvature. Its evolution is described by equation

$$\frac{d\rho}{dt} = \frac{\rho_s - \rho}{\lambda_\rho} \frac{da}{dt} + (\rho_b - \rho)\frac{d\psi_s}{dt} + (\rho_c - \rho)\frac{d\psi_c}{dt} + (\rho_h - \rho)\frac{d\psi_h}{dt} \tag{8.37}$$

which, compared with the right-hand-side of equation (8.12), contains the term that takes into account blunting or sharpening due to hydrogenization. Here ρ_h is the magnitude of the effective radius corresponding to hydrogen saturation. As no sufficient experimental data are available, we will, as in Section 8.4, unite the three last terms setting $\rho_b = \rho_c = \rho_h$ where ρ_b is the common blunt effective radius. Then the summed tip damage measure $\psi = \psi_s + \psi_c + \psi_h$ enters equation (8.37).

The last relationship closing the set of governing equations connects the generalized resistance force with the damage measures ahead of the crack tip. Not multiplying the number of material parameters whose magnitudes are hard to measure directly, we assume that the specific fracture work is given as

$$\gamma = \gamma_0[1 - \chi(\omega_s + \omega_c + \omega_h)^\alpha] \qquad (8.38)$$

at $\alpha > 0$, $\chi \geq 1$. Equation (8.38) is similar to (8.17). Then $\gamma = \gamma_0(1-\chi)$ characterizes the residual fracture toughness in the presence of a hydrogen-containing agent. In the literature evidence can be found that the residual fracture toughness is much lower than that for the virgin material. However, the lower bound of γ_h strongly depends on chemical composition (both of material and active agent), temperature, pressure, etc. In any case, the crack growth rate diagrams in the presence of hydrogen embrittlement correspond to lower magnitudes of stress intensity factors, for example, between 5 and 50 MPa $\cdot$ m$^{1/2}$, while the fracture toughness in the neutral environment may be of the order 100 MPa $\cdot$ mm$^{1/2}$.

When all environmental parameters but the tip content are fixed, one may assume that

$$\psi_h(c_t) = \psi_h^\infty[1 - \exp(-c_t/c_h)] \qquad (8.39)$$

Here the damage measure ψ_h^∞ corresponds to saturation, and $\gamma = \gamma_0(1 - \chi\psi_h^\infty)$ is the lower bound for the specific fracture work.

Taking into account equation (8.38), we obtain that for the one-parameter planar crack the generalized resistance force is

$$\Gamma = \Gamma_0[1 - \chi(\psi_s + \psi_c + \psi_h)^\alpha] \qquad (8.40)$$

with $\Gamma_0 \equiv \gamma_0$.

Obviously, there is a wide choice of analytical presentations for governing equations beginning from (8.35) to (8.40). In some aspects the proposed set of equations is "economical." This set contains the minimal amount of material parameters covering, at the same time, all main features of the physical phenomenon.

To illustrate the suggested analytical model, let us consider the case of "pure" hydrogen degradation. In other words, let only two damage measures enter the picture, mechanical ω_s and hydrogen assisted ω_h. In equations (8.36) and (8.39) we assume $\lambda_h = 10$ μm, $\psi_h^\infty = 0.5$ and treat the ratio c_∞/c_h, where c_∞ is the constant at the crack mouth, as a control parameter. In equation (8.38) we assume $\rho_s = 10$ μm, $\rho_b = \rho_h = \rho_c = \lambda_\rho = 100$ μm, and in equation (8.38) $\alpha = \chi = 1$. Numerical data (except for the newly introduced parameters) are the same as in Sections 8.5-8.8. This allows comparison of the effects of "pure" corrosion and of hydrogen embrittlement.

The numerical results are presented in Figures 8.17-8.19 for the applied stress $\sigma_\infty = 200$ MPa and initial data $a_0 = 1$ mm, $\rho_0 = 50$ μm. Initial damage is equal to zero. Time history of the hydrogen content at the tip is shown in Figure 8.17a. There lines 1, 2, 3, and 4 are drawn for the normalized hydrogen content at the crack mouth $c_\infty/c_h = 0.25$, 0.5, 0.75, and 1.0. The change of the effective radius ρ is shown in Figure 8.17b. These time histories are similar to those for stress corrosion cracking (Figure 8.4).

The behavior of the damage measures ψ_s, ψ_h and $\psi = \psi_s + \psi_h$ is depicted in Figure 8.18 for $c_t/c_h = 0.5$. An almost instantaneous effect of hydrogen diffusion is

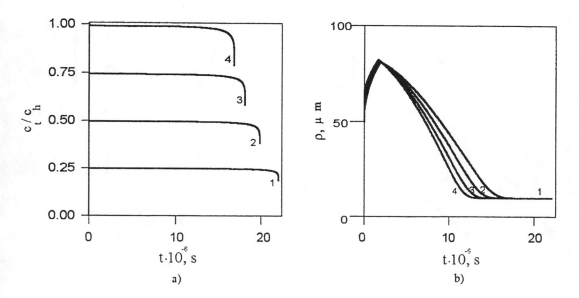

FIGURE 8.17
Evolution of hydrogen content at the crack tip (a) and effective tip radius (b)
at $c_\infty/c_h = 0.25, 0.5, 0.75, 1.0$ (lines 1, 2, 3, 4, respectively).

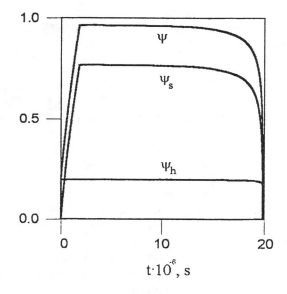

FIGURE 8.18
Evolution of tip damage measures at $c_\infty/c_h = 0.5$.

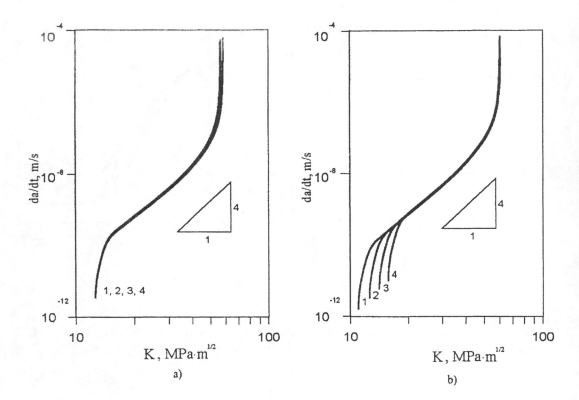

FIGURE 8.19

Crack growth rate diagrams in stress cracking for different hydrogen content (a) and different applied stresses (b); notation in text.

observed, and the embrittlement component ψ_h follows the time history of c_t. The crack growth rate diagrams are presented in Figure 8.19a. The rate da/dt counted in meters per second is plotted there in the function of the stress intensity factor K at different c_∞/c_h. The diagrams are close to merging, and the total fatigue life remains of the same order of magnitude when c_t/c_h varies in a wide range. In Figure 8.19b, to the contrary of former diagrams, various magnitudes of the applied stress σ_∞ are considered at the constant ratio $c_\infty/c_h = 0.5$. Lines 1, 2, 3, 4 are drawn for $\sigma_\infty = 175$, 200, 225 and 250 MPa, respectively. It is not surprising that the influence of hydrogen-assisted damage looks similar to the "common" stress corrosion cracking. Actually, the major difference between the two types of damage from the purely mechanical viewpoint is characterized by times needed for damage saturation. Chemical or electro-chemical damage needs a significant time to develope compared with rapid hydrogen embrittlement. It should be noted that we have not included the oxide film formation which enters classical stress corrosion cracking as a retardation factor.

8.12 Combination of Cyclic Fatigue and Hydrogen Degradation

To close the application of the general theory to embrittlement assisted fatigue, let us discuss the cycle fatigue in the presence of hydrogen embrittlement. It is evident that the set of governing equations is composed of equations considered in the above sections, and that the versatility of the theory hardly needs additional evidence. Therefore, only brief comments will be presented here. In terms of Section 8.11, equations (8.35)-(8.40) remain valid; only equations (8.35) must be complemented by the equation with respect to the cyclic component of damage ω_f:

$$\frac{\partial \omega_f}{\partial N} = \left(\frac{\Delta\sigma - \Delta\sigma_{th}}{\sigma_f} \right)^{m_f} \tag{8.41}$$

This equation is similar to (8.23) with the same interpretation of material parameters. Equation (8.5) takes into account the frequency effect on the active agent transport. A relevant component must be added in equation (8.37). In the case when only two extreme magnitudes of the characteristic tip radius are used, the "sharp" ρ_s and the "blunt" ρ_b ones, we arrive at equation

$$\frac{d\rho}{dt} = \frac{\rho_s - \rho}{\lambda_\rho} \frac{da}{dt} + (\rho_b - \rho) \frac{d}{dt} (\psi_s + \psi_f + \psi_h) \tag{8.42}$$

The cyclic load frequency enters equation (8.42). Actually, the cyclic damage rate is measured, as in equation (8.41), in $d\omega_f/dN$. At last, the cyclic damage component ψ_f must be added in equations (8.38) and (8.40).

Two types of diagrams are presented in Figure 8.20 where the growth rate is treated as da/dt and da/dN. The maximal stress intensity factor K_{max} and the range ΔK are used there as correlating parameters, respectively. The same material parameters are taken there as in Sections 8.6-8.8 and 8.10. The loading and initial conditions are: $\Delta\sigma_\infty = 200$ MPa, $R = 0.2$, $c_\infty/c_h = 1.0$, $a_0 = 1$ mm, $\rho_0 = 50$ μm. The only external parameter subjected to varying is load frequency. Lines 1, 2, 3, and 4 in Figure 8.20 are drawn for $f = 0.1, 1, 10$, and 100 Hz, respectively. The curves diverge considerably in both types of diagrams. The cause is an almost equal contribution of cyclic and hydrogen embrittlement damage to the total damage at the crack tip during most of fatigue life.

Dissatisfaction is always present when we discuss corrosion fatigue and related phenomena treating the material as linearly elastic with the fixed compliance parameters. A proper choice of characteristic tip radii and its proper interpretation make such models more agreeable from the academic viewpoint, not to mention the possibility of fitting numerous experimental data. Meantime, a similar approach is possible in the framework of the thin plastic zone model (see Section 8.10). In addition, Chapter 6 contains sufficient tools to treat corrosion fatigue, stress corrosion cracking, hydrogen embrittlement, etc., taking into account the influence of damage of any nature on the yield limit (or some equivalent limit stress relevant to the thin plastic zone model). By necessity, the effect of damage on Young's modulus and other stiffness parameters can be included in the model.

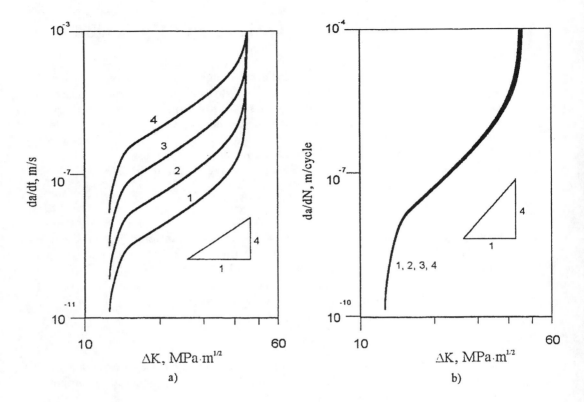

a)   b)

FIGURE 8.20
Crack growth rate diagrams in hydrogen, assisted fatigue at $c_\infty/c_h = 1.0$ referred
to time (a) and cycle number (b) at load frequencies $f = 0.1$, 1, 10, and 100 Hz
(lines 1, 2, 3, and 4, respectively).

8.13 Summary

The approach to modeling of fatigue crack growth developed in previous chapters is supplemented in this chapter with the effects of active media, which may be gaseous, vapor or liquid. The essential issue is to account for damage produced by an active agent in the vicinity of crack tips. A combination of this damage with purely mechanical damage under sustained and/or cyclic loading results in various forms of structural fatigue labelled as stress corrosion cracking, corrosion fatigue, etc.

The set of governing equations was suggested in this chapter to describe these phenomena. Compared with ordinary fatigue, the analytical models of environmentally assisted fatigue include the equations for active agent transport to the crack tip and the equations of damage accumulation due to the influence of this agent. A special emphasis was put on two items: first, the comparison of the contribution of various damage processes in crack growth; second, the choice of an appropriate form for crack growth rate diagrams. The latter problem is of special significance when two temporal variables, physical time and cycle number, are involved. A number of computational experiments were performed to assess the effects of load level, load frequency, and active agent content on the crack growth rate plotted versus various correlating load parameters. The results can be used in planning tests and their adequate interpretation in terms of solid mechanics. Some important aspects of environmentally assisted fatigue have not been discussed in this chapter explicitly, though they might be described by the suggested models with minor modifications or supplementations.

Chapter 9

FRACTURE AND FATIGUE OF FIBER COMPOSITES

9.1 Composite Materials and Composite Structures

One says composites, meaning the combinations of two or more materials with different properties resulting in materials with new, desirable and frequently advanced properties. From this viewpoint, most widely used natural and artificial materials, such as timber or concrete, ought to be called composites. However, using this term, we mean comparatively new and newly developed materials. The history of composite technology began, perhaps, from fiberglass reinforced plastics, though, for example, plywood is also a typical composite.

In most composites two kinds of components are distinguished, matrix and fillers. Usually the material of a matrix is ductile and has low strength compared with the material of fillers. Thus, usually we say matrix and reinforcement. The latter may be in the form of particles, fibers, threads, textiles, and laminae. Various combinations are used in composite technology. A typical example is a laminated shell whose plies are fabricated as unidirectional fiber composites of different orientation and may be of different composition also.

Composite materials are the subject of modern technology, and mechanics of composites is one of the rapidly developing domains of applied mechanics. This chapter is dedicated to fracture and fatigue of unidirectional fiber composites with continuous fibers. More precisely, we consider the case of ductile matrix and high-strength, however rather brittle, fibers. As matrix materials, polymers (both thermoset resins and thermoplastics) and metals are used. Fibers are supposed to be made of glass, carbon, ceramics, or metals. The main task of a matrix is to redistribute loads between fibers in the case when some of them are ruptured. Quite a different family of fiber composites is developed on the base of a ceramic matrix.

In this case the exchange of roles between the matrix and fibers takes place. The assignment of fibers is to redistribute loads in the case of local cracking of ceramics and, as a result, to increase fracture toughness and other mechanical properties characterizing structural reliability. Fracture of composites on the base of ceramics which is now on the cutting edge of research is discussed in [45, 142]. In this chapter, however, we are going to discuss more conventional materials such as glass-epoxy and carbon-epoxy unidirectional composites with the detailed analysis of mechanisms of fracture and fatigue of these materials.

Mechanics of fracture and fatigue of composite materials and composite structures is a very special section of fracture mechanics. Patterns of damage accumulation, fatigue crack initiation and propagation in composites significantly differ from those in conventional structural materials such as metallic alloys. Most composites have essentially nonhomogeneous structure and strong anisotropy. This refers both to fiber reinforced and laminate composites. Their high strength and stiffness in the direction of reinforcement are accompanied by low resistance and high compliance in the transverse direction. The composites of granular structure also have specific features. They could be macroscopically isotropic, but, being highly heterogeneous, they are sensitive to processes on the microscale. Most crack-like defects that become the origins of macroscale cracking usually are quite different from those in conventional structural materials. In particular, various kinds of delaminations: initial, acquired in the manufacturing process, and newborn ones, are typical for laminate composites. In addition, most composites are rather sensitive to thermal actions and environmental effects.

On the other side, many composite structures expose rather high structural safety in the presence of local initial or newborn defects. As an example, fiber or laminate composites with sufficiently compliant matrix may be mentioned. When a reinforcement element is ruptured, or the interface is damaged, load redistribution takes place, and damage remains localized in a comparatively small volume. Due to that, the structure keeps its integrity and a certain portion of the load-carrying capacity. The principle of load redistribution is also used in composites with brittle matrix and reinforcement elements of high ductility. The stress localization is maintained in such materials due to comparatively high ductility of the reinforcement.

The final failure of composites, as a rule, is preceded by damage accumulation on a microscale level, i.e., on the level of fibers, particles, etc. Therefore, most of the conventional notions of fracture mechanics, in particular, of the linear fracture mechanics have a limited application to composite materials. The cracks in composites, as a rule, grow in patterns rather dissimilar to those in conventional, macroscopically isotropic materials. In particular, "brush-like" failure is observed in unidirectional composites with brittle fibers and ductile matrix under the tension along fibers. Under three-point bending of specimens of such composites, longitudinal cracking is usually observed. On the other side, planar interlaminar defects in laminate composite structures, as a rule, propagate along the initial plane with damage and fracture concentrated in the interlayer. When such a delamination is situated within the bulk of a composite, its behavior is similar to that of cracks in conventional materials. The only complication originates from a strong anisotropy. However, if a delamination is situated near the surface, crack growth is frequently accompanied by local buckling of the delaminated part. Thus, microdamage, fracture and buckling phenomena become closely linked. This results in the necessity to develop special approaches to modeling even in the case of monotonous loading. This conclusion certainly refers also to delaminations growing under cyclic and/or sustained loading.

In some aspects, mathematical modeling of fracture and fatigue is more necessary for composites than for conventional materials. Experimental data on fatigue and fracture of composites are scarce compared with those for metallic alloys. Moreover, not mentioning a very wide variety of properties, each composite, as a matter of fact, is usually designed and manufactured together with a corresponding engineering structure. Thus, the properties of composites are nonseparable from the properties of the structure. On the other hand, the properties of components are frequently known beforehand, while the properties of the final product, i.e., of the composite as a part of a certain structure, need prediction, and sometimes at the earlier design stage.

Damage and fracture of composite materials are widely discussed in the literature, both from theoretical and experimental viewpoints. The early studies were performed by Rosen (1965) and Kelly (1973). Among the books and manuals, the publications edited by Broutman and Krock [42], Tamuzs and Protasov [143], Herakovich and Tarnopolsky [67], Stinchomb and Ashbaugh [138], and Martin [100] could be listed.

The theory of fatigue developed in this book is applicable, in its general part, to composite materials. Really, the principle of virtual work is valid for all kinds of materials. It is a unversal tool to analyze the states of the system cracked body-loading and predict fatigue crack propagation. The specific features of most composites, in particular, their essential nonhomogeneity require considering different damage mechanisms. For example, damage to unidirectional fiber composites consists at least of three mechanisms: rupture of single fibers, rupture of matrix, and damage of the fiber-matrix interface.

The damage processes on the microscale are strongly probabilistic. This concerns, in a larger degree, the microdamage accumulation in composite materials. If high structural safety is required, the failure must be considered as a rare event. Consequently, theoretical predictions must take into account very small probabilities. There is a lot of statistical information, obtained both in laboratories and service conditions, concerning the behavior of conventional materials and corresponding structural components. The situation with composites is not so clear. It is highly desirable to develop probabilistic models of fracture for composite structures. These models must satisfy at least two requirements: to remain consistent at small probabilities of failure, and to include the size effect of strength allowing the extrapolation upon large scales.

The size (scale) strength effect is, in fact, a deviation from the classical similitude laws observed in mechanical tests of geometrically similar specimens. It means that the strength of a specimen is affected by certain length parameters which do not enter classic equations of continuum mechanics. The sizes of particles, diameters of fibers, etc. take the part of such length parameters. The coarser the microstructure, the stronger, under other given circumstances, the size effect (see also Chapter 2).

The size effect of fracture and damage in composite materials is a consequence of their nongomogeneous structure, which is more or less of a random nature due to the scatter of mechanical properties of components, their random spacing, initial ruptures of fibers and their initial curvatures, local loss of adhesion, porosity of matrix, etc. Therefore, the size strength effect and the stochastic nature of fractures are closely connected.

The well-known model of "weakest link" by Weibull (1939) may be considered as an example of probabilistic models satisfying the above mentioned requirements. But Weibull's model as well as its generalizations relate to the ideal brittle material not including ductile effects and stress redistribution effects. These models are not applicable to most composites with polymer or metal matrices. There are, of course, successful attempts to treat experimental results using Weibull's model. As a matter of fact, these attempts only mean the fitting of statistical data with the use of two- or three-parameter Weibull's distribution.

An alternative model is the so-called "bundle-of-threads" model usually attributed to Daniels (1945). According to this model, the ultimate tensile load for a bundle of fibers is equal to the expected value of the sum of ultimate loads for separate fibers. Hence, Daniel's model takes into account both redundancy of the system and the ductile character of the final fracture. But this model, being applied to reliability problems, produces too optimistic predictions, especially in the areas of high reliability. This model, as a rule, underestimates the size strength effect.

A number of models incorporating the approaches by Weibull and Daniels were suggested in the last decades. For example, a specimen of unidirectional fiber composite was schematized as a chain whose links are of length equal to the double ineffective fiber length (the distance required for the load transfer from a ruptured fiber to the surrounding domain). Each link was considered in the frame of Daniel's model, and the set of consequent links follows Weibull's model. In some models, the possibility of rupture of neighboring fibers and the stress concentration near a single or multiple rupture are taken into account [64, 157]. These models are more flexible than those by Weibull and Daniels. They provide a better agreement with experimental results if numerical values of parameters are properly chosen.

In this chapter the mechanisms of damage and fracture of unidirectional fiber composites are considered in detail including the dispersed damage accumulation, failure due to multiple cracking, macroscale cracks formation and their propagation until the final fracture. The composites with continuous brittle fibers and high-ductile matrix such as glass-epoxy or carbon-epoxy composites are considered. The case of brittle matrix whose properties are improved by ductile fibers (both continuous and short ones) are also of interest, especially with applications to ceramics and concrete [45, 76, 142]. We do not discuss this case here. Fatigue fracture in laminate composites and, in the first line, the theory of growth for buckled delaminations in composites structures are the subject of Chapter 10.

9.2   Micromechanics of Dispersed Damage

Typical damage and fracture modes of unidirectional fiber composites under tension along fibers are schematically depicted in Figure 9.1. Dispersed damage of single elements of microstructure such as rupture and pulling-out of fibers and localized damage of fiber-matrix interface is typical for unidirectional composites (Figure 9.1a). When the density of single ruptures attain a certain critical level (Figure 9.1b), multiple fracture of the body takes place. Later on, this type of structural failure is called the loss of integrity. Another mode of structural failure is the formation of a damaged domain with significant dimensions both across and along the fiber directions (Figure 9.1c). This domain grows with the increase of load level. The growth results into structural failure analogous to quasi-brittle fracture of conventional materials. Such a mode of damage, which is certainly a kind of crack, we call brush-like crack [11]. A similar mode of damage occurs not only as a result of interaction of isolated ruptures. Due to multiple delaminations, the damage originating from an initial notch is also gradually transformed into a brush-like crack (Figure 9.1d).

The most general approach to fracture and fatigue of unidirectional composites is based on models that take microstructure into account. This approach includes, in the framework of unified model, various damage and fracture phenomena, in particular, macrocrack initiation and growth, fracture due to multiple cracking, etc. The dimension of the parameter space for such models is extremely high; therefore only simplified models give comparatively transparent results.

There are many attempts to model the process of damage accumulation in unidirectional components. Most of the suggested models, due to the statistical nature of damage, are probabilistic ones. Hereafter a model of quasi-independent dispersed microdamage will be considered [19]. This model seems to be valid both

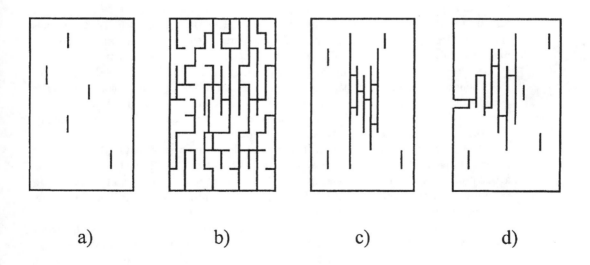

a)   b)   c)   d)

FIGURE 9.1
Damage and fracture modes of unidirectional fiber composite in tension along fibers: (a) dispersed damage; (b) multiple damage before the loss of integrity; (c) formation of a brush-like crack; (d) propagation of a brush-like crack from an initial notch.

for dispersed and localized damage. In the latter case the model is suitable for its incorporation in the theory of macrocrack initiation and growth. In general features, the model discussed hereafter is similar to that discussed in Chapter 2 in connection with fatigue crack nucleation in ordinary (quasihomogeneous) materials.

Introduce the following assumptions:

1. A body (specimen or structural member) consists of a large number of similar (in the statistical sense) microscale elements. An element becomes ruptured when a reference stress, say, the maximal tensile stress attains an ultimate stress which is inherent to the considered element. The ultimate stress s is a random variable with given cumulative distribution function $F_s(s)$.

2. A body consists of a finite number of critical domains; fracture of even a single critical domain results in the failure of the body as a whole. In two extreme cases, the critical domain coincides either with a microscale element or with the total volume of the body.

3. A critical domain becomes fractured when the number of ruptured microscale elements in its volume attains a critical level. The latter is assumed as a deterministic value. The ratio of the critical number of microscale elements to their total number is assumed as sufficiently small compared with unity.

4. The number of microscale elements in the critical volume, and their ultimate number in that volume, are large compared with unity.

Assumption 1 is used in most probabilistic models of fracture. Assumption 2 is in fact the weakest-link concept applied, however, not to smaller elements of microstructure but to macroscale volumes. It is assumed that sizes, shapes and positions of critical domains in real structures ought to be estimated from the observation of fracture of similar structures. The choice of critical volumes is to be done taking into account the geometry of a structure, load and stress-strain fields, and material properties. Introduction of an intermediate geometrical scale makes the size effect prediction more reliable. The first part of assumption 3 does not need special commentaries. The second part allows that the rupture of a single element does not influence the behavior of other elements. Hence, in the frame of the discussed model, the rupture of two or more neighboring elements as well as the growth of a macroscopic crack are not included in consideration. Last, assumption 4 is needed to substantiate the applicability of limit theorems of probability theory and the transition to asymptotic distributions.

It is easy to see that the main features of the above model enter the frames of the micromechanical damage model developed in Section 2.5. In particular, we treat a body under quasimonotonous, quasistatic loading as a set of weakly interacting random elements whose rupture follows the Laplace-Moivre scheme. Due to a large number of elements in a body, the central limit theorem is applicable.

Let the damage measure be introduced as $\omega = n/n_0$ where n_0 is the total number of elements in a certain reference domain, for example in a unity volume, and n is the number of ruptured elements in this domain. The ultimate stress s for elements is a random variable with the cumulative distribution function $F_s(s)$. Then, following equation (2.40), we present the cumulative distribution function for the damage measure ω under the applied stress σ as follows:

$$F_\omega(\omega;\sigma) = \Phi\left(\frac{\omega - F_s(\sigma)}{F_s(\sigma)[1 - F_s(\sigma)]n_0^{-1/2}}\right) \qquad (9.1)$$

Here $\Phi(\cdot)$ is the normalized function of Gaussian distribution. The simplified estimate for ω

$$\omega \approx F_s(\sigma) \qquad (9.2)$$

is applicable when the variance of ω is sufficiently small. The applicability condition on the basis of equations (9.1) and (9.2) takes the form

$$\omega n_0^{1/2} \gg 1$$

This means that, to apply equation (9.2), the number n_0 of elements has to be very large, and the damage measure ω not too small.

The difference with the model discussed in Section 2.5 is that here we consider large-scale systems consisting of a number of subsystems, and each subsystem is responsible for the safe life of the whole system. That means the weakest-link approach on the macroscale level. However, a specialization of the model is needed to apply it to unidirectional composites.

Let us consider a composite with initially continuous fibers. The fibers have circular cross section with radius r and are uniformly (at least in the statistical sense) distributed in the volume of composite. Anyway, the volume fraction v_f of fibers is constant in the considered volume. The fibers are assumed elastic with Young's modulus E_f until rupture. The matrix material is elastoplastic with Young's modulus E_m, shear modulus μ_m and ultimate stress τ_m. The stress τ_m may be interpreted as the yield stress of the matrix, or the friction stress on the damaged interface, or as a certain ultimate shear stress that takes into account all inelastic phenomena associated with damage of the matrix and interface. The fibers are supposed to be high-performance ones, and the matrix material sufficiently ductile. This is typical for most industrial unidirectional composite materials. In particular, we assume that $E_f \gg E_m$, $s_c \gg \tau_m$ where s_c is a characteristic ultimate tensile stress for fibers. We suppose that the fiber volume ratios v_f and $v_m = 1 - v_f$ are moderate, i.e., $v_f \sim v_m$.

The above model is comparatively simple when we treat the composite as non-damaged and the placing of fibers as regular. Analytical solutions were obtained for a number of problems in the framework of classical elasticity or plasticity theories. However, the problem becomes more complicated when we turn to composites with ruptured fibers, moreover, in the presence of crack-like damage. Some calculations can be performed only on the so-called "physical level of rigor." The estimates by the order of magnitudes are also sometimes acceptable. For example, the homogenized tensile stress, following the mixture rule, is

$$\sigma_\infty = v_f \sigma_f + v_m \sigma_m \qquad (9.3)$$

where σ_f and σ_m are tensile stresses in fibers and matrix, respectively. Assuming that both fiber and matrix are deformed elastically, we have

$$\sigma_f = \frac{v_f^{-1}\sigma_\infty}{[1+(v_m E_m/v_f E_f)]} \tag{9.4}$$

When $v_m E_m/v_f E_f \ll 1$, we assume $\sigma_f \approx v_f^{-1}\sigma_\infty$ and even $\sigma_\infty \sim \sigma_f$. The sign $\sim$ means equality by the order of magnitude. Sometimes one can neglect the change of applied stresses in the composite due to crack growth as well as the stress concentration near ruptured fibers. The latter neglect is justified, at least in the first approximation, because the stress concentration relaxes due to the delamination while the size effect of fiber strength becomes more significant.

A number of simplifications will be used in the treatment of probabilistic models. Among them are the limit theorems of the theory of probability, and "almost sure estimates" for macroscale parameters which are valid with the probability of the order of unity. Some deterministic variables will be identified with medians and fractiles of probabilistic distributions, say, with the fractiles corresponding to probability $1 - e^{-1} = 0.632\ldots$. Most of the above assumptions may be easily removed; but without them the final results will be less compact and transparent. In general, many estimates in the mechanics of composites are rather approximate. Even sophisticated models frequently fail to take into account the complicated distribution of loads in composites, especially in the presence of multiple damage. Hence, even crude estimates may be useful for engineering analysis not commenting on physical understanding of phenomena.

One of the main features of most high-performance fibers is their strong size effect of tensile strength. It means that the ultimate stress of fibers is a random variable whose distribution depends on the "working" length $2l$ at those points where the rupture occurs. The most suitable distribution is certainly the Weibull distribution for minima. The cumulative distribution function

$$F_s(s;l) = 1 - \exp\left[-\frac{l}{l_c}\left(\frac{s-s_{th}}{s_c}\right)^\alpha\right] \tag{9.5}$$

is used hereafter with material stress parameters $s_c > 0$, $s_{th} \geq 0$, and shape exponent $\alpha > 0$. At $s < s_{th}$ the right-hand side of (9.5) is to be put to zero. The characteristic size l_c enters (9.5). As $l_c s_c^\alpha = \text{const}$ for the same material, the same loading and environmental conditions, the choice of l_c is arbitrary. However, it is conditioned by the choice of s_c. Not multiplying the number of parameters, we might put l_c equal to the fiber cross-section radius r. Then equation (9.5) takes the form

$$F_s(s;l) = 1 - \exp\left[-\frac{l}{r}\left(\frac{s-s_{th}}{s_r}\right)^\alpha\right] \tag{9.6}$$

Certainly, the length parameter l in damage analysis is larger than r, and there is no sense to testing fiber specimens of very short length. This means that the characteristic ultimate stress s_r entering equation (9.6) is to be estimated by the extrapolation of test results of much longer fiber segments. Comparing (9.5) and (9.6), we see that $s_r = s_c(l_c/r)^{1/\alpha}$.

Numerical data concerning the parameters in equations (9.4) and (9.5) can be found in the literature on composite technology and mechanics of composites [42, 138]. We note here only that one of the most important parameters is the shape exponent α. It characterizes the scatter of tensile strength and essentially depends on the quality of fibers and by-products and on the level of manufacturing and handling. The better the quality, the less the scatter and the higher exponent α. The average magnitudes are $\alpha = 6 \ldots 8$. The larger magnitudes such as $\alpha = 10 \ldots 12$ can be achieved with the use of high-level technology. At $\alpha \gg 1$, $s_c \gg s_{th}$ the variance coefficient of the ultimate stress is

$$\text{var}[s] = \frac{\pi}{\alpha\sqrt{6}} + O\left(\frac{1}{\alpha^2}\right) \tag{9.7}$$

i.e., almost inversely proportional to α.

Equations (9.5)-(9.7) are illustrated in Figure 9.2a where the probability density functions $p_s(s) = F_s'(s)$ are shown for three shape exponents $\alpha_1 < \alpha_2 < \alpha_3$ (lines 1, 2, 3). Line 4 for matrix material is also shown in Figure 9.2a. Statistical scatter of matrix properties is treated as small compared with those of fibers. For illustration only, Figure 9.2b shows the corresponding density functions for the case of ductile fibers and brittle matrix with high scatter of tensile strengh. Such a distribution is typical for ceramics reinforced with metallic matrix. In the latter case, most cracking of the matrix occurs before the fiber ruptures.

Returning to the case of brittle fibers, consider a model of dispersed damage accumulation in more detail. The first step is to introduce the elements of microstructure. The natural way is to identify these elements with the segments of

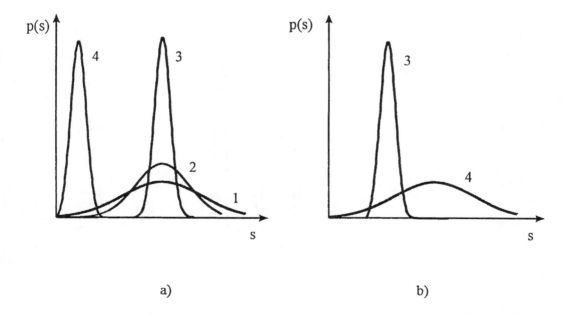

a)   b)

FIGURE 9.2
Probability density functions for ultimate stresses in case of ductile (a) and brittle (b) matrix; lines 1, 2, 3 for fibers at $\alpha_1 < \alpha_2 < \alpha_3$; line 4 for matrix.

fibers with the attached matrix material. The working length of elements is equal to $2l$, where l is the distance needed to restore the principal portion of the load on a ruptured fiber. This length, primarily introduced by Rosen (1965) as "the ineffective length," and by Bolotin (1965) as "the edge effect length," was estimated later by a number of authors with the use of more sophisticated models.

The simplest model is depicted in Figure 9.3. All fibers are subjected to the tensile stress σ_f. One of the fibers is broken. Hence some distance is needed to transfer the load from the neighboring fibers and matrix for recovering the initial stress level. Notating the axial fiber displacement $u(x)$, we define the current tensile stress in the broken fiber as $E_f \pi r^2 (du/dx)$. The load redistribution is produced by the shear stresses $\tau(x)$. If the matrix and the fiber-matrix interface are deformed elastically, one may assume $\tau = \mu_m u / h_m$, where h_m is an effective thickness of the interlayer. The tangential load per unit length of fiber is $2\pi r \tau$. Hence, the function $u(x)$ satisfies equation

$$E_f \pi r^2 \frac{d^2 u}{dx^2} - \frac{2\pi r \mu_m u}{h_m} = 0$$

with boundary conditions $E_f \pi r^2 (du/dx) \to \sigma_f$ at $x \to \infty$, $du/dx = 0$ at $x = 0$ (in the ruptured section).

The characteristic length, as it follows from the above equation, is

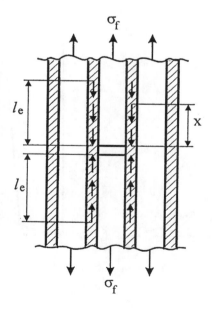

FIGURE 9.3
Evaluation of the transfer length in the vicinity of broken fiber.

$$l_e = \left(\frac{E_f r h_m}{2\mu_m} \right)^{1/2} \tag{9.8}$$

This length is supplied with the subscript e (elastic) to distinguish it from the characteristic length l_p in the plastic case. At $x = l_e$ the current stress in the broken fiber attains the magnitude $(1 - e^{-1})\sigma_f$. Noting $(rh_m)^{1/2} = g_e(v_f)r$, equation (9.8) takes the form

$$l_e = g_e(v_f)r \left(\frac{E_f}{2\mu_m} \right)^{1/2} \tag{9.9}$$

The correction factor $g_e(v_f)$ depends on the accepted model of the stress transfer. For example, formula $g_e(v_f) = (v_f^{-1/2} - 1)^{1/2}$ was suggested by Rosen (1965). For rough estimations, $g_e(v_f) \sim 1$ at moderate v_f; moreover the condition $\sigma(l_e) = (1 - e^{-1})\sigma_f$ is rather arbitrary.

Equations (9.8) and (9.9) correspond to the asumption that the interface is not damaged and the matrix remains elastic. To take into account plasticity, assume that the load transfer is performed by shear boundary stresses equal to a certain ultimate (yield) magnitude τ_m. Then instead of (9.9) we obtain

$$l_p = g_p(v_f)r \left(\frac{\sigma_f}{2\tau_b} \right) \tag{9.10}$$

Here the correction factor $g_p(v_f)$ is also of the order of unity. In particular, $g_p(v_f) = 1$ when the uniform distribution of τ upon all the concerned surface is assumed. The characteristic length similar to that given in (9.10) was introduced by Kelly (1973). Note that, due to the irreversible character of damage, the length l_p depends on the loading history. Neglecting hardening and softening phenomena, we use equation (9.10) for nonmonotonous loading under the condition that σ_∞ is taken equal to its maximal magnitude during all the preceding loading history.

Evidently, the transfer length is equal to l_e at low applied stresses, and to l_p when the stresses increase. To develop a model of damage, it is convenient to deal with a given number of microscale elements. Thus, the length of each element will be equal to $2l_e$.

As follows from the above, several types of microdamage are to be considered. We need to quantify fiber ruptures, martix ruptures, and interface damage. To describe the first type of damage, consider a reference volume M_0 of the composite with homogeneous distribution of stresses and macroscopic properties. It may or may not coincide with the critical volume responsible for the integrity of the whole body, with the domain of a structural component, specimen, etc. Let the total number of elements in M_0 be n_0. Notating with n the number of fiber ruptures in M_0, we introduce the damage measure as

$$\omega_f = \frac{n}{n_0} \tag{9.11}$$

The second damage measure ω_m may be equalized to ω_f when each fiber rupture is accompanied by the matrix cracking in the same cross section. The third measure ω_b concerns the damage of the fiber/matrix interface, i.e., the debonding of fibers and the delamination of the composite as a whole. We assume that, when the shear stresses near the fiber rupture are sufficiently high, the boundary will be damaged along the most of the transfer length. Therefore,

$$\omega_b = \omega_f \frac{l}{l_e} \tag{9.12}$$

When matrix is deformed elastically, $l = l_e$, and we put $\omega_b = 0$. Then there is no need for any damage measures except ω_f. At $l > l_e$ the measure ω_b is interpreted as the ratio of the sum of all ineffective lengths to the total length of all fibers in the considered volume. When equation (9.12) gives $\omega_b > 1$, we must put $\omega_b = 1$. Certainly, it is a coarse schematization: in fact, there is a continuous transition from the elastic state of the interface to the plastic one.

To illustrate the above notions, consider a monotonous quasistatic loading along fibers with the applied stress σ_∞. Combining equations (9.2), (9.5) and (9.11) we obtain

$$\omega_f(\sigma_\infty) = 1 - \exp\left[-\frac{l}{l_c}\left(\frac{\kappa\sigma_\infty - s_{th}}{s_c}\right)^\alpha\right] \tag{9.13}$$

Here the fiber stress σ_f is replaced, on account of (9.4), by $\kappa\sigma_\infty$. At $v_m E_m \ll v_f E_f$, $w_f \ll 1$ we have $\kappa \approx v_f^{-1}$. To take into account the stress redistribution because of fiber ruptures, we use the following equation for κ:

$$\kappa = \frac{1}{v_f(1 - \omega_f)[1 + (v_m E_m / v_f E_f)]} \tag{9.14}$$

At $\omega_f \ll 1$ equation (9.13) takes the simplified form

$$\omega_f(\sigma_\infty) \approx \frac{l}{l_c}\left(\frac{\kappa\sigma_\infty - s_{th}}{s_c}\right)^\alpha \tag{9.15}$$

The final equation for $\omega_f(\sigma_\infty)$ depends on the assumed behavior of the matrix. When the matrix is deformed elastically, the size parameter l is to be inserted according to equation (9.9). Then equation (9.15) gives

$$\omega_f(\sigma_\infty) \approx g_e(v_f)\left(\frac{E_f}{2\mu_m}\right)^{1/2}\left(\frac{\kappa\sigma_\infty - s_{th}}{s_r}\right)^\alpha \tag{9.16}$$

The fiber strength distribution is taken here according to (9.6), i.e., for $l_c = r$, $s_c = s_r$. In the case when equation (9.10) is to be used for l, we obtain instead of (9.16)

$$\omega_f(\sigma_\infty) \approx g_p(v_f)\left(\frac{\sigma_\infty}{2\tau_b}\right)\left(\frac{\kappa\sigma_\infty - s_{th}}{s_r}\right)^\alpha \tag{9.17}$$

The transition from (9.16) to (9.17) takes place at $\sigma_\infty = \sigma_p$ where

$$\sigma_p = 2\tau_b\frac{g_e(v_f)}{g_p(v_f)}\left(\frac{E_f}{2\mu_m}\right)^{1/2} \tag{9.18}$$

To demonstrate the order of magnitude for the involved variables, let us consider a conditional unidirectional composite with a ductile and compliant matrix.

Assume that $E_f = 100$ GPa, $\mu_m = 3$ GPa, $\tau_b = 50$ MPa. At $v_f = 0.5$ we may take the correction factors in (9.9) and (9.10) as $g_e(v_f) = 0.64$, $g_f(v_f) = 1$. The elastic transfer length according to (9.9) is $l_e \approx 2.6r$. We will use equation (9.11) beginning from the stress given in (9.18). For the assumed numerical data, we estimate σ_p as 130 MPa. It is a rather low stress for high-performance composites. Hence, most of the dispersed damage takes place when the fiber-matrix interface is deformed inelastically and/or damaged.

The damage measures are given in equations (9.13)-(9.17). Material parameters s_r, s_{th}, and α are of primary significance in the assessment of these measures. Because of the strong size effect of fiber strength, we must take into account the length of fiber specimens used in tests. For example, when the working length of tested specimens is 20 mm, we refer the stress s_c to half-length $l_c = 10$ mm. Using s_r as a strength parameter, we must apply the relationship $s_r = s_c(l_c/r)^{1/\alpha}$. In particular, at $s_c = 1$ GPa, $r = 10$ μm, and $\alpha = 8$ we obtain $s_r \approx 2.37$ GPa.

9.3   Microdamage Accumulation in Cyclic Loading

To extend the above model to fatigue damage (cyclic as well as static, i.e., accumulated under sustained loading), we have to treat damage accumulation as a temporal process. The preliminary notion of this approach was given in Section 2.5. We introduce the equation of damage accumulation for each microscale element

$$\frac{d\chi}{dN} = f(\sigma_f^{max}, \sigma_f^{min}, \chi) \tag{9.19}$$

where χ is the damage measure for an element. The right-hand side of this equation (9.19) may be chosen, similar to (2.47), in the form of the power-threshold law:

$$\frac{d\chi}{dN} = \frac{1}{N_c} \left(\frac{\Delta\sigma_f - \Delta\sigma_{th}}{s - s_{th}} \right)^m \tag{9.20}$$

Here s, s_{th} and $\Delta\sigma_{th}$ and m are material parameters. The characteristic cycle number N_c is chosen arbitrarily under the condition that $N_c(s - s_{th})^m = $ const for the given material, the given loading and environmental conditions. Since the fiber properties are randomly distributed, it is natural to treat s and, perhaps, $\Delta\sigma_{th}$ as random variables. In particular, the distribution given in (9.6) is convenient. Later on we assume, for simplicity, that $\Delta\sigma_{th} = s_{th}$ or that $\Delta\sigma_{th}$ is proportional to s. Then the resistance of fibers to fatigue rupture is controlled by the single random variable s.

The first damage measure ω_f, as in case of quasistatic loading, is defined as the ratio of the number of ruptured elements to their total number. Similar to the estimate given in (9.2), we state that

$$\omega_f(N) = 1 - \text{Pr}\{\chi(N) < 1\} \tag{9.21}$$

where $\chi(N)$ is the damage measure for an arbitrary chosen element attained at the cycle number N. For example, when the loading process is steady, i.e., $\Delta\sigma_\infty = $

const, $\sigma_\infty^{max} = $ const, and $\Delta\sigma_f \approx \kappa\Delta\sigma_\infty = $ const for all cycles, equation (9.20) results in

$$\chi(N) = \frac{N - N_0}{N_c}\left(\frac{\kappa\Delta\sigma_\infty - \Delta\sigma_{th}}{s - s_{th}}\right)^m$$

Here the initial condition is taken as $\chi(N) = 0$ at $N \leq N_0$. Therefore

$$\omega_f(N) = \Pr\left\{s < s_{th} + (\kappa\Delta\sigma_\infty - \Delta\sigma_{th})\left(\frac{N - N_0}{N_c}\right)^{1/m}\right\}$$

or, on account of (9.5) at $s_{th} = \Delta\sigma_{th}$,

$$\omega_f(N) = 1 - \exp\left[-\frac{l}{l_c}\left(\frac{\kappa\Delta\sigma_\infty - \Delta\sigma_{th}}{s_c}\right)^\alpha\left(\frac{N - N_0}{N_c}\right)^\beta\right] \qquad (9.22)$$

Here the exponent is introduced similar to that given in (2.9)

$$\beta = \alpha/m \qquad (9.23)$$

A more general equation for $\omega_f(N)$ is the result of substitution in (9.5), instead of s, the solution of the boundary problem for the equation (9.20) and the boundary conditions $\chi(N_0) = 0, \chi(N) = 1$:

$$[s(N) - s_{th}]^m = \frac{1}{N_c}\int_{N_0}^{N}(\kappa\Delta\sigma_\infty - \Delta\sigma_{th})^m \, d\nu \qquad (9.24)$$

When $\sigma_\infty^{max} > \sigma_p$, where σ_p is defined as in (9.18), the problem becomes more complicated. Due to irreversibility of damage, the transfer length l depends on the maximal stress σ_∞^{max} observed during the prehistory. That means that $\sigma_\infty^{max}(N)$ is to be entered in equation (9.9) as

$$\sigma_\infty^{max}(N) = \max\{\sigma_\infty(\nu), \nu \in [N_0, N]\}$$

Evidently, this involves the necessity of considering the loading process from the viewpoint of extremes of random processes. However, under steady cyclic loading when $\sigma_\infty^{max} = $ const, $\Delta\sigma_\infty = $ const for all cycles, the situation is much simpler. We can use equation (9.5), just replacing l by l_p. Then we come to equation

$$\omega_f(N) = 1 - \exp\left[-g_p(v_f)\frac{\sigma_\infty^{max}r}{2\tau_b l_c}\left(\frac{\kappa\Delta\sigma_\infty - \Delta\sigma_{th}}{s_c}\right)^\alpha\left(\frac{N - N_0}{N_c}\right)^\beta\right] \qquad (9.25)$$

under the condition that the ultimate shear stress τ_b is a deterministic constant. As to the boundary damage measure $\omega_b(N)$, it is defined by equation (9.12). More general equations were suggested in the book [19] and the survey paper [20].

9.4   Fracture Due to the Loss of Integrity

One of the fracture modes for unidirectional composites, both under monotonous and cyclic loading, is the loss of integrity due to the multiple rupture of fibers and/or damage of the fiber-matrix interface (Figure 9.1b). This mode was studied, both analytically and experimentally, by Bolotin and Tamuzs [12].

It is evident that the condition of such a type of fatigue can be formulated in terms of $\omega_f(N)$ (when the interface remains nondamaged) or in terms $\omega_f(N)$ and $\omega_b(N)$ (when the damage of the interface is significant). In the general case, we have to introduce the boundary with equation $f(\omega_f, \omega_b) = 1$ dividing the safe domain on the ω_f, ω_b plane from that where failure occurs. In the simplest case, the safety conditions are given as

$$\omega_f < \omega_f^*, \quad \omega_b < \omega_b^* \tag{9.26}$$

with two critical damage levels, ω_f^* and ω_b^*.

A question arises on the critical magnitudes ω_f^* and ω_b^* of microdamage measures. Elementary considerations show that these values are inversely proportional to the number n^* of elements neighboring a failed element, including this element, when the total failure is expected. For hexagonal arrays of fibers it is natural to suppose that $n^* = 7$, though on account of skewly positioned elements we have $n^* = 21$.

Another, perhaps more consistent, approach is to find the critical damage level following the damage history during the loading process. Consider a monotonous loading with applied stress σ_∞. The conditions of stability with respect to the loss of integrity are

$$\frac{\partial \omega_f}{\partial \sigma_\infty} > 0, \quad \frac{\partial \omega_b}{\partial \sigma_\infty} > 0 \tag{9.27}$$

The first violation of these conditions corresponds to a limit point in the space of σ_∞, ω_f and ω_b (Figure 9.4a). In fact, the ascending branch of the curve $\sigma_\infty = \sigma_\infty(\omega_f, \omega_b)$ corresponds to stable states, the descending one to unstable states. Similarly, in fatigue the conditions of stability are

$$\frac{\partial \omega_f}{\partial N} > 0, \quad \frac{\partial \omega_b}{\partial N} > 0 \tag{9.28}$$

This is illustrated in Figure 9.4b. Evidently, the conditions (9.27) and (9.28) taken with the equality sign give the magnitudes of critical damage measures for monotonous and cyclic loading, respectively.

It is expedient to compare the critical magnitudes of damage given in (9.27) and (9.28) with coarse estimates following from elementary geometrical considerations. It is also of interest to compare these magnitudes for the cases of monotonous and cyclic loading because they are expected to be very close. For simplicity, assume that the loss of integrity is controlled by ω_f only. Actually, even debonding of all fibers does not necessarily mean the total loss of load-carrying capacity in uniaxial tension.

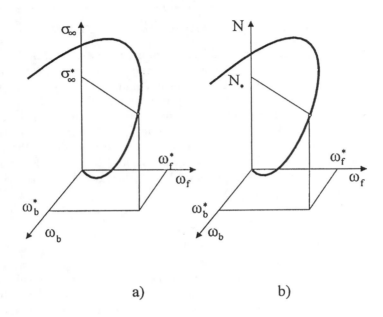

a)   b)

FIGURE 9.4
Damage measures in the function of applied stress in monotonous loading (a)
and of the cycle number in cyclic loading (b).

Some numerical examples are presented in Figures 9.5 and 9.6. The following material parameters are used for Figure 9.5: $E_f = 100$ GPa, $E_m = 10$ GPa, $\mu_m = 3$ GPa, $\tau_b = 50$ MPa, $s_r = 2$ GPa, $r = 10$ μm, $v_f = 0.8$. Figure 9.5a is drawn for $s_{th} = 100$ MPa, Figure 9.5b for $s_{th} = 250$ MPa. Lines 1-5 are depicted for $\alpha = 4, 6, 8, 10$, and 12, i.e., they relate to fibers with different strength distribution. Returning curves correspond to the fiber damage measure ω_f, monotonously ascending ones to the interface damage measure ω_b. The point of interest is that the curves are situated rather closely although the shape exponent α varies in a wide range. Due to the influence of the matrix, the scatter effect of fiber strength is not as significant as one could expect. The ultimate stress with respect to the critical condition $\partial\sigma_\infty/\partial\omega_f = 0$ varies in case Figure 9.5a from $\sigma_\infty^* = 634$ MPa at $\alpha = 4$ to $\sigma_\infty^* = 1082$ MPa at $\alpha = 12$. The corresponding critical damage measure varies from $\omega_f^* = 0.201$ to $\omega_f^* = 0.074$. In case Figure 9.5b we obtain, respectively, $\sigma_\infty^* = 716$ MPa and $\sigma_\infty^* = 716$ MPa, $\omega_f^* = 0.174$ and $\omega_f^* = 0.067$. As to the measure ω_b, its behavior is in agreement with the preliminary qualitative considerations. When the applied stress grows, this measure approaches unity. The critical stresses at $\omega_b = 1$ are practically the same as those with respect to ω_f.

Figure 9.6 is drawn for the case of cyclic loading at $\Delta\sigma_\infty = 250$ MPa, $R = 0.2$. The additional material parameters are: $\Delta\sigma_{th} = 100$ MPa in Figure 9.6a and $\Delta\sigma_{th} = 250$ MPa in Figure 9.6b. In both cases $N_0 = 0$, $N_c = 1$, $m = 6$. The same notation is used as in Figure 9.5. The abscissas, however, show the cycle number N. It is seen in Figure 9.6 that the critical damage is of the same order in both

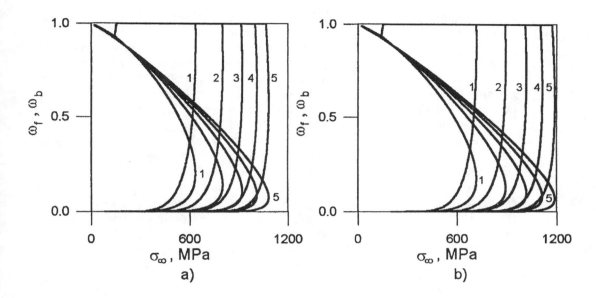

FIGURE 9.5
Damage measures versus the applied stress at $s_{th} = 100$ MPa (a) and $s_{th} = 250$ MPa (b) for shape exponents $\alpha = 4, 6, 8, 10, 12$ (lines 1-5).

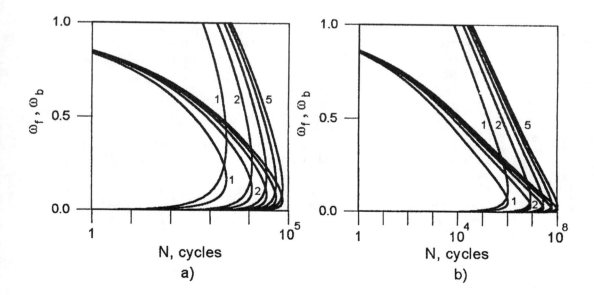

FIGURE 9.6
Damage measures versus the cycle number at $\Delta\sigma_{th} = 100$ MPa (a) and $\Delta\sigma_{th} = 250$ MPa (b) for shape exponents $\alpha = 4, 6, 8, 10, 12$ (lines 1-5).

cases, and the dependence on α is rather strong. The estimate $w_f^* = 0.1$ is suitable for preliminary analysis. Note that, as in monotonous loading, the critical state is attained when the debonding measure w_b becomes close to unity.

Complementary information is presented in Figure 9.7. There the influence of applied stresses on the fatigue damage is demonstrated at $\Delta\sigma_{th} = 100$ MPa, $R = 0.2, \alpha = m = 6$. Lines 1-5 are drawn for $\Delta\sigma_\infty = 200, 225, 250, 275, 300$ MPa. Plotting the stress range $\Delta\sigma_\infty$ against the critical cycle number, we obtain ordinary fatigue (Wöhler's) curves. A family of such curves is given in Figure 9.8. The lines 1-5 correspond to $\alpha = 4, 6, 8, 10, 12$.

The above numerical examples seem sufficiently instructive. Two conclusions might be drawn. First, the critical state of unidirectional fiber composites under tension along fibers might be identified with the limit points of $\sigma_\infty(w_f)$ and $N(w_f)$ relationships, i.e., without dealing with the debonding measure w_b. Second, the critical damage measure w_f^*, though depending on the shape exponent and the threshold stresses, does not vary significantly when these parameters vary in a wide range. In the considered examples we have observed that $0.07 \leq w_f^* \leq 0.20$ with $w_f^* = 0.1$ in the middle of the parameter domain. Hence, simplified analytical results are available both in case of monotonous and cyclic loading. Later on, to come to transparent equations, we use the critical condition of failure in the form $w_f = w_f^*$ with the given critical value w_f^*.

In the case of quasistatic monotonous loading the damage measure w_f as a function of the applied tensile stress σ_∞ is given in equation (9.13). Since $w_f^* \ll 1$, we may use the approximate equation (9.16). Then the ultimate stress resulting in the loss of integrity is estimated as

$$\kappa\sigma_\infty^* = s_{th} + s_c \left(\frac{w_f^*}{g_e(v_f)}\right)^{1/\alpha} \left(\frac{l_c}{r}\right)^{1/\alpha} \left(\frac{2\mu_m}{E_f}\right)^{1/2\alpha} \tag{9.29}$$

where the matrix is assumed to be deformed elastically. If damage of the matrix is taken into account, equation (9.16) is to be replaced with (9.17). Then instead of (9.29) we come to the equation with respect to σ_∞^*:

$$\kappa\sigma_\infty^* = s_{th} + s_c \left(\frac{w_f^*}{g_p(v_f)}\right)^{1/\alpha} \left(\frac{l_c}{r}\right)^{1/\alpha} \left(\frac{2\tau_b}{\sigma_\infty}\right)^{1/\alpha} \tag{9.30}$$

In the general case when both types of microdamage contribute to the loss of integrity, we have to use a limit condition connecting w_f and w_b such as $f(w_f, w_b, w_\infty) = 1$.

This approach remains valid when the loading is cyclic. Then instead of equations (9.15) and (9.16), the corresponding equations such as (9.22) and (9.25) are to be used. As above, we assume that $w_f^* \ll 1$. Then equation (9.22) takes the form

$$w_f(N) \approx \frac{l}{l_c} \left(\frac{\kappa\Delta\sigma_\infty - s_{th}}{s_c}\right)^\alpha \left(\frac{N - N_0}{N_c}\right)^\beta \tag{9.31}$$

The cycle number N^* at the loss of integrity is determined as

$$N^* = N_0 + N_c \left(w_f^*\right)^{1/\beta} \left(\frac{l_c}{l}\right)^{1/\beta} \left(\frac{s_c}{\kappa\Delta\sigma_\infty - s_{th}}\right)^m \tag{9.32}$$

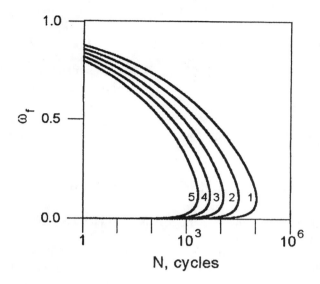

FIGURE 9.7
Damage measures in the function of the applied stress range and the cycle number at $\Delta\sigma_\infty = 100, 125, 150, 175, 200, 250$ MPa (lines 1-6).

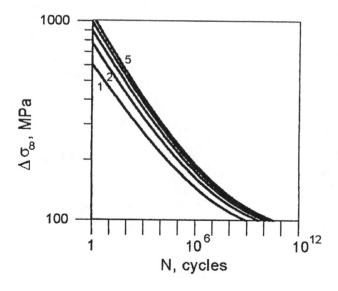

FIGURE 9.8
Wöhler's curves with respect to the loss of integrity for shape exponents $\alpha = 4, 6, 8, 10, 12$ (lines 1-5).

Depending on the matrix behavior, the fiber working length l is to be inserted in (9.32) equal to l_e or l_p.

Equation (9.32) is in fact an equation of the fatigue curve for a critical volume M_0 which is responsible for integrity of the structural member as a whole (see assumption 2 in Section 9.2). This equation looks like a deterministic one. As a matter of fact, it is valid "almost sure," precisely, with the probability $1 - e^{-1} \sim 1$. To obtain its probabilistic equivalent, we must come back to the modifications of equation (9.1) to account for fatigue damage.

Introduce the damage measure $\omega_f(N)$ similar to that in equation (9.11) as follows:

$$\omega_f(N) = \frac{1}{n_0} \sum_{k=1}^{n_0} \eta(N - N_k)$$

Here n_0 is the total number of microstructural elements in the considered domain M_0; N_k is the fatigue life of the k-th element, and $\eta(\cdot)$ is the Heaviside function. Suppose that at fixed N the number of ruptured elements follows the Bernoulli scheme and that the central theorem of probability theory is applicable. The asymptotic estimate for $\omega_f(N)$ has the form similar to (9.1)

$$F_\omega(\omega_f; N) = \Phi\left(\frac{\omega_f - F_0(N)}{F_0(N)[1 - F_0(N)]n_0^{-1/2}}\right) \tag{9.33}$$

As in (9.10), $\Phi(\cdot)$ is the normalized function of the Gaussian distribution.

The cumulative distribution function $F_0(N)$ of fatigue life for an arbitrarily chosen microstructural element can be determined when an equation of damage accumulation in elements and the distribution of their properties are given. In particular, with the use of equations (9.20) and (9.5) we obtain for the case $\Delta\sigma_{th} = s_{th}, \Delta\sigma_\infty = $ const that

$$F_0(N) = 1 - \exp\left[-\frac{l}{l_c}\left(\frac{\kappa\Delta\sigma_\infty - s_{th}}{s_c}\right)^\alpha \left(\frac{N - N_0}{N_c}\right)^\beta\right] \tag{9.34}$$

The right-hand side in (9.34) coincides with that in (9.22) which is not unexpected at all.

To obtain the cumulative distribution function $F_N(N)$ for the fatigue life of the critical volume, it is sufficient to note that

$$F_N(N) = \Pr\left\{\omega_f(N) > \omega_f^*\right\}$$

Then, using (9.33) we find:

$$F_N(N) = 1 - \Phi\left\{\frac{\omega_f^* - F_0(N)}{F_0(N)[1 - F_0(N)]n_0^{-1/2}}\right\} \tag{9.35}$$

The total number n_0 in the reference domain M_0 is, as a rule, very large compared with unity, and the variance of distributions (9.33) and (9.35) is very low. Therefore, the deterministic estimate $F_N(N) \approx \omega_f^*$ follows, which means the return to equation (9.32).

Several questions remain open after the above presentation. First, there is the question of numerical data for all material parameters. Some data can be extracted from the available results of standard tests. The magnitude of other parameters has to be evaluated in the future. The second question concerns the size effect on fatigue life. The size effect for fibers is hidden in equations for $\omega_f(N), \omega_b(N)$ and $F_0(N)$ where the ratio l/l_c or l/r is always present. The influence of the size of the domain M_0 is presented in (9.33) and (9.35) with the total number n_0 of microstructural elements in this domain. At last, the size of structural components, test specimens, etc. will enter when we estimate their fatigue life. Following assumption 2 in Section 9.2., we postulate that the cumulative distribution function $F_N(N)$ for the whole body is connected with those for its critical parts $F_N^{(1)}(N), \ldots, F_N^{(m)}(N)$ as

$$F(N) = 1 - \prod_{k=1}^{m} \left[1 - F_N^{(k)}(N) \right]$$

Thus, we observe the size effect in three capacities: at the level of fibers, at the level of critical domains, and at the level of the whole body. The first and the third effects are similar to the weakest-link effect while the second one exhibits an opposite tendency. However, the latter effect is, as a rule, negligible at $n_0 \gg 1$.

9.5 Macrocrack Initiation

The simplest model of a nucleus of a macroscopic crack is a cluster of microcracks formed around one of the ruptured elements. To realize such a model, we must consider various ways to initiate a macroscopic crack. The probabilities of all these ways must be identified and the stress concentration factors for each sequence of local ruptures must be assessed. The experience shows that this scheme is easy to realize for very simple models only, for example, when a row of single fibers with the attached matrix interlayer is considered. Even in the framework of such a model, a number of uncertainties arise concerning, for example, the evaluation of stress concentration and size effect when several neighboring fibers are ruptured in various cross sections. In addition, one must use cumbersome combinatorical calculations to build up a complete probabilistic space of events.

All these difficulties can be avoided when we limit ourselves to the objective of estimating fatigue life by the order of magnitude [11]. The simplest assumption is that a macrocrack initiates when $n_* - 1$ elements neighboring a single initially ruptured element are subjected to rupture simultaneously. The average number of nuclei in the considered domain is $\omega_f(N)n_0$. Here n_0 is the total number of elements in the reference domain M_0. Then the cumulative distribution function for the number N_* at the macrocrack initiation may be evaluated as follows:

$$F_*(N_*) = 1 - \left[1 - \omega_f^{n_* - 1}(N_*) \right]^{\omega_f(N_*)n_0}$$

As $N_* \gg 1$, $\omega_f(N_*) \ll 1$, an asymptotic approximation is natural to use following the spirit of the extreme value theory. Then we came to the Weibull type distribution

$$F_*(N_*) \approx 1 - \exp\left[-n_0 \omega_f^{n_*}(N_*)\right] \tag{9.36}$$

The drawbacks of the above heuristic model are evident. In reality, the formation of a macrocrack nucleus is not a single moment act. The fibers participating in this act are damaged in different degrees. The stress concentration and the "working" lengths of fiber segments vary from one fiber to another, etc. However, the final result given in (9.40) is very simple and intuitively understandable.

For example, let the applied loading be steady and, therefore, equation (9.21) be applicable. At $\omega_f(N) \ll 1$ we simplify this equation as follows:

$$\omega_f(N) = \frac{l}{l_c}\left(\frac{\kappa \Delta \sigma_\infty - \Delta \sigma}{s_c}\right)^\alpha \left(\frac{N - N_0}{N_c}\right)^\beta$$

Substituting in (9.36), we obtain:

$$F_*(N_*) = 1 - \exp\left[-n_0\left(\frac{l}{l_c}\right)^{n_*}\left(\frac{\kappa \Delta \sigma_\infty - \Delta \sigma_{th}}{s_c}\right)^{n_*\alpha}\left(\frac{N - N_0}{N_c}\right)^{n_*\beta}\right] \tag{9.37}$$

Equation (9.37) is similar to (9.34) for the fatigue life of a single element. Two distinctions are to be noted. First, the total number of the elements n_0 in the considered domain enters equation (9.37). When n_0 is related to the reference domain M_0 of a body, and total domain is M, n_0 is to be replaced with $n_0 M/M_0$. In the case of nonhomogeneity a natural generalization of (9.37) follows:

$$F_*(N_*) = 1 - \exp\left[-\int\limits_M\left(\frac{l}{l_c}\right)^{n_*}\left(\frac{\kappa \Delta \sigma_\infty - \Delta \sigma_{th}}{s_c}\right)^{n_*\alpha}\left(\frac{N - N_0}{N_c}\right)^{n_*\beta}\frac{dM}{M_0}\right]$$

Equation (9.37) and its generalizations describe the size effect with respect to the life until the formation of the first macroscopic crack. The exponents characterizing this effect differ from those for the fatigue life of single fibers, and this is the second point of difference between equations (9.37) and (9.34).

The measure of the size effect with respect to the cycle number at the macrocrack formation can be introduced in terms of the mean cycle number. When equation (9.37) is valid we have

$$E(N_*) = N_0 + N_c n^{-1/n_*\beta}\left(\frac{l}{l_c}\right)^{1/\beta}\left(\frac{s_c}{\kappa \Delta \sigma_\infty - \Delta \sigma_{th}}\right)^m \Gamma\left(1 + \frac{1}{n_*\beta}\right)$$

Here $n = n_0(M/M_0)$. That means that at $N_0 \ll N_c$ the ratio of the expected fatigue lives counted until the first macrocrack formation is

$$\frac{E[N_{*1}]}{E[N_{*2}]} \approx \left(\frac{M_2}{M_1}\right)^{\frac{1}{n_*\beta}} \tag{9.38}$$

A similar result can be obtained by comparison of the cycle numbers corresponding to a given probability, i.e., by comparison of the roots N_* of equation $F_*(N_*) = p$ at $M_2 \neq M_1$.

It is seen from equation (9.38) that the size effect with respect to macrocrack formation essentially depends on the number n_* of fractured elements identified with an initiating macrocrack. In thin plates, sheets and plies it may be $n_* = 3$ and more. Larger numbers, beginning from $n_* = 6$, are expected for internal cracks. The conclusion depends on the properties of fibers, matrix, and interface. However, equation (9.37) and its generalizations result in fatigue curves for fiber reinforced composites under the condition that the initiation even of a single macroscale crack is interpreted as a structural failure. For example, the condition $F_*(N_*) = p$, where $F_*(N_*)$ is given in (9.37), gives:

$$ N = N_0 + N_c n_0^{-1/n_* \beta} \left(\frac{l_c}{l} \right)^{1/\beta} \left(\frac{s_c}{\kappa \Delta \sigma_\infty - \Delta \sigma_{th}} \right)^m \left[\log \frac{1}{1-p} \right]^{1/n_* \beta} $$

Both macroscopic crack initiation and failure due to the loss of integrity are caused by the same mechanisms: fiber ruptures, matrix damage, and debonding of fibers. It is easy to estimate the conditions under which a certain type of failure becomes more probable. As it follows from (9.36), the macrocrack initiates with probability $1 - e^{-1}$ at $\omega_f \approx n_0^{-1/n_*}$. When $n_0 = 10^6$, $n_* = 6$ this gives $\omega_f \approx 10^{-1}$. The loss of integrity takes place when the damage density attains the same order of magnitude. But when $n_0 = 10^{12}$, a similar estimate gives $\omega_f \approx 10^{-2}$ for the beginning of macrocrack initiation. In the latter case, the failure due to crack propagation is much more probable than the loss of integrity under the obvious condition that no notches and other macroscale stress concentrators are present in a body.

Some other details concerning probabilistic models of fatigue crack initiation in terms of probabilistic models can be found in the book by Bolotin [19].

9.6   Types of Cracks in Unidirectional Fiber Composites

Several types of cracks are met in unidirectional fiber composites with comparatively weak matrix. We begin the discussion from the case of fracture under monotonous quasistationary tension in the fiber direction. We will take into account the size effect of fiber strength and, as far as possible, the effect of stress concentration at the crack front. In fatigue analysis, we also include in consideration the dispersed damage accumulated in the far field. Some of the listed factors will be neglected in simplified versions of the model; however, the size effect of fiber strength remains the most important factor that enters into all stages of analysis.

Under certain initial conditions, a crack begins to grow in a transverse mode similar to the ordinary mode I crack. It is schematically depicted in Figure 9.9a, where a crack in a unidirectional sheet propagates from a notch. Another mode is shown in Figure 9.9b. There the initial flaw is a narrow transversal slit. The crack starts to grow in a longitudinal mode mostly because of fiber debonding, pulling-out and ply delamination. Later on, we will say transverse and longitudinal modes (T-mode and L-mode), respectively. Some of these cracks, in the process of their development, are approaching the brush-like cracks that combine, in an equal degree, both fiber rupture and debonding. For natural conditions, when a crack formation takes place as a result of interaction of multiple damage, the brush-like mode is typical. This mode briefly notated B-mode is schematically shown in Figure 9.9c.

The type of fracture depends on many factors among which is the resistance against debonding. This resistance is characterized with the ultimate stress τ_b. When τ_b is small, fracture in the L-mode is expected. At high τ_b fracture may occur in a conventional the T-mode. Intermediate domain is occupied by the B-mode (Figure 9.10). Along with material properties, the shape of initial notches also influences on the fracture mode. It seems paradoxical, but sharp notches may develop in longitudinal cracks while a crack initiating from a blunt notch may be a source of a transverse crack. In any case, most cracks are transforming in the process of their propagation in the B-mode, and the final failure often occurs as a combination of cracking along fibers, splitting and delaminations, i.e., in the B-mode. The final failure in the L-mode is also frequently observed in laboratory experimentation (Figure 9.9d).

The virtual work of external and internal forces may be presented as follows

$$\delta W_e + \delta W_i = G_T \delta a + G_L \delta L \tag{9.39}$$

This equation is valid both for T- and/or L-modes where δa and δL are independent variations of transverse and longitudinal dimensions of the fractured domains, respectively. The virtual fracture work can be presented in a similar form as

$$\delta W_f = -(\Gamma_T \delta a + \Gamma_L \delta L) \tag{9.40}$$

The type of fracture depends on the relationship between the generalized forces. At

$$G_T = \Gamma_T, \quad G_L < \Gamma_L$$

a crack begins to grow in the T-mode. When

$$G_T < \Gamma_T, \quad G_L = \Gamma_L$$

it begins to grow in the L-mode. In the case of the B-mode, the dimensions a and L are connected. As a result, equations (9.39) and (9.40) take the form

$$\delta W_e + \delta W_i = G_B \delta a, \quad \delta W_f = -\Gamma_B \delta a \tag{9.41}$$

The equilibrium condition for B-mode cracks takes the form

$$G_B = \Gamma_B \tag{9.42}$$

A simplified model of the brush-like crack is presented in Figure 9.11. A crack in a plate or a ply under tension along fibers is treated as a three-dimensional domain with width $2a$, length $2L$ and unity thickness. The sizes of this domain

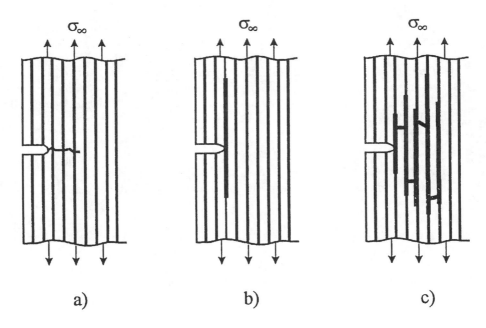

FIGURE 9.9
Cracking of unidirectional fiber composite in tension along fibers.

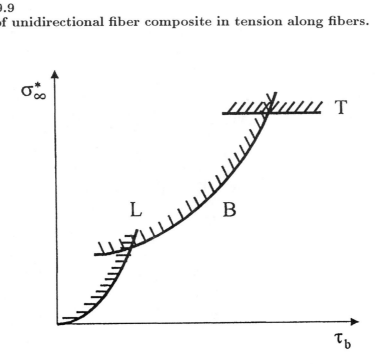

FIGURE 9.10
Influence of the ultimate debonding stress on the mode of fracture.

are small compared with the dimensions of the plate. The composite is assumed macroscopically homogeneous. In the cracked domain the material does not carry any loading. The situation is similar to that observed near a single rupture of fibers. The load $2a\sigma_\infty$ from the cracked domain is to be transferred with shear stresses distributed on the surface $2aL$. These stresses are a kind of ultimate debonding stress on the longitudinal boundary of the cracked domain. We assume the shear stresses, are uniformly distributed and equal to a material constant τ_b related to τ_b in (9.9), (9.15), etc. The transfer length is determined as

$$L = g_p(v_f)a(\sigma_\infty/\tau_b) \tag{9.43}$$

where the multiplier $g_p(v_f)$ is similar to $g_p(v_f)$ in (9.9).

The strain energy of a composite plate under the uniformly applied stress σ_∞ is

$$U = \text{const} - \frac{\sigma_\infty^2 A}{2E_r} \tag{9.44}$$

where A is the cracked area which is treated as a source of strain energy release, E_r is the effective Young modulus in the fiber direction, i.e.,

$$E_r = v_f E_f + v_m E_m \tag{9.45}$$

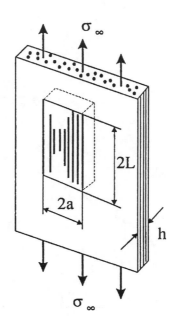

FIGURE 9.11
Schematization of the internal crack in a plate of unidirectional fiber composite in tension along fibers.

The energy in (9.44), as well as other energy variables, are referred to a unit of plate thickness and to a half of the plate, for example, to the right-hand half, $x \geq 0$ (Figure 9.11). The constant in the right-hand side of (9.44) is associated with the strain energy of the plate without the crack.

The question of how to estimate the area A is not at all simple. Half of the damaged area is approximately equal to $2aL$. Evidently, this does not include the limit case of an ordinary crack when $L \to 0$. To include this extremal case, we must increase the area by the term proportional to a^2. The contribution of this area also depends on anisotropy, in particular, on the ratio of elastic moduli, E_y/E_x. In further analysis, dealing mostly with brush-like cracks, we neglect this contribution. Then

$$A = 2aL \tag{9.46}$$

Considering (9.44) and (9.46), the generalized driving forces are

$$G_a = \frac{\sigma_\infty^2 L}{E_r}, \quad G_L = \frac{\sigma_\infty^2 a}{E_r} \tag{9.47}$$

For brush-like cracks, taking into account equations (9.39), (9.40), and (9.46), we obtain

$$G_B = \frac{2g_p(v_f)\sigma_\infty^3 a}{E_r \tau_b} \tag{9.48}$$

When, instead of (9.46), the area of energy release is determined as $A = 2aL + Y^2 a^2$ with the form-factor Y, we obtain that

$$G_a = \frac{\sigma_\infty^2 (L + Y^2 a)}{E_r}, \quad G_L = \frac{\sigma_\infty^2 a}{E_r}, \quad G_B = \frac{\sigma_\infty^2 a}{E_r}\left(\frac{2g_p(v_f)\sigma_\infty}{\tau_b} + Y^2\right)$$

9.7   Evaluation of the Fracture Work

To establish conditions of crack growth, transformation, and instability, we need an analytical approach to the evaluation of generalized resistance forces for fracture modes inherent to unidirectional fiber composites. The proposed model of damage and fracture requires more information on material properties than the models discussed in the previous chapters. It is evident that the statistical scatter and scale effect of fiber strength has the same origin both in quasistatic and cyclic loading. This origin is the microscale defects distributed along the fiber axis. Therefore, the same or closely related random variables are involved in the models of fracture under monotonous loading and in the models of fatigue. When the fatigue crack front approaches a certain bundle of fibers, they are already damaged and partially ruptured. This requires a more accurate account of differences between the far field and near field phenomena that can be done using proper conditional distributions.

The virtual fracture work for any model discussed in Section 9.6 may be presented as follows:

$$\delta W_f = -(v_f \gamma_f \delta a + v_m \gamma_m \delta a + 2\gamma_b \delta L) \tag{9.49}$$

This work is referred to a unit of plate thickness and half of the cracked domain, say for $x \geq 0$. The first term in the right-hand side of (9.49) takes into account the fracture work of fibers. The second one accounts for the matrix under the condition that any fiber and the attached matrix on the crack boundary are considered failures when even a single cross section of the fiber is ruptured. Therefore, the increments $v_f \delta a$ and $v_m \delta a$ of the fibers and matrix summed cross sections enter equation (9.49). The last term is the virtual work of fiber debonding and pulling-out. The specific work γ_b is referred to the smoothed surface of the crack front with the virtual increment $2\delta L$. Certainly, the magnitude γ_f for macrocracks may differ from that for single fibers.

The specific fracture work γ_f is a random variable. Hence, the magnitude γ_f averaged upon the newborn cross section is to be inserted in equation (9.48) accounting for the size effect of fiber strength. In the beginning, to avoid a number of additional assumptions about probabilistic properties of fibers, we postulate the following deterministic relation between the short-time ultimate tensile strength s and the specific fracture work for fibers:

$$\gamma_f = \frac{l_\gamma s^2}{E_f} \tag{9.50}$$

Here l_γ is a length parameter. Equation (9.50) follows just from dimensional considerations. Hence, assuming the probabilistic distribution for s, we easily obtain the corresponding distribution for γ_f. In particular, the average specific fracture work for a set of fibers with the working half-length l is

$$E[\gamma_f] = \frac{l_\gamma}{E_f} \int\limits_0^\infty p(s)s^2 \, ds \tag{9.51}$$

Here $p(s)$ is the probability density function corresponding to the cumulative distribution function (9.5). Introduction of (9.5) into (9.51) gives

$$E[\gamma_f] = \frac{l_\gamma}{E_f} \left[s_{th}^2 + 2s_{th}s_c \left(\frac{l_c}{l}\right)^{1/\alpha} \Gamma\left(1 + \frac{1}{\alpha}\right) + \right.$$
$$\left. + s_c^2 \left(\frac{l_c}{l}\right)^{2/\alpha} \Gamma\left(1 + \frac{2}{\alpha}\right) \right] \tag{9.52}$$

when $\Gamma(\cdot)$ is gamma function (do not mix it with the similarly notated generalized resistance force).

The size effect of fiber fracture toughness is illustrated in Figure 9.12. There the average fracture work given in (9.52) and denoted simply γ_f is plotted versus the working length l. The following numerical data are used: $s_c = 1$ GPa, $s_{th} = 100$ MPa, $E_f = 100$ GPa, $l_c = l_\gamma = 1$ mm. Lines 1–5 are drawn for the shape exponents $\alpha = 4, 6, 8, 10, 12$. At $\alpha = 4$ the average specific fracture work $E[\gamma_f]$ varies on an order of magnitude when the fiber length increases from $l = 0.1$ mm

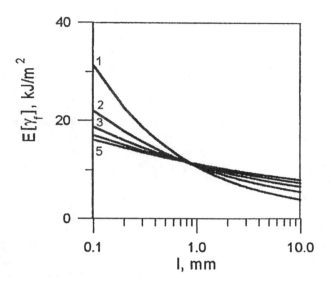

FIGURE 9.12
Average specific fracture work for the fibers as a function of the working length at $\alpha = 4, 6, 8, 10, 12$ (lines 1–5, respectively).

to 10 mm. At $\alpha = 12$ the change of $E[\gamma_f]$ is not so significant. Of course, different α correspond to different material properties.

When a part of the fibers is already ruptured under the quasistatically applied stress σ_∞, we have to estimate the conditional average for the specific fracture work:

$$E[\gamma_f | \kappa \sigma_\infty] = \frac{l_\gamma}{E_f} \int\limits_{\kappa \sigma_\infty}^{\infty} p(s) s^2 ds$$

Instead of (9.52) we obtain

$$E[\gamma_f | \kappa_t \sigma_\infty] = \frac{l_\gamma}{E_f} \left[s_{th}^2 \exp(-u_\infty) + 2 s_{th} s_c \left(\frac{l_c}{l} \right)^{1/\alpha} \Gamma\left(1 + \frac{1}{\alpha}, u_\infty\right) + \right.$$

$$\left. + s_c^2 \left(\frac{l_c}{l} \right)^{2/\alpha} \Gamma\left(1 + \frac{2}{\alpha}, u_\infty\right) \right]$$

(9.53)

Here the incomplete gamma function enters

$$\Gamma(n, u) = \int\limits_{u}^{\infty} x^{n-1} \exp(-x) dx$$

In addition, the notation is introduced for brevity

$$u_\infty = \frac{l}{l_c}\left(\frac{\kappa\sigma_\infty - s_{th}}{s_c}\right)^\alpha \tag{9.54}$$

Now we can approach a consistent interpretation of the specific fracture work γ_f in equation (9.49). Let the cracked zone length be $2L$, and the working fiber length be L. In general, the tensile stress in fibers located at the crack front is $\kappa_t\sigma_\infty$ where the stress concentration factor κ_t depends on the crack geometry, the ratio L/a in particular.

Stress concentration at the crack front in unidirectional composites with ductile matrices was studied by many authors beginning with Hedgepeth (1961) who considered stress distribution ahead of a transverse crack in one-dimensional row of fibers. Later on, it was shown that dispersed fiber damage essentially affects the stress distribution near the crack. Interaction between the initial notch and several isolated fiber fractures ahead of the notch results in the reduction of stress concentration along the crack path. This effect was studied by Tsai and Dharani [148]. An additional relaxation effect is produced by the damage of the fiber-matrix interface. When the size effect of fiber strength is high enough, the increase of the "naked," working segment of fibers results in the diminishing of their load-carrying capacity. This effect may be more significant than the increase of acting stresses in fibers due to stress concentration. When the size strength effect dominates, one may neglect the stress concentration; moreover, the general picture of composites in the presence of dispersed fiber ruptures and delaminations is very vague. Such an approach to modeling of damage in unidirectional fiber composites was suggested by Bolotin (1976).

In the presented analysis of brush-like fracture, we assume that the inherent stress concentration factor κ_0 is constant, say $\kappa_0 = 4/3$ for the first row of fibers at the crack front. Then

$$\kappa_t = \frac{\kappa_0}{v_f[1 + (E_m v_m/E_f v_f)](1 - \omega_f)} \tag{9.55}$$

The denominator in (9.55) is the same as in (9.14); it takes into account the load distribution between fibers and matrices as well as the diminishing of the net cross section due to damage. Anyway, before the crack advancement a part of the fibers are already ruptured, namely those which ultimate stress $s \le \kappa_t\sigma_\infty$. This means that the residual fracture work of fibers is determined by equation (9.53). In terms of the specific fracture work this equation takes the form

$$E[\gamma_f|\kappa_t\sigma_\infty] = \gamma_{th}\exp(-u_t) + 2(\gamma_{th}\gamma_c)^{1/2}\left(\frac{l_c}{L}\right)^{1/\alpha}\Gamma\left(1 + \frac{1}{\alpha}, u_t\right) +$$

$$+ \gamma_c\left(\frac{l_c}{L}\right)^{2/\alpha}\Gamma\left(1 + \frac{2}{\alpha}, u_t\right) \tag{9.56}$$

where u_t is determined as in (9.54) at $l = L, \kappa = \kappa_t$. The notation is also used for the threshold fracture work γ_{th} and its characteristic magnitude γ_c:

$$\gamma_{th} = l_\gamma s_{th}^2/E_f, \quad \gamma_c = l_\gamma s_c^2/E_f \qquad (9.57)$$

Equation (9.56) looks too cumbersome. In the paper [38] a simplified version of this equation is used. As no sufficient information is available about the relationship between the ultimate stress and the specific fracture work, we replace equation (9.51) as follows:

$$\gamma = \frac{l_\gamma(s - s_{th})^2}{E_f}$$

Then only the last term remains in the right-hand side of (9.56). Evidently, the same result will be obtained at $s_{th} = 0$. Another simplification is to neglect the difference between conditional and unconditional distributions. Actually, when $\kappa_t\sigma_\infty \ll \sigma_c$, and $\alpha \gg 1$, the difference between $E[\gamma_f]$ and $E[\gamma_f|\kappa_t\sigma_\infty]$ is small. At high stresses and high strength scatter, the discrepancy between conditional and nonconditional mean values becomes rather large. There is a way to take into account this discrepancy using the concept of damage measure; moreover, such an approach is in the spirit of the theory of fatigue crack growth. In monotonous loading, one may replace equation (9.56) with its simplified version, for example,

$$E\left[\gamma_f|\kappa_t\sigma_\infty, l\right] = \gamma_c \left(\frac{l_c}{l}\right)^{2/\alpha} \Gamma\left(1 + \frac{2}{\alpha}\right) \exp\left[-\frac{l}{l_c}\left(\frac{\kappa_t\sigma_\infty - \sigma_{th}}{s_c}\right)^\alpha\right] \qquad (9.58)$$

The exponential multiplier in (9.58) is equal to the relative portion of fibers at the crack front which are not yet fractured due to quasistatic tension.

9.8   Stability with Respect to Brush-Like Fracture

Consider fracture of a wide plate or a ply of unidirectional fiber composite under monotonous quasistationary tension in the fiber direction. Assume that a plate or ply contains a brush-like crack similar to the central mode I crack in ordinary materials (see Figure 9.11). The origin of this crack is arbitrary: it may be a transformed T-mode crack as well as a result of the coalescence of several local fiber ruptures associated with the damage of matrix and debonding of fibers. As above, we schematize the crack as a rectangular domain with the length $2L$ along fibers, the width $2a$ across fibers. The thickness of the crack domain is equal to the plate thickness which is taken as equal to unity.

The crack does not propagate when $G_B < \Gamma_B$ where G_B is given in (9.48). As to Γ_B, we have to use equation (9.49) together with the relationship (9.43) between the half-length L of the cracked domain, and its half-width a. Then equation (9.49) gives

$$\Gamma_B = v_f\gamma_f + v_m\gamma_m + 2\gamma_b g_p(v_f)\left(\frac{\sigma_\infty}{\tau_b}\right) \qquad (9.59)$$

where the fiber fracture work, notated briefly as γ_f, is to be determined by equation (9.56) or, maybe, (9.58). At $G_B = \Gamma_B$ an equilibrium state (in Griffith's sense) is attained. This state is unstable. Actually, it is seen from (9.47) that $\partial G_B / \partial a > 0$ for all a. On the other side, equations (9.56) and (9.58) distinctly show that $\partial \Gamma_B / \partial L < 0$. Therefore, $\partial \Gamma_B / \partial a < 0$ for all a. Thus, the driving force is increasing during the crack propagation, while the resistance force is decreasing (by the way, due to the size effect of fiber strength). This means that, when a brush-like crack attains an equilibrium state, the total failure occurs if, of course, the loading conditions are not subjected to a favourable change during this process.

Let us analyze the equilibrium condition (9.42). After elementary transformations we come to equation

$$a = \frac{E_r \tau_b}{2 g_p(v_f) \sigma_\infty^3} \left(v_f \gamma_f + v_m \gamma_m + 2 \gamma_b g_p(v_f) \frac{\sigma_\infty}{\tau_b} \right) \tag{9.60}$$

The size a of the crack domain enters both sides of (9.60). The solution of this equation gives the critical crack size a^*.

The relationship between a^* and σ_∞ in the critical state depends on the ratio between material parameters. We are interested in the power exponent p in the relationship $a^* \sim \text{const} \cdot \sigma_\infty^{-p}$. For example, when the fracture toughness of the composite is controlled only by fibers, we obtain that

$$p = \frac{3 + 2/\alpha}{1 + 2/\alpha} \tag{9.61}$$

Then at $\alpha = 2$ we have $p = 2$, i.e., we obtain, as in classical fracture mechanics, that $a^* \sim \text{const} \cdot \sigma_\infty^{-2}$. When α increases, the right-hand side in (9.61) tends to $p = 3$. Hence in the case of dominating of fiber fracture toughness we have $2 \leq p \leq 3$ for all $\alpha \geq 2$. When the resistance to debonding becomes high, the last term in the brackets of (9.60) dominates. Then $a^* \sim \text{const} \cdot \sigma_\infty^{-2}$ as in the classical case.

Some results of direct computations are presented in Figure 9.13. Figure 9.13a illustrates the effect of the shape exponent α on the critical size a^* as a function of the applied stress σ_∞. Lines 1-5 are drawn for $\alpha = 4$, 6, 8, 10, 12. Other material parameters $E_f = 100$ GPa, $E_m = 10$ GPa, $v_f = 0.8$, $\gamma_c = 10$ kJ/m^2, $\gamma_{th} = 0.1$ kJ/m^2, $\gamma_m = \gamma_b = 1$ kJ/m^2, $\tau_b = 50$ MPa, $l_\gamma = 1$ mm. As is seen in Figure 9.13a, the size effect is not significant, even in the area of low stresses when the critical crack size is measured in millimeters. Note that the slope exponent is close to $p = 2$ for all considered α.

Another item to discuss is the effect of the fiber fracture toughness. Figure 9.13b is plotted for $\alpha = 6, \gamma_b = 0.2$, 0.4, 0.6, 0.8, 1.0 kJ/m^2 (lines 1-5). Other material parameters are the same as in the above example. The effect of the fiber-matrix interface quality is less significant.

In the author's earlier publication [11] an internal brush-like crack in a massive bulk of composite was considered. The crack was modeled as a circular cylindrical domain with radius a and half-length L. Elementary considerations similar to those resulting in equation (9.43) lead to the relationship between L and a:

$$L = g_p(v_f) a \frac{\sigma_\infty}{2 \tau_b}$$

Instead of (9.44) we obtain

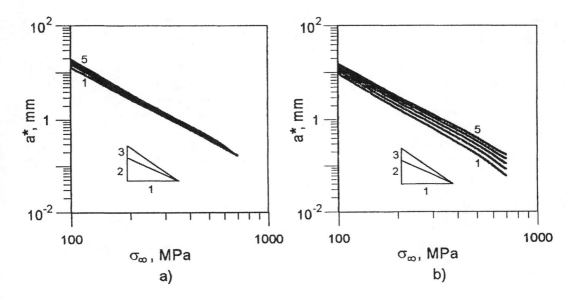

FIGURE 9.13
Critical size of the brush-like crack as a function of applied stress: (a) at shape exponent $\alpha = $ 4, 6, 8, 10, 12 (lines 1-5); (b) at specific fracture work $\gamma_b = $ 0.2, 0.4, 0.6, 0.8, 1.0 kJ/m^2 (lines 1-5).

$$U = \text{const} - \frac{\pi \sigma_\infty^2 a^2 L}{E_r}$$

The generalized driving force is

$$G_B = \frac{3\pi g_p(v_f)\sigma_\infty^3 a^2}{2 E_r \tau_b} \tag{9.62}$$

The generalized resistance force, instead of equation (9.59), is determined as

$$\Gamma_B = 2\pi a \left(v_f \gamma_f + v_m \gamma_m + 2\gamma_b g_p(v_f)\frac{\sigma_\infty}{\tau_b} \right) \tag{9.63}$$

Combining (9.62) and (9.63), we come to the equation with respect to the critical crack radius a^*:

$$a^* = \frac{4 E_r \tau_b}{3 g_p(v_f)\sigma_\infty^3} \left(v_f \gamma_f + v_m \gamma_m + 2\gamma_b g_p(v_f)\frac{\sigma_\infty}{\tau_b} \right)$$

It is evident that this equation differs from (9.60) only in the multipliers before the brackets.

9.9   Fatigue Brush-Like Crack Propagation

When the applied loading is cyclic, we have to include in consideration several mechanisms: damage accumulation due to cyclic loading in the far field; the same in the near field, for example, in the row of fibers located directly on the crack front; instantaneous, quasistatic fracture due to maximal stresses in the far field; and the same for the first-row fibers. In addition, the influence of microdamage on the bulk compliance of the composite cannot be neglected, either. To begin with the latter factor, we just replace the effective Young's modulus E_r given in (9.22) as follows:

$$E_r = v_f(1 - \omega_f)E_f + v_m(1 - \omega_m)E_m \qquad (9.64)$$

Here ω_f and ω_m are damage measures in the far field. In the framework of the microstructural model we connect them with the probabilities of single fiber ruptures. In particular, following equation (9.12), we assume that

$$\omega_f = \omega_m = 1 - \exp\left[-\frac{l}{l_c}\left(\frac{\kappa\Delta\sigma_\infty - \Delta\sigma_{th}}{s_c}\right)^\alpha \left(\frac{N - N_0}{N_c}\right)^\beta\right]$$

where, depending on the matrix behavior, we insert the length l either as given in (9.8) or in (9.9).

To evaluate the generalized resistance force, we have to take into account all the types of damage, far field and near field, produced by cycle loading or by single overloading. Not to get too deep into arising difficulties, let us assume that

$$\Gamma_B = (v_f\gamma_f + v_m\gamma_m)(1 - \psi_i - \psi_f) + 2\gamma_b g_p(v_f)(\sigma_\infty^{max}/\tau_b)(1 - \psi_b) \qquad (9.65)$$

This equation is similar to other formerly used equations for generalized resistance forces. Equation (9.65) makes a distinction between various types of resistance as well as between various types of damage. The quantities $v_f\gamma_f + v_m\gamma_m$ and γ_b relate to nondamaged material. The damage measure ψ_i takes into account the instantaneous rupture of fibers and matrix produced by maximal stresses. We assume that this measure is equal to the probability of failure for fibers with working length L loaded with stress $\kappa_t\sigma_\infty$:

$$\psi_i = 1 - \exp\left[-\frac{a\sigma_\infty}{l_c\tau_b}\left(\frac{\kappa_t\sigma_\infty - s_{th}}{s_c}\right)^\alpha\right]$$

The measure ψ_f, as usual, takes into account the purely fatigue damage that is divided into the far field damage and near field damage. For the sake of simplicity, we will not individualize the far field damage with respect to each fiber describing the far field damage with the averaged measure $\omega_f(N)$. The model of this damage is discussed in detail in Section 9.3. Under stationary loading, the far field damage measure $\omega_f(N)$ is given in equation (9.24) and its generalizations. As to the near field damage, we use the quasistationary approximation to evaluate its contribution into the total damage.

The equation of continuous growth for brush-like cracks has the form

$$G_B(\sigma_\infty^{max}, a) = \Gamma_B(\psi_i, \psi_f, \psi_m, \psi_b) \tag{9.66}$$

where the driving force is taken according to (9.47), and the resistance force according to (9.65). The right-hand side in (9.66) contains four damage measures, ψ_i, ψ_f, ψ_m and ψ_b . Two latter ones are connected with ψ_f.

To evaluate the damage measure ψ_f, we return to equations (9.20) and (9.21). The crack front arrives at a certain row of fibers at the cycle number N and leaves, due to crack propagation, at the cycle number $N + \Delta N$. The damage measure for a single fiber is

$$\chi_t(N) = \omega_f(N) + \int\limits_{N}^{N+\Delta N} \left(\frac{\kappa_t \Delta \sigma_{th} - \Delta \sigma_{th}}{s - s_{th}} \right)^m d\nu$$

and the damage measure for the assembly of all the first-row fibers, as in (9.21), is

$$\psi_f(N) = \Pr\left\{ s \leq s_{th} + \frac{1}{(1 - \omega_f)^{1/m}} \left[\int\limits_{N}^{N+\Delta N} (\kappa_t \Delta \sigma_\infty - \Delta \sigma_{th})^m d\nu \right]^{1/m} \right\}$$

Hence, taking into account the cumulative distribution function (9.5), we obtain

$$\psi_f(N) = 1 - \exp\left\{ -\frac{1}{(1 - \omega_f)^\beta} \frac{a\sigma_\infty^{max}}{l_c \tau_b} \times \right.$$

$$\left. \times \left[\int\limits_{N}^{N+\Delta N} \left(\frac{\kappa_t \Delta \sigma_\infty - \Delta \sigma_{th}}{s_c} \right)^m d\nu \right]^\beta \right\} \tag{9.67}$$

where the working half-length is inserted according to (9.43), and the notation (9.23) for exponent β is used.

When a crack is propagating slowly, we may neglect the change of the integrand in (9.67) during the segment $[N, N + \Delta N]$. Then we come to equation similar to (9.22):

$$\psi_f(N) = 1 - \exp\left\{ -\left(\frac{\Delta N}{1 - \omega_f} \right)^\beta \frac{a\sigma_\infty^{max}}{l_c \tau_b} \left(\frac{\kappa_t \Delta \sigma_\infty - \Delta \sigma_{th}}{s_c} \right)^\alpha \right\} \tag{9.68}$$

or, in the case of small damage,

$$\psi_f(N) = \left(\frac{\Delta N}{1 - \omega} \right)^\beta \frac{a\sigma_\infty^{max}}{l_c \tau_b} \left(\frac{\kappa_t \Delta \sigma_\infty - \Delta \sigma_{th}}{s_c} \right)^\alpha \tag{9.69}$$

Equation (9.69) is similar to the equations used in Chapter 4 and later on for quasistationary approximation. To realize the same approach to this case, we have to note that the average cycle number required for a crack to advance at one row of fibers is

$$\Delta N = \lambda_p \left(\frac{da}{dN}\right)^{-1} \tag{9.70}$$

Here the size λ_p, being similar to the process zone size, corresponds to one row of fibers. At square placing of fibers we have $v_f \lambda_p^2 = \pi a^2$, i.e.,

$$\lambda_p = r(\pi/v_f)^{1/2}$$

In any case, λ_p has the order of $v_f^{-1/2} r$.

The damage measure for the matrix may be assumed equal to $\psi_i + \psi_f$, and that for the interface determined as in (9.12), namely

$$\psi_b = (\psi_i + \psi_f)v_b \tag{9.71}$$

In the case when equation (9.71) gives $\psi_b > 1$, we must put $\psi_b = 1$.

Some numerical examples based on equations (9.48), (9.66)-(9.69) and (9.71) are presented in Figures 9.14 and 9.15. The numerical data are $E_f = 100$ GPa, $E_m = 10$ GPa, $\mu_m = 3$ GPa, $\gamma_m = \gamma_b = 0.1$ kJ/m^2, $\tau_b = 50$ MPa, $r = 10$ μm, $l_c = l_\gamma = 1$ mm, $s_c = 1$ GPa, $\Delta\sigma_{th} = 100$ MPa, $m = \alpha = 6$, $N_0 = 0$, $N_c = 10^6$. The numerical values of γ_c and γ_{th} are in accordance with equations (9.50). The initial crack size is $a_0 = 0.1$ mm.

Crack growth histories are depicted in Figure 9.14 at $R = 0.2$ and $\Delta\sigma_\infty = 175, 200, 225, 250$, and 275 MPa (lines 1, 2, 3, 4, and 5, respectively). The cycle number at the start of crack growth varies from $N_* = 54.8 \cdot 10^3$ at $\Delta\sigma_\infty = 175$ MPa to $N_* = 0.63 \cdot 10^3$ at $\Delta\sigma_\infty = 275$ MPa. The cycle number until the final failure varies, respectively, from $N^* = 535 \cdot 10^3$ to $N^* = 2.98 \cdot 10^3$. The crack size at final failure also varies on a wide scale. The shorter fatigue life at higher applied stresses correlates with the shorter cracks at failure. At higher stresses we have a higher contribution of the instantaneous fiber damage ψ_i into the total damage and, as a result, a significant decrease of the generalized resistance force.

The crack growth rates da/dN are plotted versus cycle number in Figure 9.14b. There is a peculiarity in all the curves that is not observed when we look on the crack size histories. Namely, the loss of smoothness is visible in the middle part of the fatigue life. The origin of this phenomenon is the interaction between various microdamage mechanisms. It is illustrated in Figure 9.15. Figure 9.15a shows the histories of the summed fiber damage $\psi_i + \psi_f$ during all the fatigue life. The stage of initial damage accumulation is distinctive in all curves. The stage of crack propagation is accompanied by the gradual decreasing of fiber damage. The points where the loss of smoothness occur are also here. Looking at Figure 9.15b, we mark that the crack growth starts when most fibers at the crack front are debonded. Later, when the cracked domain in the fiber directions becomes longer, the interface damage measure ψ_b begins to drop, and this process proceeds until the final failure. Switching of this mechanism results in temporary moderation of the diminishing of fiber damage.

It follows from the above analysis that parameters γ_b and τ_b of the interface toughness are of significance for the fatigue analysis in the B-mode. As an example, the influence of the specific fracture work γ_b is illustrated in Figures 9.16 and 9.17. Lines 1, 2, 3, 4, 5 are plotted for $\gamma_b = 0.2$, 0.4, 0.6, 0.8, and 1 kJ/m^2, respectively.

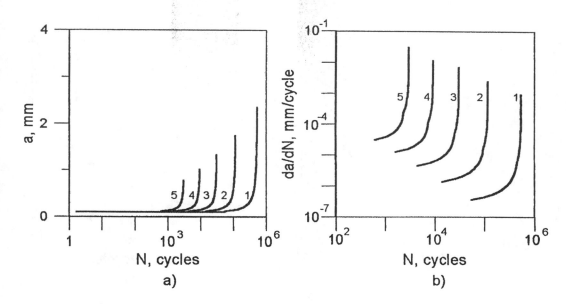

FIGURE 9.14
Crack size (a) and growth rate (b) as a function of the cycle number at $\Delta\sigma_\infty =$ 175, 200, 225, 250, and 275 MPa (lines 1, 2, 3, 4, and 5, respectively).

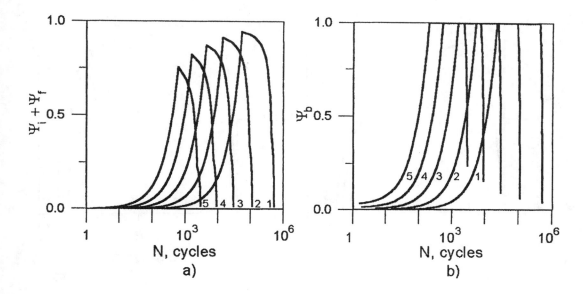

FIGURE 9.15
Fiber rupture measure (a) and debonding measure (b) as a function of the cycle number; notation as in Figure 9.14.

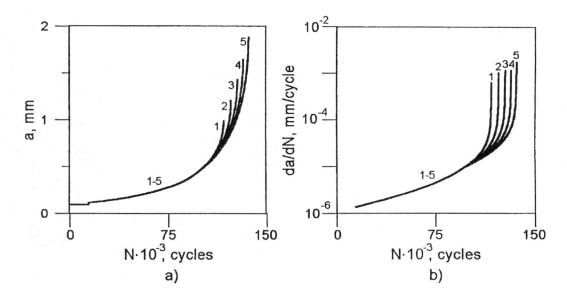

FIGURE 9.16
Crack size (a) and growth rate (b) as a function of the cycle number at $\gamma_b =$ 0.2, 0.4, 0.6, 0.8, and 1.0 kJ/m^2 (lines 1, 2, 3, 4, and 5, respectively).

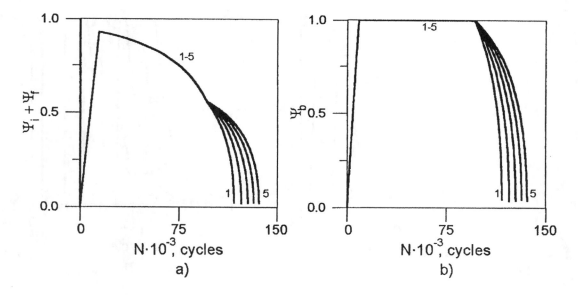

FIGURE 9.17
Fiber damage measure (a) and debonding measure (b) as a function of the cycle number; notation as in Figure 9.16.

The loading is given as $\Delta\sigma_\infty = 200$ MPa at $R = 0.2$; other parameters are the same as in the above example.

Figure 9.16a shows the histories of the crack size at various magnitudes of γ_b. The picture is completely understandable: higher γ_b correspond to higher fatigue life and longer critical crack sizes. However, the difference in fatigue life is not significant varying from $N^* = 103 \cdot 10^3$ at $\gamma_b = 0.2$ kJ/m^2 to $N^* = 115 \cdot 10^3$ at $\gamma_b = 1$ kJ/m^2.

The histories of crack growth rates are given in Figure 9.16b. At the initial stage of crack growth, the fibers at the crack front become debonded (within the assumed model) during the first cycle of loading. Therefore, the influence of debonding on the crack growth rate is negligible during this stage. Later on the situation changes, and we observe splitting of curves. Line 1 is drawn for $\gamma_b = 0.2$ kJ/m^2, i.e., for very weak interface. When γ_b increases, temporary moderating of crack growth occurs.

To illustrate the phenomenon, the damage measures $\psi_i + \psi_f$ and ψ_b are plotted in Figure 9.17 for the same numerical data. The comments are the same as for Figure 9.15. The picture might not look agreeable for experimenters and specialists dealing with composite technology. However, two remarks must be made. First, it is a coarse model of real composites going, nevertheless, into intimate details of the composite behavior. Second, turning to numerical results, we see that the discussed phenomenon does not strongly affect the final results. Note that a uniform scale for cycle numbers is used in Figures 9.16 and 9.17 on contrary to Figures 9.14 and 9.15 where the cycle number is plotted in the log scale. At last, comparing, for example, lines 1 and 5 in Figure 9.16,a we must take into account that they correspond to the fracture toughnesses of the interface varying on a wide range.

9.10 Growth   Rate   Diagrams   for   Brush-Like Cracks

Turning to growth rate diagrams, we have to choose the load parameters which will be most suitable to present diagrams in a compact form. The stress intensity factor certainly is a poor candidate. Actually, the situation ahead of brush-like cracks in composites significantly differs from that at the tips of cracks in common materials. To substantiate the choice of the correlating load parameters, let us use the quasistationary approximation.

Part of the way for developing approximate equations has been covered by equations (9.69) and (9.70). Substituting these equations in (9.66), we present the generalized resistance force as an explicit function of the rate da/dN. After that the governing equation (9.66) can be solved with respect to da/dN. The final equation is rather cumbersome. To simplify it, we neglect the term containing the damage of interface. During part of the fatigue life this term vanishes because of the equality $\psi_b = 1$ (see Figures 9.15b and 9.17b). During the remaining part $\psi_b < 1$. However, usually γ_b is small compared with γ_f. This makes the contribution of the interface in generalized resistance forces small compared with that of fibers.

The simplified version of equation (9.66) has the form $G = \Gamma_0(1 - \psi_i - \psi_f)$ where $\Gamma_0 = v_f\gamma_f + v_m\gamma_m$ is the fracture work for fibers. Replacing $\psi_f(N)$ according to (9.69), we obtain

$$\left(\frac{\Delta N}{1-\omega_f}\right)^\beta \frac{a\sigma_\infty^{max}}{l_c \tau_b} \left(\frac{\kappa_t \Delta\sigma_\infty - \Delta\sigma_{th}}{s_c}\right)^\alpha = 1 - \frac{G}{\Gamma_0} - \psi_i$$

or, after the use of equation (9.70),

$$\frac{da}{dN} = \frac{\lambda_p}{1-\omega_f}\left(\frac{a\sigma_\infty^{max}}{l_c \tau_b}\right)^{1/\beta}\left(\frac{\kappa_t \Delta\sigma_\infty - \Delta\sigma_{th}}{s_c}\right)^m \left(1 - \frac{G}{\Gamma_c} - \psi_i\right)^{-1/\beta} \quad (9.72)$$

The structure of equation (9.72) is similar to that of equations (4.43), (4.44), etc. An uncommon participation of the far field damage in (9.72) reflects its special character in unidirectional fiber composites under tension along fibers. In contrast to ordinary fatigue cracks with their essentially localized stress and damage fields, brush-like cracks in composites cover large domains and propagate with wide fronts across the load direction. Nevertheless, at $m = \alpha$ (which is not very far from experimental data), we have $\beta = 1$. Then equation (9.72) approaches the form typical to common materials. Another peculiarity of equation (9.72) is entering of σ_∞^{max} in such a way that the complex $\lambda_p(\sigma_\infty^{max}/\tau_b)$ could be interpreted in terms of the process zone measured along fibers. In any case, the model of brush-like fatigue cracks is a very special case of mechanics of fatigue.

Equation (9.72) allows choosing appropriate loading parameters for the growth rate diagram. For simplicity, consider the ordinary case when $\kappa_t \Delta\sigma_\infty \gg \Delta\sigma_{th}$. At fixed σ_∞^{max} the rate given in (9.72) becomes proportional to $\Delta\sigma_\infty^m a^{1/\beta}$. As $\beta = \alpha/m$, this signifies the proportionality to $(\Delta\sigma_\infty^m a^{1/\alpha})^m$. Thus, the controlling parameter here is $\Delta\sigma_\infty^m a^{1/\alpha}$. In the case when the stress rate R is fixed, σ_∞^{max} varies with $\Delta\sigma_\infty$ as $\Delta\sigma_\infty(1-R)^{-1}$. A similar consideration gives, as the controlling parameter, the magnitude of $\Delta\sigma_\infty^m a^{1/(\alpha+1)}$. The difference between the two cases is not significant when $\alpha \gg 1$.

It is not unexpected that the stress intensity factor $K = Y\sigma_\infty(\pi a)^{1/2}$ and its range $\Delta K = Y\Delta\sigma_\infty(\pi a)^{1/2}$ are not acceptable as loading parameters in unidirectional composites, especially when they are fractured in the B-mode. However, in the special case, namely at $\alpha = 1$, we come to the controlling parameter $\Delta\sigma_\infty a^{1/2}$. This observation was done by Bolotin (1981). The reason for this peculiarity is the following: the case $\alpha = 1$ corresponds to the exponential distribution of ultimate stresses, i.e., to Poisson's distribution of random flaws along fibers. A composite with such fibers may exhibit some features similar to those of polycrystalline alloys with stochastic microstructure. Note that the usual magnitudes of the shape exponent for real fibers are $\alpha = 6$ and higher.

It is of interest to compare equation (9.72) with the corresponding equation obtained by Bolotin [11, 19] with the use of "almost sure" probabilistic approach. Let a brush-like crack be propagating almost surely when the probability of damage for most fibers on the crack front is of the order of unity. This probability, as a matter of fact, is given in (9.67). Therefore, using equations (9.69) and (9.70) and putting $\psi_f = 1 - e^{-1} \sim 1$, we come to equation

$$\frac{da}{dN} = \frac{\lambda_p}{1-\omega_f}\left(\frac{a\sigma_\infty^{max}}{l_c \tau_b}\right)^{1/\beta}\left(\frac{\kappa_t \Delta\sigma_\infty - \Delta\sigma_{th}}{s_c}\right)^m \quad (9.73)$$

This equation differs from (9.72) only by the absence of the last multiplier in the right-hand side. When $G \ll \Gamma_0$, $\psi_i \ll 1$, this approach gives a good approximation.

The crack growth rate diagrams are presented in Figure 9.18 using, as control parameters, $\Delta\sigma_\infty a^{1/\alpha}$ and $\Delta\sigma_\infty a^{1/(\alpha+1)}$. The same numerical data and the same notation are used as in Figures 9.14 and 9.15. In particular, $\alpha = m = 6$. Except for the final stage, the rate diagrams are positioned closely while the stress level varies in a wide range. In Figure 9.18b, where da/dN is plotted as a function of $\Delta\sigma_\infty a^{1/(\alpha+1)}$ a compact placing is also observable. As follows for equation (9.72), the slope exponent is approximately equal to $m(1 + 1/\alpha) = 7$. However, it is not sufficiently close to the slope shown in Figure 9.18. Perhaps, the cause is the influence of other material parameters entering the picture in complex interactions. Note that these diagrams, as well as the following ones, are obtained by the direct solution of the problem; equation (9.72) is used here only as a reasoning of how to find an appropriate form for growth rate diagrams. Just to demonstrate inconsistency of the stress intensity factor presentation, the rate da/dN is plotted in Figure 9.19 as a function of $\Delta\sigma_\infty(\pi a)^{1/2}$. Lines 1-5 corresponding to different $\Delta\sigma_\infty$ disagree completely: the discrepancy between lines 1 and 5 is measured in two orders of magnitude.

The influence of the stress ratio R is demonstrated in Figure 9.20. Lines 1, 2, 3, 4, and 5 are obtained as $R = 0.1$, 0.2, 0.3, 0.4, and 0.5. As in common fatigue, the curves are subjected to significant divergence when we approach the final failure. The origin of this divergence is obvious. However, using $\Delta\sigma_\infty a^{1/(\alpha+1)}$ as a control parameter, we obtain more compact diagrams (Figure 9.20b).

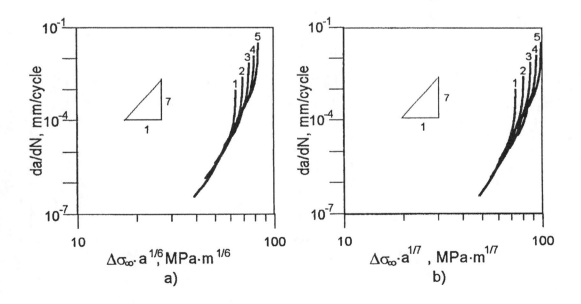

FIGURE 9.18
Growth rate diagrams for brush-like cracks; lines 1-5 correspond to $\Delta\sigma_\infty = 175$, 200, 225, 250, and 275 MPa.

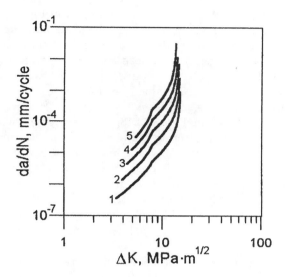

FIGURE 9.19
An attempt to present the growth rate diagram for a brush-like crack in conventional form; notation as in Figure 9.18.

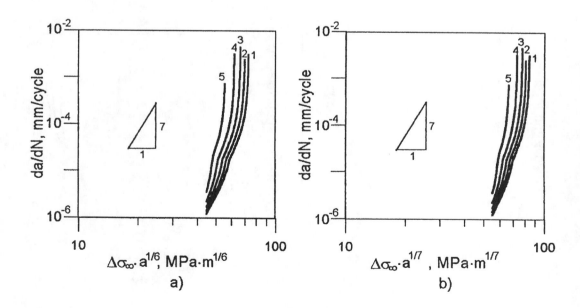

a)

b)

FIGURE 9.20
Growth rate diagrams for different stress ratios; lines 1-5 correspond to $R = 0.1, 0.2, 0.3, 0.4, 0.5$.

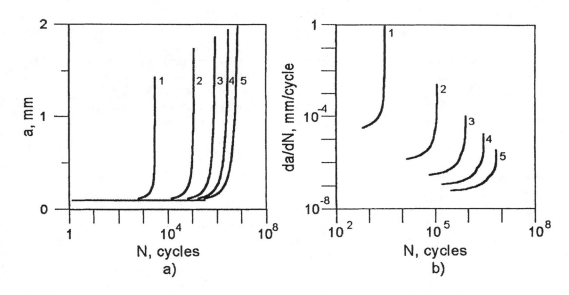

FIGURE 9.21
Crack size (a) and growth rate (b) as a function of the cycle number for different shape exponents; lines 1, 2, 3, 4, 5 correspond to $\alpha = 4, 6, 8, 10, 12$.

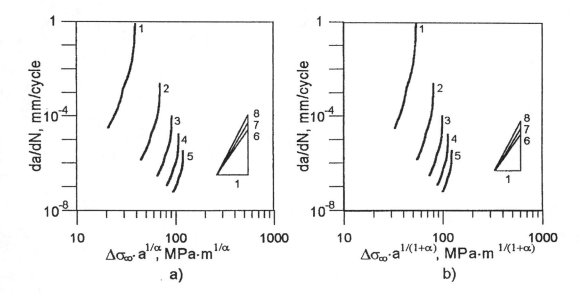

FIGURE 9.22
Growth rate diagrams for different shape exponents; notation as in Figure 9.21.

The last figure of this section demonstrates the influence of the shape exponent α on fatigue crack propagation in the B-mode. The same numerical data are used as in Figure 9.20. The only difference is that α varies as $\alpha = 4, 6, 8, 10$, and 12 (lines 1, 2, 3, 4, and 5, respectively) and the loading regime is fixed at $\Delta\sigma_\infty = 200$ MPa, $R = 0.2$. Figure 9.21 shows the histories of a and da/dN for different α . Corresponding growth rate diagrams as functions of $\Delta\sigma_\infty a^{1/\alpha}$ and $\Delta\sigma_\infty a^{1/(\alpha+1)}$ are presented in Figure 9.22. Certainly, different α refer to composites with fibers of different quality, precisely, of different scattering of their strength. At $\alpha = 4$ we have low quality of fibers; the case $\alpha = 12$ corresponds to rather fair quality. Thus, lines 1-5 in Figures 9.21 and 9.22 refer to different composite materials.

9.11 Summary

Mechanics of fracture and fatigue of unidirectional fiber components under tension along fibers was discussed here on the basis of probabilistic microstructure models. The process of damage accumulation in fibers, matrix and interface was studied until the final failure due to multiple cracking which results into loss of integrity.

The process of crack growth in composites was studied with the emphasis on brush-like cracks as cracked domains containing a set of crack ruptures, matrix crackings, and interface debondings. The role of various material parameters on macroscale crack nucleation and growth until the loss of integrity was analyzed. The problem of load parameters most appropriate for plotting of growth rate diagrams was stated. With the use of the quasistationary approximation, load parameters correlating with the propagation of brush-like cracks were found, and the consistency of this choice was confirmed in numerical experimentation. Comparison with probabilistic "almost sure" estimates for the crack growth rate was also performed.

Chapter 10

FRACTURE AND FATIGUE IN LAMINATE COMPOSITE STRUCTURES

10.1 Interlaminar Defects in Composite Materials

Interlaminar cracks and crack-like defects are as typical for composite structures as fatigue are cracks for common metal structures.

Two kinds of delaminations are to be distinguished depending on their position in a structural member. Delaminations situated within the bulk of the material (Figure 10.1a) are rather similar to common fatigue cracks. In part, the edge delaminations in thick members may be attributed to this kind, too. Delaminations situated near the surface of a structural member (Figure 10.1b) are a special kind of crack-like defects. The growth of near surface delaminations, as a rule, is accompanied by their buckling. Dealing with near-surface delaminations, we have to take into account not only their growth and interlaminar damage but also their stability, considered from the viewpoint of the theory of elastic stability. In addition, local instability and crack growth may produce global instability of structural components such as columns, plates, and shells under compression. Hence, the joint analysis of damage, fracture, local buckling and global stability is frequently required to predict load-carrying capacity and fatigue life of composite structures with near-surface delaminations. Not only complete delaminations but also multiple cracking without separation of layers (Figure 10.1c) are typical for composite structures. These crack-like flaws also affect load-carrying capacity and fatigue life of structural components.

Although the intensive studies in mechanics of delaminations were initiated in the 80's, the historical background begins much earlier. In this context, the paper by Obreimoff (1930) ought to be cited. It was dedicated to the assessment of surface

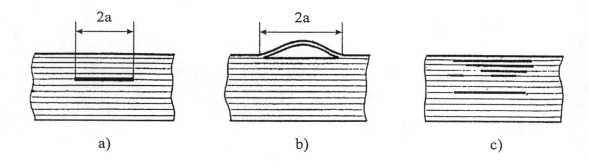

FIGURE 10.1
Interlaminar cracks in laminate composites; (a) internal crack; (b) near-surface delamination; (c) multiple cracking.

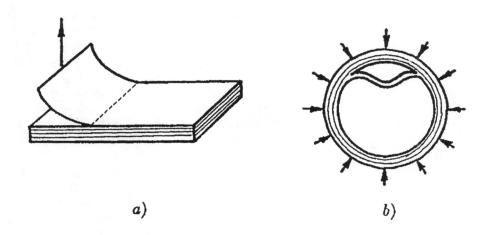

FIGURE 10.2
Classical problems of interlaminar cracking: (a) Obreimoff's problem; (b) Kachanov's problem.

energy in splitting of mica (Figure 10.2a). Being not a composite but an extremely anisotropic natural laminate, mica is similar to some modern composite materials. With direct reference to composite structures, the problem of delaminations was primarily considered by Kulkarni and Frederik (1973) and Kachanov (1975). In particular, Kachanov considered a fiberglass tube under compression with a delamination situated near the internal surface (Figure 10.2b). Now the literature in this field numbers several dozens of papers. Surveys of publications can be found in [16, 19, 51, 132, 139].

The high strength of most laminated and fibrous composites in the direction of reinforcement is accompanied by low resistance against interlaminar shear and transverse tension. Therefore, the interlaminar cracks can originate both at the fabrication stage and at the stages of transportation, storage and service. Instabilities of the manufacturing process, imperfections of various natures, and thermal and chemical shrinkage of components are the sources of initial delaminations. Among other causes of interlaminar cracking are various additional, not accounted for in design, loads and actions: local forces, thermal actions, low-energy surface impacts. Holes, notches and connections also enter as sources of delaminations in the absence of proper design and fabrication.

In recent years, a study of multiple cracking of laminates under surface impact was performed independently by a number of authors. In particular, in the paper by Bolotin, Murzakhanov and Shchugorev [24] experimental results are presented for three types of composites: organic fiber/epoxy, graphite/epoxy and glass-textile/epoxy laminates. Specimens on the rigid foundation as well as beam specimens were tested under low-energy impact. Impactors with flat and spheroidal heads up to 15 kg of mass and up to 600 J of initial energy were applied.

To evaluate the level of multiple cracking, the residual fracture work in ply-after-ply peeling was measured. An example is presented in Figure 10.3, where the longitudinal distribution of the specific fracture work γ in mode I is presented. Specimens were placed on the rigid foundation, and a flat-head impactor with a diameter equal to 20 mm was used. After the impact no separation of layers was observed; however, the fracture work under the impact area diminishes significantly (see curves 1, 2, 3 in Figure 10.3 plotted for three increasing levels of impact energy). Figure 10.4 illustrates the damage distribution along the depth for a beam specimen subjected to the three-point impact. The specimens were dyed after the impact, split into layers, and the summed cracked area A was measured. Both multiple cracking under the impact area and significant delamination near the middle surface were observed.

An analytical-numerical study of cracking under low-energy impact was performed by Bolotin and Grishko [27]. The composite was modelled as a multilayered solid with alternating elastic and elasto-plastic layers [9]. Very special properties were attributed to elasto-plastic layers which imitate the matrix and the interface. The model includes changes of the downloading modulus and secondary moduli due to damage as well as the existence of the ultimate tensile strain

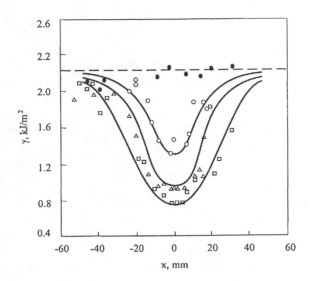

FIGURE 10.3
Decreasing of the residual specific fracture work under impacts with various energy levels.

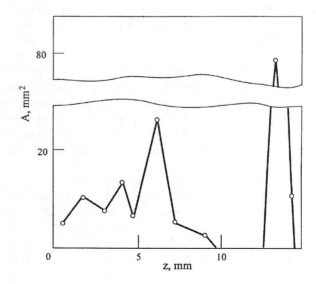

FIGURE 10.4
Summed cracked area distribution along the specimen thickness.

that corresponds to quasibrittle interlayer rupture. Wave propagation in a laminate plate under a rectangular-shaped pressure impulse was studied with numerical simulation to illustrate the consequent cracking of interlayers. Various patterns of damage were observed in numerical simulation: cracking under the impacted surface; splitting near the back side produced from wave reflection; multiple cracking through almost the whole thickness of the specimen.

10.2 Interlaminar Fracture Work

Three items must be discussed before going into analytical modeling of interlaminar cracking: the evaluation of driving generalized forces; the modeling of damage accumulation in interlayers under cyclic and/or sustained loading; the influence of damage on the resistance to interlaminar crack propagation. All these problems are complicated by strong nonhomogeneity and anisotropy of laminate composites. Usually the fractography of interlaminar cracking is rather complicated, in particular, when the boundary of two layers with dissimilar properties is concerned. Even in the case of similar layers we have some difficulties in evaluating the specific fracture work because of the anisotropy of contact surfaces.

Consider, for example, a delamination in a plate of laminate composite under compression in the plane of the layers. For simplicity we assume that the boundary S of the delamination is smooth and given in polar coordinates: $r = r(\varphi)$. The specific fracture work varying along the boundary is a function of φ. On the other side, the magnitude of the specific fracture work depends on the direction of crack growth in each point of boundary S. It is natural to expect that small advancements of the boundary occur in the direction of the external normal $\mathbf{n}$ to S (Figure 10.5). For example, when we deal with a unidirectional fiber composite with fibers oriented along x-axis, the work γ depends on the angle θ between the normal $\mathbf{n}$ and the x-axis. At $\varphi = \theta = 0$ the interlayer fracture follows the pattern shown in Figure 10.6a and at $\varphi = \theta = \pi/2$ the pattern shown in Figure 10.6b. The specific fracture work for intermediate angles varies between the maximal value γ_1 (at $\theta = 0$) and minimal value γ_2 (at $\theta = \pi/2$).

From the above considerations it follows that the specific fracture work is a kind of tensorial value. From the first look, this contradicts the common interpretation of work as a scalar. However, though the fractured areas in Figures 10.6a and 10.6b are equal, the amount of work spent for their formation is essentially different. The reason is the complex fractographic picture. In addition to anisotropy of mechanical properties, we meet multiple cracking, damage of interface, pulling-out and rupture of single fibers, etc. The picture becomes more complicated when a crack propagates between two dissimilar layers (hereafter we suppose that the neighboring layers have the same properties).

As the specific fracture work is characterized with one numerical value when the material is isotropic, and with two values in the case of orthotropy, it is natural to suppose that in the general case the specific fracture work is a tensor of second rank. Following this idea, we assume that the specific fracture work in mode I is given by the matrix

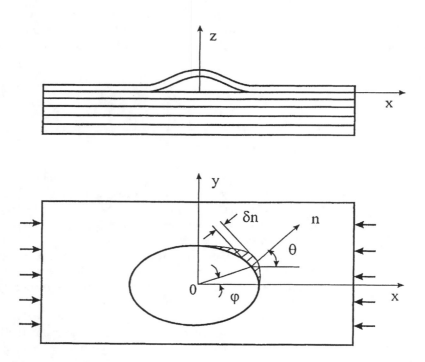

FIGURE 10.5
Delamination in laminate composite plate under compression.

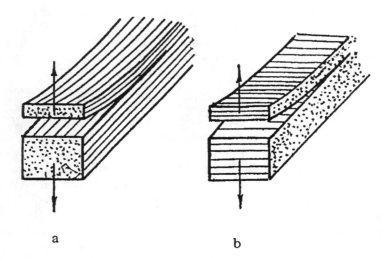

a   b

FIGURE 10.6
Dependence of the specific fracture work on the direction of crack growth.

$$\gamma = \begin{pmatrix} \gamma_{11} & \gamma_{12} \\ \gamma_{21} & \gamma_{22} \end{pmatrix} \tag{10.1}$$

This matrix is symmetrical and definitely positive. In principal axes this matrix takes the form

$$\gamma = \begin{pmatrix} \gamma_1 & 0 \\ 0 & \gamma_2 \end{pmatrix} \tag{10.2}$$

where γ_1 and γ_2 are magnitudes of the specific fracture work in x_1- and x_2-directions. We assume that the components of matrix γ vary in the rotation of the crack front as follows:

$$\gamma'_{jk} = \sum_{j=1,2} \sum_{k=1,2} v_{j\alpha} v_{k\beta} \gamma_{\alpha\beta}, \quad v_{j\alpha} = \cos(x'_j, x_\alpha) \tag{10.3}$$

Here x_1, x_2 are old axes, x'_1, x'_2 are new ones, and γ'_{jk} are the components of γ in new axes. According to this equation, the elements of γ are subjected to transformations similar to those for a second rank tensor in orthogonal Cartesian coordinates. However, this transformation does not mean change of reference axes but the change in the direction of crack propagation.

Let the external normal to the crack front be inclined to the x_1-axis at the angle θ. At $\theta = 0$ and $\theta = \pi/2$ we have $\gamma = \gamma_1$ (Figure 10.7a), and $\gamma = \gamma_2$ (Figure 10.7b). When $0 < \theta \le \pi/2$, the diagonal element γ_θ and nondiagonal one η_θ are:

$$\gamma_\theta = \gamma_1 \cos^2 \theta + \gamma_2 \sin^2 \theta$$

$$\eta_\theta = \frac{\gamma_1 - \gamma_2}{2} \sin 2\theta \tag{10.4}$$

The magnitude η_θ is taken in (10.4) with opposite sign (similar to shear stresses in the elementary stress analysis or cross products in the elementary dynamics). Such a change gives automatically the sign of the additional rotation angle $\Delta\theta$. It is illustrated in Figure 10.7. At $\gamma_1 > \gamma_2$ the crack front rotates in the x_2-direction. Thus, $\Delta\theta > 0$ as is shown in Figure 10.7c. In the opposite case shown in Figure 10.7d, we obtain $\Delta\theta < 0$. The sign of η_θ defined by equation (10.4) is subjected to the same change.

The fracture work in an oblique crack propagation consists of two parts. The first one is the work performed on the crack extension in the normal direction. Notating the normal component of small crack advancement with Δn, we obtain that this part is equal to $-\gamma_\theta b \Delta n$. Here b is the width of the plate (Figure 10.8). The origin of the second part is the crack extension in the tangential direction. The corresponding newly cracked area is equal to $\Delta t/2$, where $\Delta t \approx b \cdot \Delta\psi$. It is natural to connect this part of the cracked area with η_θ. The question is which multiplier is to be used for the product $\eta_\theta \Delta t/2$. We suppose it is equal to unity. Then the total increment of the fracture work is

$$\Delta W_f = -\left(\gamma_\theta b \Delta n + \frac{1}{2} \eta_\theta b \Delta t \right) \tag{10.5}$$

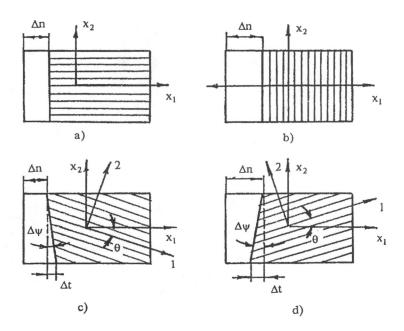

FIGURE 10.7
Interlaminar fracture of composite: (a) and (b) in principal direction of tensor
γ; **(c) and (d) in oblique directions.**

The validity of equations (10.4) and (10.5) may be checked experimentally, by the direct measurement of the work spent on oblique interlaminar cracking of laminates with significant orthotropy.

The scheme of testing is shown in Figure 10.8. A cantilever specimen with the principal axes turned with respect to the specimen axis is subjected to fracture in the mode I. The fracture work ΔW_f and the crack size increments Δn and Δt are measured. The principal values γ_1 and γ_2 are estimated by direct tests of "straight" specimens. Then equations (10.4) and (10.5) are applied for "skew" specimens, and computed results are compared with experimental ones.

However, as the work ΔW_f depends on the crack advancement, it is expedient to modify the procedure. As above, the principal values γ_1 and γ_2 are to be first evaluated by testing at $\Delta t \equiv 0$. Then, using equation (10.4), the magnitudes of γ_θ and η_θ can be found. The substitution of these magnitudes in equation (10.5) gives the amount of work ΔW_f which is to be compared with the results of testing of "skew" specimens.

The experimental results for a unidirectional organic fiber/epoxy composite are presented in Figure 10.9. The sandwich-type specimens were used with external metallic plates to maintain cracking in mode I. The average principal values of tensor γ are $\gamma_1 = 1.45 \, \text{kJ/m}^2$ and $\gamma_2 = 0.77 \, \text{kJ/m}^2$. The scatter of the experimental data in Figure 10.9 is rather large. Moreover, the discrepancy is observed between experimental results and theoretical ones (the latter are drawn by solid curves). However, the general character of functions and the order of magnitudes are in

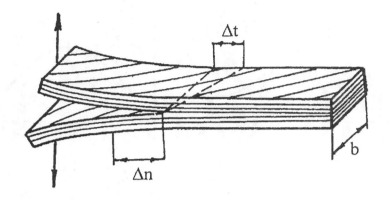

FIGURE 10.8
Scheme of tests to determine specific fracture work in oblique interlaminar cracking.

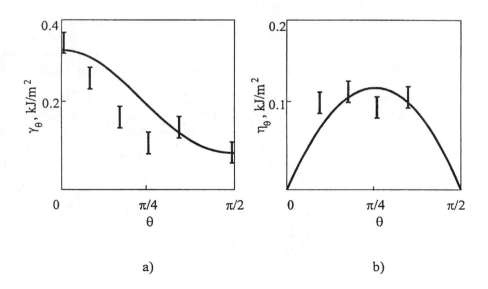

FIGURE 10.9
Components of specific fracture work for organic fiber/epoxy laminate.

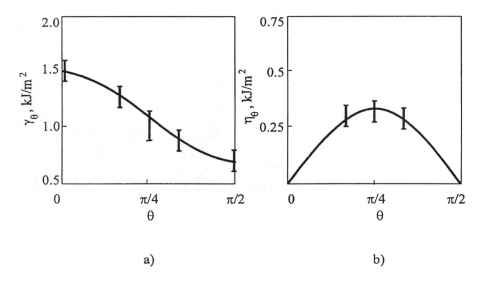

a)   b)

FIGURE 10.10
The same as in Figure 10.9 for textile glass/epoxy laminate.

agreement with theoretical prediction.

Similar data for an orthotropic textile glass/epoxy laminate composite are given in Figure 10.10. The specimens were cut from the same composite plate at various angles to the principal axes. The principal values are $\gamma_1 = 0.32\,\mathrm{kJ/m^2}$ and $\gamma_2 = 0.8\,\mathrm{kJ/m^2}$. Qualitatively, the conclusions are the same as for the unidirectional composite. Though there is doubt concerning the tensorial nature of the specific fracture work or, at least, the rank of the tensor, we can speak, at least at the current state of study, of the agreement between predictions and laboratory experimentations.

Evidently, a similar presentation can be suggested for modes II and III, for example, in the case of shear cracking. As to the mixed mode interlaminar cracks, the situation here is more complicated. The simplest case is when a mixed mode crack propagates between two similar orthotropic layers with coinciding principal axes. We have to make a distinction between $\gamma_I^j, \gamma_{II}^j$, and γ_{III}^j, where $j = 1, 2$. The specific fracture work matrix is composed of three matrices similar to (10.2)

$$
\gamma = \begin{pmatrix}
\gamma_I^1 & 0 & & & & \\
0 & \gamma_I^2 & & & & \\
& & \gamma_{II}^1 & 0 & & \\
& & 0 & \gamma_{II}^2 & & \\
& & & & \gamma_{III}^1 & 0 \\
& & & & 0 & \gamma_{III}^2
\end{pmatrix}
$$

Other elements of this 6×6 matrix are equal to zero. For obliquely propagating cracks we obtain similar to (10.2)

$$\gamma = \begin{pmatrix} \gamma_I^{11} & \gamma_I^{12} & & & & \\ \gamma_I^{21} & \gamma_I^{22} & & & & \\ & & \gamma_{II}^{11} & \gamma_{II}^{12} & & \\ & & \gamma_{II}^{21} & \gamma_{II}^{22} & & \\ & & & & \gamma_{III}^{11} & \gamma_{III}^{12} \\ & & & & \gamma_{III}^{21} & \gamma_{III}^{22} \end{pmatrix} \tag{10.6}$$

The elements of the latter matrix can be calculated using equation (10.3). When the direction of crack propagation is known, the magnitudes of the specific fracture work similar to those given in (10.4) form a (6×1)-vector with components $\gamma_I(\theta), \eta_I(\theta), \gamma_{II}(\theta), \eta_{II}(\theta), \gamma_{III}(\theta), \eta_{III}(\theta)$. The virtual fracture work for a stripe of the unity width is

$$\delta W_f = -[\gamma_I(\theta)\delta n_I + \frac{1}{2}\eta_I \delta t_I + \gamma_{II}(\theta)\delta n_{II}+$$

$$+\frac{1}{2}\eta_{II}\delta t_{II} + \gamma_{III}(\theta)\delta n_{III} + \frac{1}{2}\eta_{III}\delta t_{III}] \tag{10.7}$$

The peculiarity of equation (10.7) is that different, generally independent, virtual displacements are used for each fracture mode. It is not so extravagant from the physical viewpoint, as it might seem. Actually, the fractographic picture of interlaminar cracking in composites is very complicated. In particular, the tensile and shear damage have different patterns and, as a result, their fronts do not necessarily coincide. Moreover, to stay in the framework of analytical mechanics, we have to introduce independent generalized coordinates for each fracture mode (Section 10.4).

When the properties of the neighboring layers are dissimilar and/or in the general case of anisotropy, the interlaminar fracture modes interact. It means that the vacant places in matrices may be occupied with elements describing this interaction. To study these effects experimentally, a rather sophisticated experimentation is necessary, and its results could be vague due to the influence of secondary effects. Eventual complications in fracture at the boundary between two dissimilar materials are noted even for isotropic materials, not to mention laminate composites. We do not discuss this problem whose analysis is far beyond the frames of this book.

10.3 Generalized Resistance Forces

In many applications, a simplified approach is expedient to evaluate the resistance generalized forces. For example, when the delamination front is a smooth curve (Figure 10.5), the oblique component of the fracture work is negligible, and we obtain

$$\delta W_f = -\int_S \gamma_\theta(s)\delta n(s)ds \tag{10.8}$$

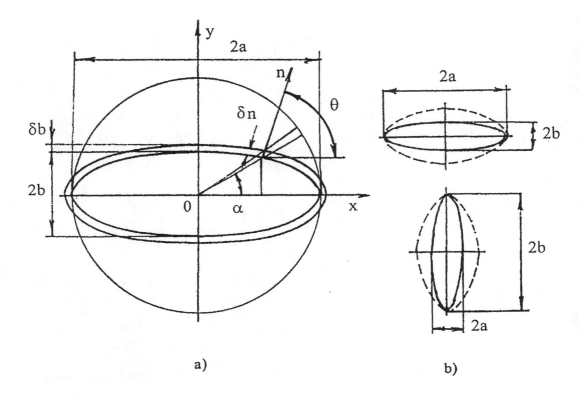

a)   b)

FIGURE 10.11
Elliptical delamination: (a) variation of dimensions; (b) growth of elongated delaminations.

Here $\gamma_0(s)$ is the specific fracture work in normal crack advancement; $\delta n(s)$ is the virtual advancement. The integral in (10.8) is taken upon the entire front length S. Let us use the polar coordinates r, φ. Omitting the terms of higher order, we have $\delta n(s)ds = r\delta r d\varphi$ where δr is the virtual polar radius increment. Equation (10.8) takes the form

$$\delta W_f = -\int_{\varphi_1}^{\varphi_2} \gamma_\theta(\varphi) r \delta r(\varphi) d\varphi \tag{10.9}$$

where the segment $[\varphi_1, \varphi_2]$ corresponds to the complete crack front. The magnitude of $\gamma_\theta(\varphi)$ referred to the principal axes x_1 and x_2 is given in (10.4).

Let us demonstrate how to calculate the generalized resistance forces for an elliptical interlaminar crack. The semi-axes of the ellipse a and b take the part of G-coordinates. Coordinates of the points of the ellipse are given by the elliptical angle α, i.e., $x = a\cos\alpha, y = b\sin\alpha$ (Figure 10.11a). Then the arc element ds and the angle θ between the external normal n and the x-axis are

$$ds = (a^2 \sin^2 \alpha + b^2 \cos^2 \alpha)^{1/2}, \quad \tan \theta = k \tan \alpha$$

where notation $k = a/b$ is used. Consider the identities

$$(a + \delta a) \cos(\alpha + \delta \alpha) = x + \delta n \cos \theta$$

$$(b + \delta b) \sin(\alpha + \delta \alpha) = y + \delta n \sin \theta$$

These relationships give the following formula for the variation δn:

$$\delta n = \frac{b \delta a \cos^2 \alpha + a \delta b \sin^2 \alpha}{(a^2 \sin^2 \alpha + b^2 \cos^2 \alpha)^{1/2}} \tag{10.10}$$

Substituting the above formulas in the general equation (10.8), we obtain

$$\Gamma_a = 4b \int_0^{\pi/2} \gamma[\theta(\alpha)] \cos^2 \alpha \, d\alpha$$
$$\Gamma_b = 4a \int_0^{\pi/2} \gamma[\theta(\alpha)] \sin^2 \alpha \, d\alpha \tag{10.11}$$

Suppose that the axes of the ellipse are directed along the principal axes of tensor γ_{jk}. Then, according to (10.4), $\gamma(\theta) = \gamma_1 \cos^2 \theta + \gamma_2 \sin^2 \theta$. Equation (10.11) takes the form

$$\Gamma_a = 4b \int_0^{\pi/2} \frac{(\gamma_1 + \gamma_2 k^2 \tan^2 \alpha) \cos^2 \alpha \, d\alpha}{1 + k^2 \tan^2 \alpha}$$

$$\Gamma_b = 4a \int_0^{\pi/2} \frac{(\gamma_1 + \gamma_2 k^2 \tan^2 \alpha) \sin^2 \alpha \, d\alpha}{1 + k^2 \tan^2 \alpha}$$

These integrals can be calculated elementarily. The final result is

$$\Gamma_a = \frac{\pi b [\gamma_1 (2k + 1) + \gamma_2 k^2]}{(k + 1)^2}$$

$$\Gamma_b = \frac{\pi a [\gamma_1 + \gamma_2 k (k + 2)]}{(k + 1)^2} \tag{10.12}$$

When the material is transversally isotropic, then $\gamma_1 = \gamma_2 = \gamma$, and we obtain

$$\Gamma_a = \pi \gamma b, \quad \Gamma_b = \pi \gamma a. \tag{10.13}$$

When $\gamma_1 \neq \gamma_2$ but the ellipse is elongated, it is easy to get simple estimates. For example, if $a \ll b$, equation (10.12) results in $\Gamma_a = \pi \gamma_1 b, \Gamma_b = \pi \gamma_2 a$. This means

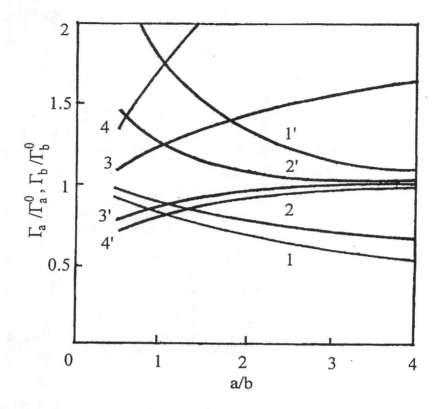

FIGURE 10.12
Generalized resistance forces for elliptical delamination at different γ_2/γ_1 (notation in text).

that $\Gamma_a \gg \Gamma_b$ at γ_1 and γ_2 of the same order of magnitude. Under other equal circumstances, a crack will grow in x-direction. This growth is controlled mainly by the fracture toughness in this direction. Oppositely, when $a \gg b$ we come to the estimate $\Gamma_a \approx \pi\gamma_2 b$, $\Gamma_b \approx \pi\gamma_2 a$. Then the crack growth is governed mainly by the fracture toughness in the y-direction. In both cases the elongated cracks have a tendency to approach the shape closest to a circular one (if, certainly, the loading and resistance factors are near to isotropic). This conclusion is illustrated in Figure 10.11b.

Some numerical results are presented in Figure 10.12. There the normalized resistance forces Γ_a/Γ_a^0 and Γ_b/Γ_b^0 are plotted as a function of normalized parameters $k = a/b$ and γ_2/γ_1. Here $\Gamma_a^0 = \pi\gamma_1 b, \Gamma_b^0 = \pi\gamma_2 a$. Curves 1, 2, 3, 4 are plotted for Γ_a/Γ_a^0 at $\gamma_2/\gamma_1 = 0.25, 0.5, 2, 4$, and curves 1', 2', 3', 4' for Γ_b/Γ_b^0 at the same ratios γ_2/γ_1. At $\gamma_2/\gamma_1 = 1$ we come to equalities $\Gamma_a = \Gamma_a^0$, $\Gamma_b = \Gamma_b^0$ that correspond to equation (10.13).

10.4 Internal Fatigue Cracks in Composites

Internal delaminations are rather similar to cracks in ordinary structural materials, and they are usually treated in terms of conventional fracture mechanics. This concerns, partially, the edge delaminations in thick laminated members. Conditions of stability with respect to the growth of delaminations are formulated in terms of energy release rates and path-independent integrals as well as, in the case of linear elasticity, in terms of stress intensity factors. In particular, the interlaminar fracture energy per unit of the new surface is widely used to characterize the toughness of composites to the growth of delaminations. Usually, we distinguish the energy release rates G_I, G_{II} and G_{III} for tensile, transverse shear and anti-plane shear, respectively, as well as the corresponding critical magnitudes G_{IC}, G_{IIC} and G_{IIIC}. However, the patterns of interlaminar fracture are complex due to strong anisotropy, interface friction, etc. In a larger degree the complexity is connected with the mixed-mode type of fracture that is present in most practical cases. In this section an analytical approach is discussed about delaminations between two layers with similar properties, without going into complications originating from dissimilarities of neighboring layers. Moreover, we consider delaminations oriented along the principal elastic axes. Then we may discuss the interlaminar fracture in terms of particular energy release rates G_I, G_{II}, G_{III} and their combinations.

Consider a planar interlaminar crack in a composite modelled as an orthotropic elastic solid. Let a crack of length $2a$ be situated in the x, y plane, and the principal elastic planes parallel to the planes of the reference system x, y, z (Figure 10.13). The strain energy release rates for particular fracture modes are as follows [134]:

$$G_I = K_I^2 (a_{11} a_{22})^{1/2} \eta, \quad G_{II} = K_{II}^2 a_{11} \eta, \quad G_{III} = \frac{K_{III}^2}{2(c_{44} c_{55})^{1/2}} \quad (10.14)$$

Here K_I, K_{II} and K_{III} are stress intensity factors in "pure" fracture modes, a_{jk} are the elements of the compliance matrix, c_{jk} are the elements of the stiffness matrix, and η is a nondimensional factor:

$$\eta = \frac{1}{\sqrt{2}} \left[\left(\frac{a_{22}}{a_{11}} \right)^{1/2} + \frac{2(a_{12} + a_{66})}{2 a_{11}} \right]$$

According to the global energy balance criterion, the quasibrittle fracture begins when the strain energy release rate G attains its critical magnitude G_C:

$$G = G_C \quad (10.15)$$

According to its nature, the energy release rates should be the additive variables. Thus

$$G = G_I + G_{II} + G_{III} \quad (10.16)$$

Combining equations (10.14)-(10.16), we obtain the limit condition that should connect K_I, K_{II} and K_{III} at the beginning of crack propagation:

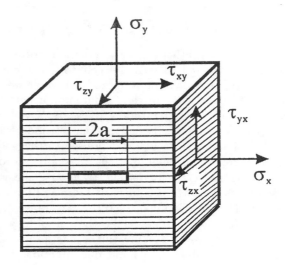

FIGURE 10.13
Internal crack in macroscopically orthotropic composite.

$$\frac{K_I^2}{K_{IC}^2} + \frac{K_{II}^2}{K_{IIC}^2} + \frac{K_{III}^2}{K_{IIIC}^2} = 1 \qquad (10.17)$$

Critical magnitudes K_{IC}, K_{IIC} and K_{IIIC} entering equation (10.17) should be connected due to the additive properties of G. For example, for the plane-stress state

$$\frac{K_{IIC}}{K_{IC}} = \left(\frac{E_x}{E_y}\right)^{1/2} \qquad (10.18)$$

The right-hand side of (10.18) contains Young's moduli E_x and E_y in x and y-directions.

Equations (10.17) and (10.18) are easy to verify experimentally. The experiments show that these equations essentially contradict experimental results. Generally, although the energy is an additive variable, the strain energy release rates are not additive for interlaminar fracture. In other words, equation (10.16) is invalid for strongly nonhomogeneous and anisotropic composites though it may be applicable for ordinary metal alloys. The cause is the complicated fractographic patterns in composites that include multiple tensile and shear crackings, both in the interlayers and on the boundaries, as well as breakage, tearing, and buckling of fibers.

Semi-empirical equations are used in the design of composite structures such as

$$\left(\frac{G_I}{G_{IC}}\right)^{m_I} + \left(\frac{G_{II}}{G_{IIC}}\right)^{m_{II}} + \left(\frac{G_{III}}{G_{IIIC}}\right)^{m_{III}} = 1 \qquad (10.19)$$

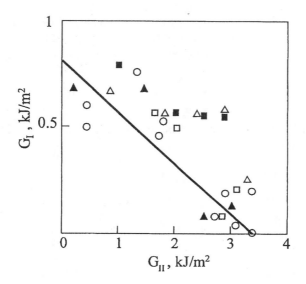

FIGURE 10.14
Relationship between the critical magnitudes of generalized driving forces for textile/glass/epoxy laminate.

Here m_I, m_{II}, and m_{III} are empirical exponents that are not necessarily equal to unity. Similarly, the empirical critical magnitudes G_{IC}, G_{IIC}, and G_{IIIC} are not analytically connected as one might conclude from (10.14) and (10.15). Since equation (10.19) is a kind of interpolation, it might fit experimental data quite satisfactorily. An example taken from the paper [24] is presented in Figure 10.14. There the relationship between G_I and G_{II} is drawn for the mass-production glass-textile/epoxy laminate. The scatter of experimental data is high due to the low quality of the material. However, the general tendency is evident. The average critical magnitudes for "pure" modes are $G_{IC} = 0.74 \, \text{kJ/m}^2$ and $G_{IIC} = 3.60 \, \text{kJ/m}^2$. That signifies a strong anisotropy of fracture toughness.

Equation (10.19) and similar relationships do not look attractive from the academic viewpoint. It seems more adequate to keep the summed energy release rate G as a generalized driving force, assuming that the critical magnitude G_C (later on $G_C \equiv \Gamma$) depends on the fracture mode. In fact, the fractographic picture of delaminations depends on which mode dominates in a mixed-mode fracture. In general, all three modes contribute to the damage near the fronts of delaminations.

An analytical approach to the interlaminar fracture of laminates is based on the principle of virtual work. Let us follow a rather nonconventional idea. Assume that the crack front is given with very close generalized Griffith's coordinates $a_1; \ldots, a_m$ with independent variations $\delta a_1, \ldots, \delta a_m$. In fact, the front of the shear damage might be a little ahead of the opening (tensile) damage front, etc. Let "pure" modes be associated with variations $\delta a_1, \delta a_2$, and δa_3 (hereafter Arabic figures are used for the numbers of modes). When any couple of damage fronts coincide, we introduce variations $\delta a_4, \delta a_5$, and δa_6. When all three modes contribute almost equally to damage, we label the variation δa_7. It ought to be stressed that all coordinates correspond to the material points located in the process zone whose size is small compared with $a_1, \ldots, a_7$. It means that G-coordinates take very close magnitudes. Assuming the variations independent and applying the principle of virtual work

$$\sum_{j=1}^{7} (G_j - \Gamma_j)\delta a_j \leq 0$$

we conclude that a delamination begins to propagate if one of the following conditions is attained:

$$G_1 = \Gamma_1, \quad G_2 = \Gamma_2, \quad G_3 = \Gamma_3$$

$$G_1 + G_2 = \Gamma_{12}, \quad G_1 + G_3 = \Gamma_{13}, \quad G_2 + G_3 = \Gamma_{23} \tag{10.20}$$

$$G_1 + G_2 + G_3 = \Gamma_{123}$$

Here Γ_1, Γ_2 and Γ_3 are generalized resistance forces in "pure" modes; $\Gamma_{12}, \Gamma_{13}, \Gamma_{23}$, and Γ_{123} are those in mixed modes. Then the forces may be associated with the principal values of submatrices entering the 6×6 matrix in (10.6). The nondiagonal submatrices that are not filled in (10.6) must be included to describe the interaction between fracture modes.

The limit surface in the space of G_1, G_2, G_3 according to equations (10.20) is piece-planar (Figure 10.15a). But it is not so bad compared, say, with the smooth surface (Figure 10.15b) given by equation (10.19). Opposite to equation (10.14) which is not more than a result of interpolation, equations (10.20) are in agreement with the principles of mechanics of solids (although it is based on the unconventional assumption of the existence of several damage fronts).

Turning to fatigue cracks, we have to introduce the relevant interlaminar measures, the equations describing damage accumulation and its effect on the generalized resistance forces. Analogous to (10.15), let us introduce damage measures $\omega_1, \ldots, \omega_7$ for "pure" and mixed fracture modes. The growth of these measures under cyclic loading is controlled by interlaminar stresses, in particular, with their range within a cycle. The interlaminar stresses can be found using the theory of multilayered structures [9, 122]. The resistance forces $\Gamma_1, \ldots, \Gamma_7$ depend on the magnitudes of damage at the fronts which, in general, do not coincide, being macroscopically very close.

Continuous fatigue crack growth takes place when at least one of the relationships given in (10.20) is satisfied. It means that one of the "pure" modes may dominate even under mixed mode loading. As a result, the macroscopic crack growth rate is defined as follows:

$$\frac{da}{dN} = \max\left\{ \frac{da_1}{dN}, \ldots, \frac{da_7}{dN} \right\}$$

A part of the variables in brackets is usually equal to zero, and all the variables are equal to zero for nonpropagating cracks. It is easy to see that we approach a transparent (though cumbersome) generalization of the theory of fatigue crack

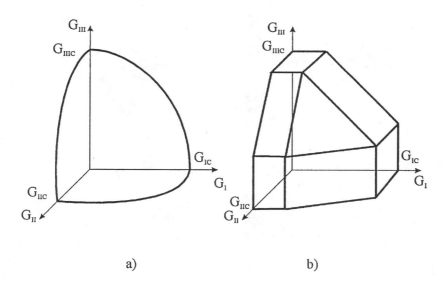

a)   b)

FIGURE 10.15
Limit surface for mixed mode fracture: (a) piece-planar; (b) smooth.

growth in conventional materials. We do not go further in this subject; turning to a more interesting type of fracture, namely, near-surface delaminations.

10.5 Buckling and Stability of Near-surface Delaminations

When a delamination is situated near the surface of a structural member, its behavior under loading is often accompanied by buckling. This is typical for members under compression, under surface heating, and sometimes for components under tension (due to Poisson's effect). Examples of delaminations are shown in Figure 10.16. Case (a) corresponds to the delamination propagating in a member under tension. In case (a) we say of an open delamination that growth is accompanied by the formation of longitudinal cracks. Delaminations in the components under compression are shown in Figure 10.16, too. In case (b) the delamination is open and subjected to buckling. In case (c) the delamination is also buckled, but is considered as closed. An edge delamination is presented in Figure 10.16d. In the last case a secondary crack may appear during the buckling of the delamination. One-dimensional bending approximation is applicable in cases (a) and (b). In other cases we have to treat delaminations with the use of the theory of plates and shells. In case (c) the delamination may be considered as elliptical in the plane, in case (d) as semi-elliptical.

Various versions of the energy approach were used to predict the stability of

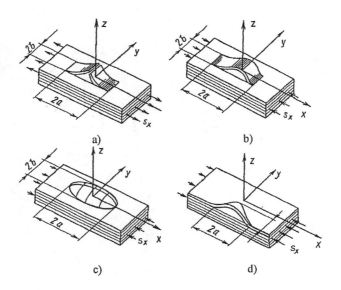

FIGURE 10.16
Near-surface delaminations: (a) open in tension; (b) open and buckled in compression; (c) closed buckled in compression; (d) edge buckled in compression.

delaminations with respect to their growth: the strain energy release rate approach, the path-invariant integral approach, the strain energy density approach. In the case of a single-parameter delamination in elastic structural members all these methods produce either identical or numerically close predictions.

The certainly simplest problems are those depicted in Figure 10.16a. The crack size is given with a single G-coordinate a. The thickness h and the width b of the cracked domain are fixed. The plate is assumed to be orthotropic and subjected to uniformly applied strain ε_∞ (not necessary in compression). The direction of straining coincides with one of the principal directions of orthotropy. The generalized driving force is

$$G = -\frac{\partial U}{\partial a} \tag{10.21}$$

where U is the potential energy of the system, i.e.,

$$U = \text{const} - \frac{1}{2\eta_x} E_x \varepsilon_\infty^2 abh \tag{10.22}$$

Here b is the width of delamination; $h \ll a$ is its thickness. The energy U as well as other variables refer to a half of the plate and, respectively, to one of the delamination fronts. Hereafter $\eta_x = 1 - \nu_{xy}\nu_{yx}$ in the plane-strain state, where ν_{xy} and ν_{yx} are Poisson's ratios, and $\eta_x = 1$ in the plane-stress state. Equations (10.21) and (10.22) result in

$$G = \frac{E_x bh\varepsilon_\infty^2}{2\eta_x} \tag{10.23}$$

The virtual fracture work, evidently, is $\delta W_f = -\gamma b \delta a$, where γ is the specific work of an interlaminar fracture. Hence

$$\Gamma = \gamma b \qquad (10.24)$$

The delamination does not propagate when $G < \Gamma$, and begins to propagate, more precisely, attains an equilibrium state when $G = \Gamma$. The critical strain ε_0 is defined equalizing the right-hand sides in (10.23) and (10.24):

$$\varepsilon_0 = \left(\frac{2\eta_x \gamma}{E_x h} \right)^{1/2} \qquad (10.25)$$

Evidently, this state is neutral because neither G nor Γ depends on a. The instability region is depicted in Figure 10.17a. To produce the proceeding cracking, it is necessary to increase the loading level.

Another simple problem is a beam-like initially flat but buckled delamination in a plate under compression (Figure 10.16b). Various approaches to this problem were discussed in [13, 16, 32, 57, 132, 139]. When the normal displacements are small, i.e., the maximal deflection $f \ll a$, we may assume that the membrane stresses and strains in the delamination are equal to those attained at the beginning of buckling. Hence, we take the membrane strain as

$$\varepsilon_*(a) = \frac{\pi^2}{12} \left(\frac{h}{a} \right)^2 \qquad (10.26)$$

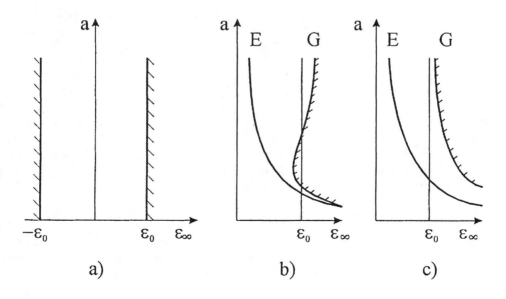

FIGURE 10.17
Stability diagrams for simplest delamination problems: (a) open delamination; (b) closed beam-like delamination in compression; (c) circular isotropic delamination under uniform compression.

The total energy is

$$U = \text{const} - \frac{E_x abh(\varepsilon_\infty^2 - \varepsilon_*^2)}{2\eta_x} + \frac{E_x bh^3}{24\eta_x} \int\limits_0^a \left(\frac{d^2 w}{dx^2}\right)^2 dx \qquad (10.27)$$

The first term, as in (10.22), is equal to the initial amount of energy, the second one is equal to the strain energy release due to crack propagation, and the last one is the energy of buckling. Here $w(x)$ is normal displacement at $|x| \leq a$. The shape of the delamination is close to that in Euler's buckling. Therefore,

$$w(x) = f \cos^2\left(\frac{\pi x}{2a}\right) \qquad (10.28)$$

and equation (10.27) takes the form

$$U = \text{const} - \frac{E_x abh}{2\eta_x}(\varepsilon_\infty - \varepsilon_*)^2 \qquad (10.29)$$

The generalized driving force, according to (10.21) and (10.29), is

$$G = \frac{E_x bh}{2\eta_x}\left(\varepsilon_\infty^2 + 2\varepsilon_\infty \varepsilon_* - 3\varepsilon_*^2\right) \qquad (10.30)$$

Equalizing the right-hand sides of (10.24) and (10.30), we come to the equation that connects the critical magnitudes of ε_∞ and a:

$$\varepsilon_\infty^2 + 2\varepsilon_\infty \varepsilon_*(a) - 3\varepsilon_*^2(a) = \varepsilon_0^2. \qquad (10.31)$$

Here the notation (10.25) is used for the critical strain of an open delamination. The stability diagram following from equations (10.26) and (10.31) is shown in Figure 10.17b. Line E is the boundary of Euler's buckling, and line G is the boundary of the initiation of fracture. The states of the system corresponding to the descending part of the line G are unstable, and those corresponding to the ascending part are stable (in Griffith's sense).

Multiparameter delaminations were considered primarily by Bolotin [13]. The simplest among them is the elliptical delamination that is assumed to remain elliptical during its growth. The semi-axes a and b play here the role of Griffith's generalized coordinates (Figure 10.11). In the case of a near-surface delamination whose behavior does not affect the strain-stress field in the main bulk of the structural member, we present the strain energy as

$$U = \text{const} - U_1(a_1, \ldots, a_m) \qquad (10.32)$$

Here the constant represents the total energy in the absence of the delamination, and the term $U_1(a_1, \ldots, a_m)$ depends on Griffith's coordinates $a_1, \ldots, a_m$ as well as on the loading parameters. The generalized driving forces are

$$G_j = -\frac{\partial U_1}{\partial a_j}, \quad (j = 1, \ldots, m). \qquad (10.33)$$

Consider a closed initially flat delamination in a plate under compression similar to one depicted in Figure 10.16b. Let the thickness of the delamination h be sufficiently small to neglect effects of the delamination behavior on the stress-strain distribution in the bulk of the plate. The material of the plate is considered as orthotropic with the principal elasticity axes parallel to the reference axes x and y. The principal axes of the nominal stress-strain field are assumed parallel to these axes, too. Denote the applied stresses s_x, s_y and the applied strains e_x, e_y. Stresses and strains in the delaminated part are denoted σ_x, σ_y and $\varepsilon_x, \varepsilon_y$, respectively. The influence of components τ_{xy} and γ_{xy} in the delaminated part will be neglected. This is equivalent to the assumption that the principal axes of stresses and strains at all points of the delamination are parallel to the reference axes. The domain occupied with the delamination in the x, y plane is denoted Ω, and its boundary S. The membrane strain energy is defined as

$$U_0 = \text{const} - \frac{h}{2(1 - \nu_{xy}\nu_{yx})} \iint_{\Omega} \left[E_x(e_x^2 - \varepsilon_x^2) + \right.$$

$$\left. + 2\nu_{xy}E_x(e_x - \varepsilon_x)(e_y - \varepsilon_y) + E_y(e_y^2 - \varepsilon_y^2) \right] dx dy.$$

$$(10.34)$$

The potential strain energy of bending (and, maybe, of torsion, also) is accumulated in the delamination only. As the initial state of the delaminated part is assumed as planar, we obtain the following equation for the energy of bending:

$$U_b = \frac{h^2}{24(1 - \nu_{xy}\nu_{yx})} \iint_{\Omega} \left[E_x \left(\frac{\partial^2 w}{\partial x^2} \right)^2 + 2\nu_{xy}E_x \frac{\partial^2 w}{\partial x^2} \frac{\partial^2 w}{\partial y^2} + \right.$$

$$\left. + E_y \left(\frac{\partial^2 w}{\partial y^2} \right)^2 + (1 - \nu_{xy}\nu_{yx})G_{xy} \left(\frac{\partial^2 w}{\partial x \partial y} \right)^2 \right] dx dy.$$

$$(10.35)$$

Here $w(x, y)$ is the buckling shape function within the delaminated area.

The critical buckling strains ε_x^* and ε_y^* can be found approximately using the variational approach. Consider the quadratic functional of the theory of elastic stability

$$V = \frac{h}{2} \iint_{\Omega} \left[s_x \left(\frac{\partial w}{\partial x} \right)^2 + s_y \left(\frac{\partial w}{\partial y} \right)^2 \right] dx dy \qquad (10.36)$$

where stresses in the middle surface of the delamination are taken equal to the applied stresses s_x and s_y (positive in compression). To go further, we have to assume the buckled shape $\varphi(x, y)$ of the delamination with the characteristic deflection f:

$$w = f\varphi(x, y; a_1, \ldots, a_m) \qquad (10.37)$$

Substituting (10.37) in (10.34) and (10.36) and equalizing the results, we obtain one of the required equations. To obtain the remaining equations, we use relations between displacements at the delamination boundary obtained in two ways: expressing them through deformation of the main bulk and through deformation of the delaminated part of the plate. In [13] the following relations were suggested to complete the set of equations to find the critical strains ε_x^* and ε_y^*:

$$\frac{1}{2}\iint\limits_{\Omega} \left(\frac{\partial w}{\partial x}\right)^2 dx\,dy = (e_x - \varepsilon_x^*)\Omega;$$

$$\frac{1}{2}\iint\limits_{\Omega} \left(\frac{\partial w}{\partial y}\right)^2 dx\,dy = (e_y - \varepsilon_y^*)\Omega; \tag{10.38}$$

Evidently, both sides in (10.38) have the meaning of length reduction of chords of the delamination averaged over the area Ω.

10.6 Elliptical Delamination

Consider a delamination that grows keeping the elliptical shape in the plane (Figure 10.11). Then the semi-axes a and b enter as G-coordinates. It is convenient to assume the buckling mode (10.37) as follows:

$$w(x,y) = f\left(1 - \frac{x^2}{a^2} - \frac{y^2}{b^2}\right)^2. \tag{10.39}$$

The right-hand side, evidently, satisfies the boundary conditions of clamping along the boundary S. Substituting (10.39) in (10.35) and (10.36) we obtain

$$U_b = \frac{4\pi f^2}{a^3 b^3}\left[D_x b^4 + \frac{2}{3}(D_{xy} + 2D_t)a^2 b^2 + D_y a^4\right]$$

$$V = \frac{\pi f^2 h}{3ab}(s_2 b^2 + s_1 a^2) \tag{10.40}$$

Here D_x and D_y are bending stiffnesses of the delamination, D_{xy} is mixed stiffness, and D_t is torsional stiffness:

$$D_x = \frac{E_x h^3}{12(1 - \nu_{xy}\nu_{yx})}, \quad D_y = \frac{E_y h^3}{12(1 - \nu_{xy}\nu_{yx})}$$

$$D_{xy} = \frac{\nu_{xy} E_x h^3}{12(1 - \nu_{xy}\nu_{yx})}, \quad D_t = \frac{G_{xy} h^3}{12}$$

The relationship between the critical strains $e_x^*(a,b)$ and $e_y^*(a,b)$ is to be found from condition $U_b = V$. On account of (10.35) and (10.36) this condition takes the form

$$(e_x^* + \nu_{xy}e_y^*)E_x b^2 + (e_y^* + \nu_{yx}e_x^*)E_y a^2 = \frac{h^2}{a^2 b^2}H(a,b) \tag{10.41}$$

where the following notation is used

$$H(a,b) = \frac{12(1 - \nu_{xy}\nu_{yx})}{h^2}\left[D_x b^4 + \frac{2}{3}(D_{xy} + 2D_t)a^2 b^2 + D_y a^4\right]$$

To close the set of equations with respect to e_x^*, e_y^*, and f, we use equations (10.38). Substituting the buckling mode given in (10.39), we obtain

$$f^2 = 3a^2(e_x - e_x^*) = 3b^2(e_y - e_y^*) \tag{10.42}$$

The combination of (10.41) and (10.42) results in the explicit equations for the critical strains

$$e_x^* = \frac{1}{H_1(a,b)}\left[E_y(a^2 + \nu_{yx}b^2)(e_x a^2 - e_y b^2) + \left(\frac{h}{a}\right)^2 H(a,b)\right]$$

$$e_y^* = \frac{1}{H_1(a,b)}\left[E_x(b^2 + \nu_{xy}a^2)(e_y b^2 - e_x a^2) + \left(\frac{h}{b}\right)^2 H(a,b)\right]$$

where the following notation is used

$$H_1(a,b) = E_x b^4 + 2\nu_{xy}E_x a^2 b^2 + E_y a^4$$

Then, returning to equation (10.42), we find the deflection f in the center of the delamination:

$$f^2 = \frac{3a^2 b^2}{H_1(a,b)}\left[E_x e_x(b^2 + \nu_{xy}a^2) + E_y e_y(a^2 + \nu_{yx}b^2) - \frac{h^2}{a^2 b^2}H(a,b)\right] \tag{10.43}$$

Combining equations (10.34), (10.40), and (10.43), we find the summed energy $U(a,b,e_x,e_y)$ and the generalized driving forces

$$G_a = -\frac{\partial U}{\partial a}, \quad G_b = -\frac{\partial U}{\partial b} \tag{10.44}$$

As to the generalized resistance forces, Γ_a and Γ_b, we have to return to equations (10.12).

The condition of nonpropagation of the delamination is $G_a < \Gamma_a, G_b < \Gamma_b$. When one of the inequalities is violated, the delamination begins to propagate in the corresponding direction. To illustrate the situation in detail, consider a very special case, a circular isotropic delamination under uniform compression. Let $a = b$, $E_x = E_y = E$, $\nu_{xy} = \nu_{yx} = \nu$, $e_x = e_y = \varepsilon_\infty$. Then equation (10.42) takes the form

$$f^2 = 3a^2(\varepsilon_\infty - \varepsilon_*)$$

with the critical strain

$$\varepsilon_*(a) = \frac{4}{3(1+\nu)}\frac{h^2}{a^2} \tag{10.45}$$

The formula for the energy of the system is given with a very simple equation similar to (10.29):

$$U = \text{const} - \frac{E\pi a^2 h}{1-\nu}(\varepsilon_\infty - \varepsilon_*)^2$$

The generalized driving force $G = -\partial U/\partial a$ is defined as

$$G = \frac{2E\pi ah}{1-\nu}(\varepsilon^2 - \varepsilon_*^2) \tag{10.46}$$

Virtual fracture work is $W_f = -\gamma[\pi(a + \partial a) - \pi a^2]$. Hence

$$\Gamma = 2\pi\gamma a \tag{10.47}$$

Note that the generalized forces G and Γ refer to the whole circle while in (10.44) and (10.13) they refer to one of two axes. Therefore, for example, Γ defined in (10.47) is equal to the double Γ in (10.13) at $a = b$.

The critical relationship between the radius of delamination and applied strain follows from the equality $G = \Gamma$. Taking into account equations (10.46) and (10.47) we obtain the critical radius

$$a^* = \frac{2h(\varepsilon_\infty^2 - \varepsilon_0^2)^{1/4}}{[3(1+\nu)]^{1/2}} \tag{10.48}$$

Here, in contrast to (10.25), the following notation is introduced:

$$\varepsilon_0 = \frac{\gamma(1-\nu)}{Eh}$$

A delamination under in-plane compression remains flat at $\varepsilon_\infty < \varepsilon_*$, where ε_* is defined in (10.45). At $\varepsilon_* < \varepsilon_\infty < \varepsilon_\infty^*$, where ε_∞^* is the root $\varepsilon_\infty(a)$ of equation (10.48), the delamination is buckled but nonpropagating. The G-equilibrium is attained at $\varepsilon_\infty = \varepsilon_\infty^*(a)$, and this equilibrium is unstable. In contrast with the buckled beam-like delamination (Figure 10.17b), no stable branches of the equilibrium curve exist for circular delaminations. The stability diagram for a circular delamination is shown in Figure 10.17c.

A number of more complicated problems were considered in the same manner, among them: initially buckled delaminations (Bolotin [13]); edge semi-elliptical delaminations with secondary cracks (Murzakhanov and Shchugorev [107]); elliptical delaminations in cylindrical shells (Kislyakov [77], Nefedov [108]); and in spherical shells (Nesin [109]).

An illustrative example is shown in Figure 10.18. A laminate cylindrical shell of transversally isotropic material is subjected to longitudinal loading with the applied strain ε_∞. The delamination whose dimensions are small in comparison with the radius of the shell R has an elliptical form with semi-axes a and b. The delamination is situated near the internal surface. The semi-axes a and b are plotted along the horizontal axes, and the applied compression strain ε_∞ is plotted along the vertical

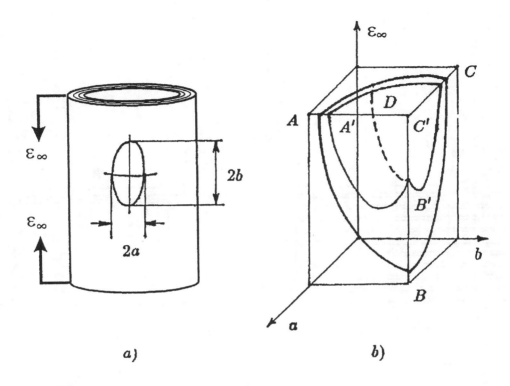

FIGURE 10.18
Elliptical delamination in cylindrical shell: (a) scheme of loading; (b) stability chart.

axis. Curve ABC is the trace of the buckling boundary. Curve A′B′C′ is the trace of the similar surface corresponding to one of the conditions $G_a = \Gamma_a, G_b = \Gamma_b$. At point B″ and at similar points of surface A′B′C′, stability of the equilibrium state transfers instability.

10.7 Interlaminar Damage in Cyclic Loading

Modeling of growth of delaminations in composite structures is similar to modeling of fatigue cracks in ordinary materials. The direction of growth is prescribed: it is the interlayer situated on the prolongation of the initial delamination. One of the mechanisms controlling this growth is the stress-strain field in the interlayer and the damage produced by interlayer cyclic stresses.

At least two types of stresses have to be included in the model: the normal stress σ_m directed perpendicular to the interface plane, and the shear stress τ_m in the

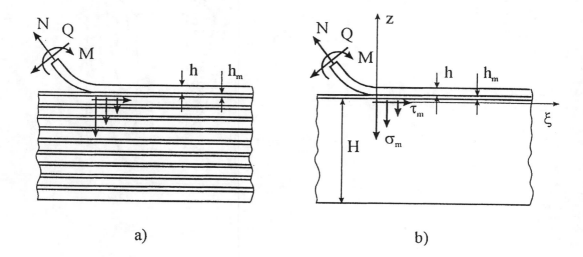

a)   b)

FIGURE 10.19
Interlaminar stresses in the vicinity of delamination: (a) multilayered model;
(b) three-layered model.

interface, i.e., acting along the expected growth direction (Figure 10.19). Therefore, at least two damage measures must be introduced: the measure ω_σ of mode I microcracking in the interlayer, and the measure ω_τ of mode II microcracking. Then we assume the specific work of debonding along the interlayer to be presented in the form:

$$\gamma = \gamma_0 f(\omega_\sigma, \omega_\tau) \qquad (10.49)$$

Here γ_0 is the specific fracture work for nondamaged material, and function $f(\omega_\sigma, \omega_\tau)$ is similar to the corresponding function in (5.24). Generally, γ_0 is varying along the contour of the delamination, $\gamma_0 = \gamma_0(s)$.

The simplest equations for damage accumulations are similar to (5.23):

$$\frac{\partial \omega_\sigma}{\partial N} = \left(\frac{\Delta \sigma_m - \Delta \sigma_{th}}{\sigma_d} \right)^{m_\sigma}, \quad \frac{\partial \omega_\tau}{\partial N} = \left(\frac{|\Delta \tau_m| - \Delta \tau_{th}}{\tau_d} \right)^{m_\tau} \qquad (10.50)$$

Material parameters σ_d, τ_d, $\Delta \sigma_{th}$, $\Delta \tau_m$, m_σ, and m_τ depend, in general, on temperature, frequency, as well as on the stress ratios R_σ and R_τ of extremal stresses

10-20

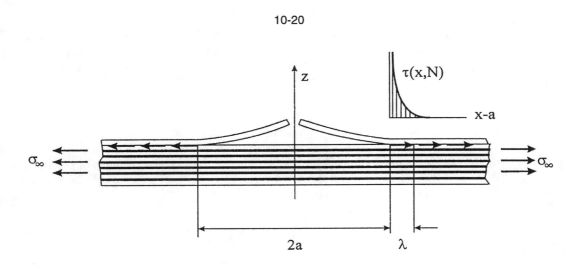

FIGURE 10.20
Open delamination under cyclic loading.

within a cycle.

The measures ω_σ and ω_τ depend on the position of the boundary point given with the arc s, on the distance ξ from the boundary counted along the normal (Figure 10.5) and, certainly, on the cycle number N. To predict the growth of delaminations, we deal mostly with the damage at the front, i.e., at $\xi = 0$. We denote these damage measures $\psi_\sigma(s, N)$ and $\psi_\tau(s, N)$. Then, to evaluate the generalized resistance forces, we use equation (10.49) at $\omega_\sigma = \psi_\sigma$, $\omega_\tau = \psi_\tau$. In particular, assuming that

$$\gamma = \gamma_0 \left[1 - (\psi_\sigma + \psi_\tau)^\alpha\right] \tag{10.51}$$

at $\alpha > 0$, we minimize the number of material parameters. Then $\psi = \psi_\sigma + \psi_\tau$ represents the total interlaminar damage directly at the delamination front.

The next problem is to evaluate the interlaminar stresses. Laminate composites are, in fact, multilayered structures that consist of a large number of layers with various mechanical properties. When a delamination is thin compared with the structural member under consideration, we may neglect the influence of delamination on the stiffness of the main bulk.

In general, the evaluation of interlaminar stresses is a rather complicated problem. Here we discuss a very simple case when a composite plate with an open delamination is subjected to cyclic tension/compression given with the applied strain ε_∞ (Figure 10.20). Considering the equilibrium of a small segment of the upper layer, we come to the equation with respect to the longitudinal displacement $u(x, N)$ of this layer:

$$E_x h \frac{d^2 u}{dx^2} - \frac{G_m}{h_m} u = \frac{G_m \varepsilon_\infty}{h_m}(x - a)$$

Here h and h_m are thicknesses of delamination and interlayer, respectively, E_x is the elastic modulus of the upper layer, and G_m is the effective shear modulus of the interlayer. In general, the stiffness G_m/h_m could be measured in direct shear tests. Boundary conditions are: $du/dx = 0$ at $|x| = a$ and $du/dx \to \varepsilon_\infty$ at $|x| \to \infty$. Hence

$$u(\xi) = \varepsilon_\infty \lambda_\tau \exp\left(-\frac{\xi}{\lambda_\tau}\right) + \varepsilon_\infty \xi$$

where $\xi = x - a$, and $\lambda_\tau = (h h_m E_x/G_m)^{1/2}$ is the characteristic length of the edge effect in interlaminar shear. Normal stresses σ_m in this case are negligible while the tangential stresses are connected with $u(\xi)$ as $\tau_m = (G_m/h_m)(u - \varepsilon_\infty \xi)$. Thus,

$$\tau_m = \frac{G_m \varepsilon_\infty \lambda_\tau}{h_m} \exp\left(-\frac{\xi}{\lambda_m}\right) \tag{10.52}$$

The condition of (non)propagation of the delamination, as usual, is taken as $G \gtrless \Gamma$. The generalized driving force for this problem is given in (10.23). The generalized resistance force differs from (10.24) as follows:

$$\Gamma = \gamma_0 b \left(1 - \psi_\tau^\alpha\right)$$

The tip damage ψ_τ is to be found as the solution of the second equation in (10.50). The cycle number N_* at the start of growth satisfies equation

$$\omega_0(a_0) + \int_0^{N_*} \left[\frac{|\Delta \tau_m(a_0, \nu)| - \Delta \tau_{th}}{\tau_d}\right]^m d\nu =$$

$$= \left\{1 - \left[\frac{\varepsilon_\infty^{max}(N_*)}{\varepsilon_0}\right]^2\right\}^{1/\alpha}$$

Here $\omega_0(x)$ is the initial damage measure; ε_0 is the critical strain (10.25). The subscript at m_τ is temporarily omitted. The equation with respect to the current size $a(N)$ of the growing delamination takes the form:

$$\omega_0[a(N)] + \int_{N_*}^{N} \left\{\frac{|\Delta \tau[a(\nu), \nu]| - \Delta \tau_{th}}{\tau_d}\right\}^m d\nu =$$

$$= \left\{1 - \left[\frac{\varepsilon_\infty^{max}(N)}{\varepsilon_0}\right]^2\right\}^{1/\alpha}$$

To obtain a more transparent result, consider the case when $|\Delta\tau_m| \gg \Delta\tau_{th}$. Using the quasistationary approximation, we come to the differential equation with respect to $a(N)$:

$$\frac{da}{dN} = \frac{\lambda_\tau}{m}\left(\frac{\Delta\varepsilon_\infty}{\varepsilon_d}\right)^m \left\{\left[1 - \left(\frac{\varepsilon_\infty^{\max}}{\varepsilon_0}\right)^2\right]^{1/\alpha} - \omega_{ff}\right\}^{-1} \qquad (10.53)$$

Here the notation ε_d is used for the characteristic resistance strain $\varepsilon_d = \tau_d h_m/(G_m\lambda_\tau)$.

The delamination length does not enter the right-hand side of (10.53). For the same composite, the delamination thickness h may be considered as the only characteristic length subject to varying in (10.53). Then the right-hand side in (10.53) is proportional to $(\Delta\varepsilon_\infty)^m h^{(m+1)/2}$. This means that the part of the stress intensity factor $K = Y\sigma_\infty(\pi a)^{1/2}$ is taken by the complex $\varepsilon_\infty h^{(m+1)/2m}$ or, in terms of applied stresses, by $\sigma_\infty h^{(m+1)/2m}$. At $m \gg 1$ we come to the complex whose dimension is the same as that of stress intensity factors.

10.8 Application of Mechanics of Multilayered Structures

Mechanics of multilayered structures (see the book by Bolotin and Novichkov [9]) was developed with special objectives to apply its results to laminate composites as well as to structural members (beams, plates, and shells) of these composites. Though the equations governing deformation of thin delaminations can be developed without using this theory, it seems expedient to discuss here its simplest version to show how to estimate interlaminar stresses in more complex situations.

For simplicity, let all the layers be planar, their materials be linear elastic and isotropic. Thus, the structure is an elastic plate composed of an arbitrary number of "rigid" layers and intermediate "soft" layers (Figure 10.20,a). The first kind of layers imitate the reinforcement, the second ones the matrix and the interface. In the simplest case, we treat "rigid" layers as classical thin plates following the Kirchhof-Love model. As to the "soft" layers, we take into account their transversal and shear deformations. We assume that the total picture may be described by the displacements $u_k(x, u), v_k(x, u), w_k(x, u)$ of "rigid" layers. As to the straining of "soft" layers, it is controlled by the displacements of neighboring "rigid" layers. This model was primarily suggested by Bolotin (1965). Evidently, such a model is a generalization of popular models for sandwich plates developed, however, with the special aim of applying it to modeling of laminate composite structures.

Equations of the theory in the above version are

$$A_k\left(\frac{\partial^2 u_k}{\partial x^2} + \frac{1-\nu_k}{2}\frac{\partial^2 u_k}{\partial y^2} + \frac{1+\nu_k}{2}\frac{\partial^2 v_k}{\partial x \partial y}\right) +$$

$$+ B_k\left(u_{k+1} - u_k + c'_k\frac{\partial w_k}{\partial x} + c''_k\frac{\partial w_{k+1}}{\partial x}\right) -$$

$$-B_{k-1}\left(u_k - u_{k-1} + c'_{k-1}\frac{\partial w_{k-1}}{\partial x} + c''_{k-1}\frac{\partial w_k}{\partial x}\right) + q_{xk} = 0,$$

$$A_k\left(\frac{\partial^2 v_k}{\partial y^2} + \frac{1-\nu_k}{2}\frac{\partial^2 v_k}{\partial x^2} + \frac{1+\nu_k}{2}\frac{\partial^2 u_k}{\partial x \partial y}\right) +$$

$$+B_k\left(v_{k+1} - v_k + c'_k\frac{\partial w_k}{\partial y} + c''_k\frac{\partial w_{k+1}}{\partial y}\right) - \qquad (10.54)$$

$$-B_{k-1}\left(v_k - v_{k-1} + c'_{k-1}\frac{\partial w_{k-1}}{\partial y} + c''_{k-1}\frac{\partial w_k}{\partial y}\right) + q_{yk} = 0,$$

$$D_k\Delta\Delta w_k - C_k(w_{k+1} - w_k) + C_{k-1}(w_k - w_{k-1}) -$$

$$-B_k c'_k\left[\frac{\partial}{\partial x}(u_{k+1} - u_k) + \frac{\partial}{\partial y}(v_{k+1} - v_k) + c'_k\Delta w_k + c''_k\Delta w_{k+1}\right] -$$

$$-B_{k-1}c''_{k-1}\left[\frac{\partial}{\partial x}(u_k - u_{k-1}) + \frac{\partial}{\partial y}(v_k - v_{k-1}) + c'_{k-1}\Delta w_{k-1} + c''_{k-1}\Delta w_k\right] = q_{zk}$$

$$(10.55)$$

The stiffnesses of layers are

$$A_k = \frac{E_k h_k}{1-\nu_k^2}, \quad B_k = \frac{\tilde{G}_k}{\tilde{h}_k}, \quad C_k = \frac{\tilde{E}_k}{\tilde{h}_k}, \quad D_k = \frac{E_k h_k^3}{12(1-\nu_k^2)}$$

where E_k and ν_k are Young's modulus and Poisson's ratio of the k-th "rigid" layer, $\tilde{E}_k$ and $\tilde{G}_k$ are Young's and shear moduli of the k-th "soft" layers. The thicknesses of these layers are denoted h_k and $\tilde{h}_k$ (Figure 10.20). In addition, the distances between the middle planes of neighboring layers are introduced

$$c'_k = \frac{h_k + \tilde{h}_k}{2}, \quad c''_k = \frac{h_{k+1} + \tilde{h}_k}{2} \qquad (10.56)$$

The first equation (10.54) describes longitudinal straining in the x-direction, the second one in the y-direction; equation (10.55) describes bending. Notation q_{xk}, q_{yk} and q_{zk} is used in (10.54) and (10.55) for external loads in corresponding directions applied to the k-th "rigid" layer. The total number of equations in (10.54) and (10.55) is equal to $3n$ where n is the number of "rigid" layers.

The natural boundary equations at $x = $ const are

$$A_k\left(\frac{\partial u_k}{\partial x} + \nu_k\frac{\partial v_k}{\partial y}\right) = N_k, \quad D_k\left(\frac{\partial^2 w_k}{\partial x^2} + \nu_k\frac{\partial^2 w_k}{\partial y^2}\right) = M_k,$$

$$A_k \frac{1-\nu_k}{2}\left(\frac{\partial u_k}{\partial y} + \frac{\partial v_k}{\partial x}\right) = S_k,$$

$$D_k\left[\frac{\partial^3 w_k}{\partial x^3} + (2-\nu_k)\frac{\partial^3 w_k}{\partial x \partial y^2}\right] - B_k c'_k \left(u_{k+1} - u_k + c'_k\frac{\partial w_k}{\partial x} + c''_k\frac{\partial w_{k+1}}{\partial x}\right) \quad (10.57)$$

$$-B_{k-1}c''_{k-1}\left(u_k - u_{k-1} + c'_{k-1}\frac{\partial w_{k-1}}{\partial x} + c''_{k-1}\frac{\partial w_k}{\partial x}\right) = Q_k + \frac{\partial T_k}{\partial x}$$

where N_k, S_k, Q_k, M_k and T_k are the longitudinal, tangential and transversal forces, bending and torsion moments, respectively.

The interlaminar stresses $\tilde{\tau}_k \equiv \tau_{xz}$ and $\tilde{\sigma}_k \equiv \sigma_z$ modelled as stresses in "soft" layer, are:

$$\tilde{\sigma}_k = C_k(w_{k+1} - w_k)$$

$$\tilde{\tau}_k = B_k\left(u_{k+1} - u_k + c'_k\frac{\partial w_k}{\partial x} + c''_k\frac{\partial w_{k+1}}{\partial x}\right) \quad (10.58)$$

When the delamination is thin, more precisely, at $h \ll a$, $h \ll H$ (H is the total plate thickness), we may consider the displacements of the plate as controlled by external loads. Let $k = 0$ for all variables concerning the main bulk of the plate, and $k = 1$ for the near-surface layer subjected to delamination. Then instead of (10.54) and (10.55) we obtain

$$A\left(\frac{\partial^2 u}{\partial x^2} + \frac{1-\nu}{2}\frac{\partial^2 u}{\partial y^2} + \frac{1+\nu}{2}\frac{\partial^2 v}{\partial x \partial y}\right) -$$

$$-\frac{G_m}{h_m}\left(u - u_0 + \frac{h+h_m}{2}\frac{\partial w}{\partial x} + \frac{H+h_m}{2}\frac{\partial w_0}{\partial x}\right) + q_x = 0$$

$$A\left(\frac{\partial^2 v}{\partial y^2} + \frac{1-\nu}{2}\frac{\partial^2 v}{\partial x^2} + \frac{1+\nu}{2}\frac{\partial^2 u}{\partial x \partial y}\right) -$$

$$\quad (10.59)$$

$$-\frac{G_m}{h_m}\left(v - v_0 + \frac{h+h_m}{2}\frac{\partial w}{\partial y} + \frac{H+h_m}{2}\frac{\partial w_0}{\partial y}\right) + q_y = 0$$

$$D\Delta\Delta w + \frac{E_m}{h_m}(w - w_0) - \frac{G_m}{h_m}\frac{h+h_m}{2}\left[\frac{\partial}{\partial x}(u - u_0)+\right.$$

$$\left. + \frac{\partial}{\partial y}(v - v_0) + \frac{h+h_m}{2}\Delta w + \frac{H+h_m}{2}\Delta w_0\right] = q_z$$

Here the subscript $k = 1$ is omitted. In particular, the notation $\tilde{h}_1 \equiv h_m$ is used for the effective thickness of the interlayer. Stiffnesses E_m/h_m and G_m/h_m can be found by direct experimentation, and this allows the estimation of the thickness h_m.

Boundary conditions take the form

$$A\left(\frac{\partial u}{\partial x} + \nu\frac{\partial v}{\partial y}\right) = N, \quad D\left(\frac{\partial^2 w}{\partial x^2} + \nu\frac{\partial^2 w}{\partial y^2}\right) = M,$$

$$A\frac{1-\nu}{2}\left(\frac{\partial u}{\partial y} + \frac{\partial v}{\partial x}\right) = S,$$

$$D\left[\frac{\partial^3 w}{\partial x^3} + (2-\nu)\frac{\partial^3 w}{\partial x\partial y^2}\right] - \frac{G_m}{h_m}\frac{h+h_m}{2}\left(u - u_0 + \frac{h+h_m}{2}\frac{\partial w_k}{\partial x} + \frac{H+h_m}{2}\frac{\partial w_0}{\partial x}\right) =$$

$$= Q + \frac{\partial T}{\partial x}$$

$$(10.60)$$

Instead of (10.58) we obtain the equation for interlaminar stresses

$$\sigma_m = \frac{E_m}{h_m}(w - w_0),$$

$$\tau_m = \frac{G_m}{h_m}\left(u - u_0 + \frac{h+h_m}{2}\frac{\partial w}{\partial x} + \frac{H+h_m}{2}\frac{\partial w_0}{\partial x}\right) \qquad (10.61)$$

Of course, equations (10.59)-(10.61) are easy to develop from direct consideration of the thin layer attached to a thick plate with an intermediate thin layer subjected to transversal and shear straining. In particular, one could abstain from modeling the main member as a classical plate. To do that, it is sufficient to introduce displacements on the surface of the main bulk, more precisely, on the middle plane of the interlayer. In the context of equations (10.59)-(10.61) it means that the substitution is performed:

$$u_s = u_0 - \frac{H+h_m}{2}\frac{\partial w_0}{\partial x}, \quad v_s = v_0 - \frac{H+h_m}{2}\frac{\partial w_0}{\partial y}, \quad w_s = w_0$$

However, the problem is not to be oversimplified. Some equations of some sandwich plates and shells are not irreproachable (they are not invariant with respect to rigid rotations). From this viewpoint, equations (10.59) deal with characteristic distances such as $(h + h_m)/2$ and $(H + h_m)/2$ more accurately.

In one-dimensional problems, such as depicted in Figure 10.20b, the analytical solution for interlayer stresses is available. Actually, instead of equations (10.59) we have:

$$E_x h \frac{d^2 u}{d\xi^2} - \frac{G_m}{h_m} \left(u + \frac{h + h_m}{2} \frac{\partial w}{\partial \xi} \right) = -\frac{G}{h_m} u_s$$

$$D_x \frac{d^4 w}{d\xi^4} + \frac{E_m}{h_m} w - \frac{G_m}{h_m} \frac{h + h_m}{2} \frac{d}{d\xi} \left(u + \frac{h + h_m}{2} \frac{dw}{d\xi} \right) = \quad (10.62)$$

$$= \frac{E_m}{h_m} w_s - \frac{G_m}{h_m} \frac{h + h_m}{2} \frac{du_s}{d\xi}$$

Here E_x and D_x may be interpreted in various ways, for example, as $E_x = E(1 - \nu^2)^{-1}$, $D_x = (Eh^3/12)(1 - \nu^2)^{-1}$ for isotropic delamination in the plane-strain state. The characteristic equation at $h_m \ll h$ takes the form

$$\left(r^2 - r_\tau^2 \right) \left(r^4 - 3r_\tau^2 r^2 + r_\sigma^4 \right) - 3r_\tau^4 r^2 = 0 \quad (10.63)$$

where the notation used is

$$r_\sigma = \left(\frac{E_m}{D_x h_m} \right)^{1/4}, \quad r_\tau = \left(\frac{G_m}{E_x h h_m} \right)^{1/2}$$

Parameters r_σ and r_τ are connected with the characteristic lengths of edge effects in interlaminar transversal deformation $\lambda_\sigma = r_\sigma^{-1}$, and in shear $\lambda_\tau = r_\tau^{-1}$. The roots of equation (10.63) may be calculated analytically or numerically. Denote the negative real root r_1, the complex roots with the negative real parts r_2 and r_3. Then

$$u = A_1 \exp \rho_1 \xi + (A_2 \cos \rho_3 \xi + A_3 \sin \rho_3 \xi) \exp \rho_2 \xi + u_s(\xi)$$

$$w = B_1 \exp \rho_1 \xi + (B_2 \cos \rho_3 \xi + B_3 \sin \rho_3 \xi) \exp \rho_2 \xi + w_s(\xi)$$

where $\rho_1 = r_1$, $\rho_2 = \text{Re}\, r_{2,3}$, $\rho_3 = \text{Im}\, r_{2,3}$. The coefficients A_k and B_k are to be found from boundary conditions stated at the front of the delamination, i.e., at $\xi = 0$. Then the interlayer stresses are to be calculated according to equations (10.61):

$$\sigma_m = \frac{E_m}{h_m} (w - w_s), \quad \tau_m = \frac{G_m}{h_m} \left(u + \frac{h + h_m}{2} \frac{\partial w}{\partial x} - u_s \right)$$

10.9   Growth of Delamination in Cyclic Compression

If a structural member with a near-surface delamination is subjected to axial cyclic compression, the interlaminar damage takes place only in the case when the delamination is buckled (Figure 10.21). The problem of buckling was considered in Section 10.5. In particular, the generalized driving force for a buckled delamination

10-21

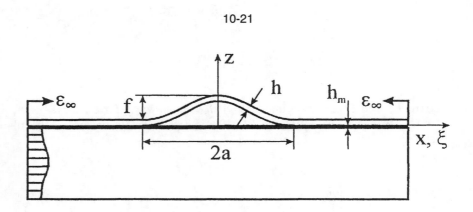

FIGURE 10.21
Buckled delamination under cyclic loading.

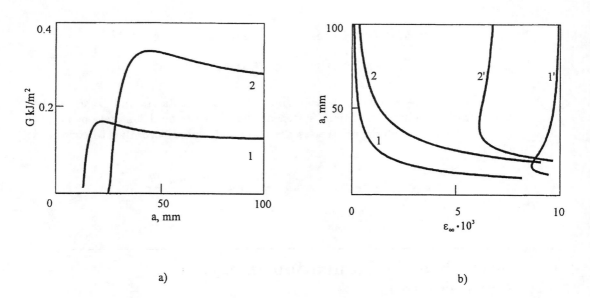

a)   b)

FIGURE 10.22
Generalized driving force (a) and boundaries of instability regions (b) at $h = 1$ and 2 mm.

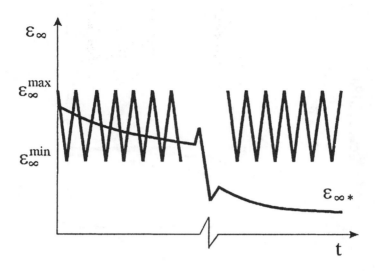

FIGURE 10.23
Patterns of delamination growth as a function of maximal strain (a) and cycle number (b).

is given in equation (10.30). This equation is illustrated in Figure 10.22a where the driving force relating to the unity width is plotted against the delamination half-length a. The following numerical data are used: $E_x = 10$ GPa, $\varepsilon_\infty = 5 \cdot 10^{-3}$. Line 1 is drawn for the delamination thickness $h = 1$ mm, line 2 for $h = 2$ mm. The driving force becomes equal to zero before buckling occurs. The thicker the delamination, the longer the delaminated section of the upper layer. However, the driving force for thicker delaminations attains its maximum soon after buckling.

There is a rather wide gap between the boundaries of buckling and instability in Griffith's sense. It is illustrated in Figure 10.22b drawn for the same data as Figure 10.22a. In addition, it is assumed that $\gamma_0 = 0.5\,\text{kJ/m}^2$. Lines 1 and 2 correspond to the boundary buckling at $h = 1$ and 2 mm, lines 1' and 2' to the boundary of unstable crack propagation along the structural member. In the domain between lines 1 and 1', 2 and 2' we expect the growth of delamination if the structural member is subjected to cyclic loading.

In general, three patterns of growth are to be distinguished. They are schematically illustrated in Figure 10.23. In case 1, continuous growth takes place after the initiation stage. In case 2, consecutively, we observe the initiation stage, a comparatively short stage of continuous growth, a jump-wise propagation to cross the instability domain, and the return to continuous growth. In case 3, the continuous process terminates shortly resulting in the total spreading of the delamination. Evidently, special initial data and loading conditions must be chosen to realize patterns 2 and 3. This is seen in Figure 10.22b where the domain in discussion is situated between instability boundaries 1 and 1', 2 and 2'.

Later on, we consider the most typical situation when the pattern labelled in

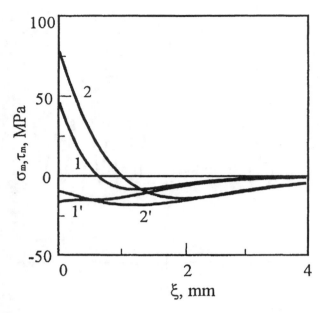

FIGURE 10.24
Interlaminar stresses σ_m and τ_m at the delamination front (a) and ahead of the front at $a = 50$ mm.

Figure 10.23 as case 1 takes place. A delamination does not propagate at $G < \Gamma$, and begins to propagate in a continuous pattern when the equality $G = \Gamma$ is primarily attained. Thus, the equation of the continuous growth is

$$G\left(\sigma_\infty^{\max}, a\right) = \Gamma\left(\psi_\sigma, \psi_\tau\right) \tag{10.64}$$

The most laborious part of computations is the evaluation of the interlaminar stresses σ_m and τ_m. The solution of equation (10.59) is to be found satisfying the boundary conditions (10.60). The buckling mode is taken according to (10.28). Some numerical examples are presented in Figures 10.24. They are obtained for $E_x = 10$ GPa, $G_m = 1$ GPa, $h_m = 0.1$ mm. Figure 10.24a shows the stresses σ_m and τ_m at the delamination front as a function of the size a. Lines 1 and 2 are drawn for σ_m at $h = 1$ and 2 mm, respectively. Lines $1'$ and $2'$ are drawn for τ_m. The distribution of these stresses ahead of the front is shown in Figure 10.24b drawn for $a = 50$ mm. The same notation is used as in Figure 10.24a.

The histories of delamination growth are presented in Figure 10.25. In addition to the above numerical data, it is assumed that in equations (10.50) $\sigma_d = 500$ MPa, $\tau_d = 250$ MPa, $\Delta\sigma_{th} = 10$ MPa, $\Delta\tau_{th} = 5$ MPa, $m_\sigma = m_\tau = 2$. The exponent in (10.50) is taken as $\alpha = 1/2$. Loading is given with the extremal strains $\varepsilon_\infty^{\max} = 5 \cdot 10^{-3}$, $\varepsilon_\infty^{\min} = 3 \cdot 10^{-3}$. Note that at $a = 50$ mm the critical strains in Euler's and Griffith's sense are $\varepsilon_{\infty *} = 3.29 \cdot 10^{-3}$ and $\varepsilon_\infty^* = 9.69 \cdot 10^{-3}$. Hence, during a part of the cycle, at least in the beginning, the process of growth is accompanied by closure of the delamination. This situation is illustrated in Figure 10.26 where the cyclic and critical strains are schematically depicted. Of course, this is closure in the literal sense, not the same as the crack closure effect in ordinary fatigue.

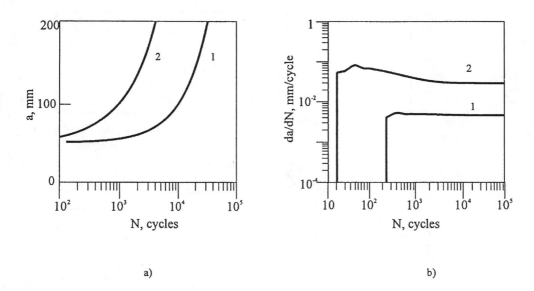

a)   b)

FIGURE 10.25
Histories of delamination size (a) and growth rate (b); notation as in Figure 10.22.

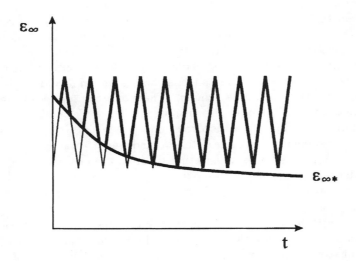

FIGURE 10.26
Closure effect for buckled delaminations.

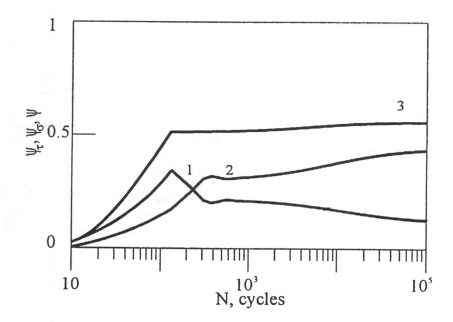

FIGURE 10.27
Histories of damage measures ψ_σ, ψ_τ, **and** $\psi = \psi_\sigma + \psi_\tau$ **(lines 1, 2, and 3).**

The crack size versus the cycle number is plotted in Figure 10.25a, and the crack growth rate in Figure 10.25b. Lines 1 correspond to $h = 1$ mm, lines 2 to $h = 2$ mm. Near the growth start the rate da/dN is high due to the damage accumulated during the initiation stage. Then the rate becomes practically constant. The cause is that the contribution of normal interlaminar stresses in the total interlaminar damage diminishes in the process of cracking while the contribution of shear stresses approaches a fixed level. This phenomenon is understandable; the delamination becomes less rigid during its growth. Therefore, the forces and moments in the right-hand side of (10.60) are diminishing.

Figure 10.27 illustrates the evolution of damage in the process of cracking. Line 1 corresponds to ψ_σ, line 2 to ψ_τ, and line 3 to $\psi = \psi_\sigma + \psi_\tau$. Note that the summed damage measure is rather far from the critical level. In this example the initial magnitude of ψ is approximately equal to half of $\psi_* = 1$. In particular, at $h = 1$ mm we have $G \approx 0.2\,\mathrm{kJ/m}^2$ while the assumed initial fracture work is $\gamma_0 = 0.5\,\mathrm{kJ/m}^2$.

Under certain conditions, the process of growth terminates when the delamination attains a certain length. This is typical for higher magnitudes of the resistance threshold stresses $\Delta\sigma_{th}$ and $\Delta\tau_{th}$. As an example, the histories of growth are presented in Figure 10.28. The same numerical data are used as for Figure 10.25. The only difference is that it is assumed $\Delta\sigma_{th} = 25$ MPa, $\Delta\tau_{th} = 10$ MPa (lines 1) and $\Delta\sigma_{th} = 25$ MPa, $\Delta\tau_{th} = 15$ MPa (lines 2). In the second case we have $|\Delta\tau_m| < \Delta\tau_{th}$

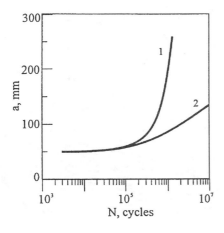
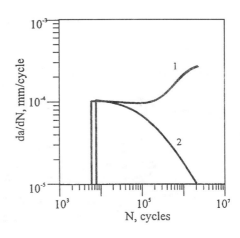

a)   b)

FIGURE 10.28
Histories of delamination size (a) and growth rate (b) at $\Delta \tau_{th} = 10$ and 15 MPa
(lines 1 and 2).

at all considered $a \geq a_0 = 50$ mm. Therefore no shear damage occurs in the interlayer. All the damage refers to mode I microcracking in the interlayers. When $a(N)$ increases, the stress range $\Delta \sigma_m(N)$ decreases. At $\Delta \sigma_m \leq \Delta \sigma_{th}$ no further damage accumulation occurs, and we observe the arrest of delamination growth.

10.10 Growth of Delaminations in Cyclic Bending

Let a composite beam or plate with a delamination be subjected to cyclic bending (Figure 10.29). The problem is similar to that considered in Section 10.9. Really, the damage and cracking are controlled by the applied strain $\varepsilon_\infty(t)$ which may be connected to the bending moment $M(t)$ as $\varepsilon_\infty \approx 6M/E_x bH^2$. The most important difference compared with the axial compression is the behavior of the buckled delamination.

Assume the buckling mode of the delamination as follows:

$$w = f \cos^2 \frac{\pi x}{2a} + \kappa \frac{x^2}{2}, \quad |x| \leq a \tag{10.65}$$

Here f is the deflection in the center of the delamination; κ is the curvature of the main structural member, $\kappa \approx 12M/E_x H^3$. The right-hand side in (10.65) satisfies the boundary conditions when small; however, finite deflections are considered. In fact, $w = \frac{1}{2}\kappa a^2$, $\partial w/\partial x = \kappa a$ at $x = \pm a$.

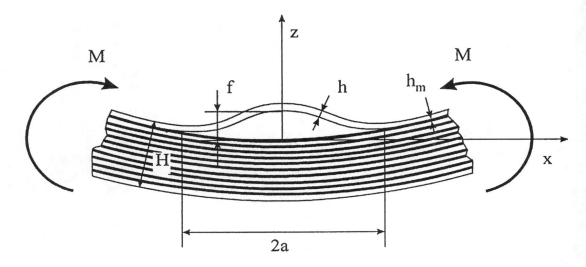

FIGURE 10.29
Buckled delamination under cyclic bending.

To estimate the deflection f, let us use the same approach as in Section 10.7. Equalizing the shortening distances between the tips measured in two ways, along the main member and along the delamination, we obtain

$$\varepsilon_\infty a = \frac{1}{2} \int_0^a \left(\frac{\partial w}{\partial x} \right)^2 dx + \varepsilon_*(a)a \tag{10.66}$$

The terms of higher order are omitted, and the membrane postbuckling strain in the delamination is taken equal to Euler's critical strain (10.26). Substitition (10.65) in (10.66) gives the equation

$$f^2 - \frac{8a^2 \kappa f}{\pi^2} + \frac{8a^4 \kappa^2}{3\pi^2} - \frac{16a^2}{\pi^2} (\varepsilon_\infty - \varepsilon_*) = 0 \tag{10.67}$$

where the applied strain ε_∞ and the applied curvature κ are entering independently. Thus, most of the further results are applicable both to bending and to bending accompanied by axial loading of the main member. At $\kappa \to 0$ equation (10.67) results in the common approximate formula for postcritical buckling. To receive the case of pure bending, it is sufficient to put $\varepsilon_\infty = \frac{1}{2}\kappa H$. Then one of the roots of equation (10.67)

$$f = \frac{4\kappa a^2}{\pi^2} \left\{ 1 \pm \left[1 - \frac{\pi^2}{6} + \frac{\pi^2}{\kappa^2 a^2} (\varepsilon_\infty - \varepsilon_*) \right]^{1/2} \right\}$$

corresponds to an ascending branch of the $f(\kappa)$ curve, another one to the descending branch. Two critical curvatures, the upper one κ_* and the lower one κ_{**}, enter the

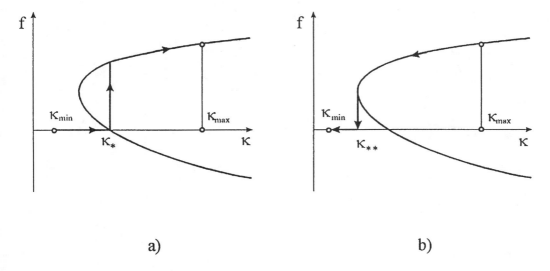

a)   b)

FIGURE 10.30
Cyclic straining accompanied by snap-through behavior: (a) loading; (b) downloading.

picture (Figure 10.30). Because of this, bending of a structural member may be accompanied by snap-through phenomena of its delaminated section.

To calculate the generalized driving force, we use an equation similar to (10.27). As a result we obtain

$$U = \text{const} - E_x abh \left(\varepsilon_\infty - \varepsilon_*^2 - \frac{h^2}{12}\kappa^2 - \frac{\pi^2}{8a^2}\varepsilon_* f^2 \right)$$

where f is the larger root of equation (10.67). Then, taking into account that $G = -\partial U/\partial a$, we obtain:

$$G = \frac{E_x h}{2} \left[\varepsilon_\infty^2 + 3\varepsilon_* - \frac{h^2}{12}\varepsilon_* f^2 + \frac{4(\varepsilon - 2\varepsilon_*)\varepsilon_*}{1 - (4\kappa a^2/\pi^2 f)} \right] \qquad (10.68)$$

A detailed parametrical analysis of the delamination growth in structural members subjected to bending was performed by Bolotin and Nefedov [38]. The analysis was based on equations (10.50), (10.51), (10.64), (10.65), and (10.68). The influence of loading and initial conditions as well as of material parameters was studied. Special attention was given to the influence of the interlayer stiffness characterized by parameters E_m/h_m and G_m/h_m.

As an example, the size of the delamination and the growth rate are plotted in Figure 10.31 versus the cycle number for interlayers with three various stiffness parameters. The same numerical data are used as for Figure 10.25. The principal difference is the change of E_m and G_m. Lines 1, 2, 3 are drawn for $E_m/E_x = 0.05, 0.075$ and 0.1, respectively. As one could expect, the higher the interlayer stiffness, the higher the interlaminar stresses and damage accumulation rates and,

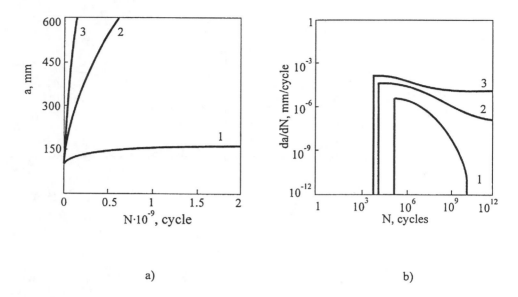

a)   b)

FIGURE 10.31
**Influence of interlayer stiffness on the growth of delaminations; lines 1, 2, 3
correspond to $E_m/E_x = 0.05, 0.075, 1.0$.**

as a result, the shorter the fatigue life. In our case, we observe three patterns: the
arrest of growth at lower stiffness (line 1), the gradual decreasing of the growth rate
with the cycle number (line 2), and the apparent tendency to a constant growth
rate (line 3).

A number of related problems were considered in the papers by Bolotin [13],
Bolotin and Nefedov [38], Murzakhanov and Shchugorev [107], Nesin [109] and
Shchugorrev [131]. Stability and load-carrying capacity of columns, plates and
shells containing delaminations, were considered in [10]. A survey of publications
is given in the papers [132, 139], where references to the Western literature can be
found.

10.11 Summary

Several aspects of fracture and fatigue of laminate composite structures were discussed in this chapter beginning with the properties of fracture toughness in strongly
anisotropic materials to the patterns of growth of delaminations with the special
attention to near-surface delaminations. Laminate composites and composite structures are considered in this chapter in the framework of the theory of multilayered
structures with the interpretation of more compliant interlayers as intermediate
"soft" layers.

Some items considered here are rather unconventional. Among them is the attempt to describe the interlaminar fracture work in laminate composite in tensorial terms. Another item is the modeling of growth of delaminations under cyclic loading taking into account buckling of the delaminated domain. In addition, two different types of interlaminar damage were included in the model, and the comparative contributions of these types were studied.

A number of related problems were left out of the discussion. Among them are the problems of fatigue growth of elliptical delaminations in plates, spherical and cylindrical shells. Some of them were briefly discussed in the text, and the references to literature are given.

REFERENCES

[1] T.L. Anderson: Fracture Mechanics. Fundamentals and Applications. CRC Press, Boca Raton, 1991

[2] S.N. Atluri, Ed.: Computational Methods in Mechanics of Fracture. North-Holland, Amsterdam, 1986

[3] S.N. Atluri: Structural Integrity and Durability. Tech Science Press, Forsyth, 1997

[4] N.V. Banichuk: Determination of the form of curvilinear crack by small parameter technique, Izvestiya AN SSSR, Mekhanika Tverdogo Tela, N 2, 130-137 (1970) (in Russian)

[5] J.A. Bannatine, J.J.Comer, and J.L.Handrock: Fundamentals of Metal Fatigue Analysis. Prentice-Hall, Englewood Cliffs, 1990

[6] Z.P. Bažant: Mechanics of distributed cracking. Applied Mechanics Review, 39, 675-705 (1986)

[7] V.V. Bolotin: Statistical Methods in Structural Mechanics. Holden Day, San Francisco, 1968 (Russian editions of 1961 and 1985)

[8] V.V. Bolotin: On safe dimensions of cracks under random loading, Izvestiya AN SSSR, Mekhanika Tverdogo Tela, N 1, 124-130 (1980) (in Russian)

[9] V.V. Bolotin and Yu.N. Novichkov: Mechanics of Multilayered Structures, Mashinostroyeniye, 1980 (in Russian)

[10] V.V. Bolotin, Z.Kh. Zebelyan and A.A.Kurzin: Stability of compressed structural members with delaminations, Problemy Prochnosti, N 7, 3-8 (1980) (in Russian)

[11] V.V. Bolotin: Stochastic models of fracture in unidirectional fiber composites, Mekhanika Kompozitnykh Materialov, N 3, 404-420 (1981)

[12] V.V. Bolotin and V.P. Tamuzs: On length distribution of ruptured fibers in unidirectional composites, Mekhanika Kompozitnykh Materialov, N 6, 1107-1110 (1982) (in Russian)

[13] V.V. Bolotin: Delamination defects in structures of composite materials, Mekhanika Kompozitnykh Materialov, N 2, 239-255 (1984) (in Russian)

[14] V.V. Bolotin: A unified approach to damage accumulation and fatigue crack growth, Engineering Fracture Mechanics, 22, 387-398 (1985)

[15] V.V. Bolotin: A unified model of fatigue and fracture with applications to structural reliability, in Random Vibration, Status and Recent Developments (S.H.Crandall Festschrift), Elsevier, Amsterdam, 47-58 (1986)

[16] V.V. Bolotin: Fracture and fatigue of composite plates and shells, in Inelastic Behavior of Plates and Shells, IUTAM Symposium Rio de Janeiro (J.Bevilacqua, R.Feijoo, and R.Valid, Eds.), Springer, Heidelberg, 131-161 (1986)

[17] V.V. Bolotin: Mechanics of stress corrosion cracking, Izvestiya AN SSSR, Mekhanika Tverdogo Tela, N 4, 20-26 (1987) (in Russian)

[18] V.V. Bolotin: Stability and growth of fatigue cracks, Izvestiya AN SSSR, Mekhanika Tverdogo Tela, N 4, 133-140 (1988) (in Russian)

[19] V.V. Bolotin: Prediction of Service Life for Machines and Structures. ASME Press, New York, 1989 (Russian editions of 1984 and 1990)

[20] V.V. Bolotin: Reliability of composite structures, in Handbook of Composites, Vol. 2, Structures and Design (C.T.Herakovich and Yu.M.Tarnopolsky, Eds.), North-Holland, Amsterdam, 264-349 (1989)

[21] V.V. Bolotin: Mechanics of fatigue fracture, in Nonlinear Fracture Mechanics, CISM Course 314, M.P. Wnuk, Ed., Springer, Wien-New York, 1-69 (1990)

[22] V.V. Bolotin and S.G. Minokin: Static fatigue crack growth in visco-elastic media, Izvestiya AN SSSR, Mekhanika Tverdogo Tela, N 1, 128-138 (1991) (in Russian)

[23] V.V. Bolotin: On generalized forces in analytical fracture mechanics, in Novozhilovsky Zbornik (N.S.Solomenko, Ed.), Sudostroyeniye, St.Peterburg, 161-170 (1992) (in Russian)

[24] V.V. Bolotin, G.Kh. Murzakhanov and V.N. Shchugorev: Fracture toughness and fatigue life of composite laminate structures under combined loading, in Mechanics of Composite Structures (V.D.Protasov, Ed.), Mashinostroyeniye, 61-101 (1992) (in Russian)

[25] V.V. Bolotin and B.V. Minakov: Crack growth and fracture in creep conditions, Izvestiya RAN, Mekhanika Tverdogo Tela, N 3, 147-150 (1992) (in Russian)

[26] V.V. Bolotin: Random initial defects and fatigue life prediction, in Stochastic Approaches to Fatigue, CISM Course 334, K. Sobczyk, Ed., Springer-Verlag, Wien, 126-163 (1993)

[27] V.V. Bolotin and A.A. Grishko: Numerical modeling of fracture of laminate composites under impact, Izvestiya RAN, Mekhanika Tverdogo Tela, N 3, 151-160 (1993) (in Russian)

[28] V.V. Bolotin and V.M. Kovekh: Numerical simulation of fatigue crack growth in a media with microdamage, Izvestiya RAN, Mekhanika Tverdogo Tela, N 2, 132-142 (1993) (in Russian)

[29] V.V. Bolotin: Mechanics of fatigue crack growth as a synthesis of micro- and macromechanics of fracture, in Handbook of Fatigue Crack Propagation in Metallic Structures, A. Carpinteri, Ed., Elsevier, Amsterdam, 883-911 (1994)

[30] V.V. Bolotin, B.V. Minakov and V.P. Chirkov: Influence of initial conditions on the start and growth of fatigue cracks, Izvestiya RAN, Mekhanika Tverdogo Tela, N 1, 73-79 (1994) (in Russian)

[31] V.V. Bolotin and V.L. Lebedev: Mechanics of fatigue crack growth in media with microdamage, Prikladnaya Matematika i Mekhanika, 59(2), 307-317 (1995) (in Russian)

[32] V.V. Bolotin: Stability Problems in Fracture Mechanics. John Wiley, New York, 1996

[33] V.V. Bolotin: Meandering propagation of fatigue cracks through solids with randomly distributed properties, in Advances in Nonlinear Stochastic Mechanics, A.Naess and S. Krenk, Eds., Kluwer, Dordrecht, 47-58 (1996)

[34] V.V. Bolotin: Brush-like modes of fatigue fracture in unidirectional fibre composites, International Journal of Non-Linear Mechanics, 31, 823-836 (1996)

[35] V.V. Bolotin and V.M. Kovekh: Effect of microdamage on fatigue crack propagation, Prikladnaya Matematika i Mekhanika, 60, 1029-1038 (1996) (in Russian)

[36] V.V.Bolotin and V.L. Lebedev: Analytical model of fatigue crack growth retardation due to overloading, International Journal of Solids and Structures, 33, 1229-1242 (1996)

[37] V.V.Bolotin, V.L. Lebedev, G.Kh. Murzakhanov, and S.V. Nefedov: A model of low-cycle fatigue for a penny-shape crack, Izvestiya RAN, Mekhanika Tverdogo Tela, N 2, 132-142 (1996) (in Russian)

[38] V.V. Bolotin and S.V. Nefedov: Growth of thin delaminations in laminate composite beams under cyclic bending, Mechanics of Composite Materials and Structures, 3, 275-295 (1996)

[39] V.V. Bolotin and A.A. Shipkov: A model of environmentally assisted fatigue crack growth, Prikladnaya Matematika i Mekhanika, 62(2), 314-323 (1998) (in Russian)

[40] H.E. Boyer, Ed.: Atlas of Fatigue Curves. ASM International, Metals Park, 1986

[41] D. Broek: The Practical Use of Fracture Mechanics. Kluwer, Dordrecht, 1988

[42] L.J. Broutman and R.H. Krock: Composite Materials. Vol 5. Fracture and Fatigue. Academic Press, New York, 1974

[43] H.H. Bryan and K.K.Ahuja: Review of crack propagation under unsteady loading. AIAA Journal, 31, 1077-1089 (1993)

[44] B. Budiansky and J.W. Hutchinson: Analysis of closure in fatigue crack growth, Journal of Applied Mechanics, ASME, 45, 267-276 (1978)

[45] B. Budiansky, J.W. Hutchinson and A.G. Evans: Matrix fracture in fiber-reinforced ceramics, Journal of Mechanics and Physics of Solids, 34, 167-189 (1986)

[46] O. Buxbaum: Betriebsfestigkeit. Sichere und wirtschaftliche Bemessung schwingbruchgefährdeter Bauteile. Stahleisen, Düsseldorf, 1986 (in German)

[47] A. Carpinteri, Ed.: Handbook on Fatigue Crack Propagation in Metallic Structures, Elsevier, Amsterdam, 1994

[48] J.-L. Chaboche: Continuum damage mechanics, Journal of Applied Mechanics, ASME 55, 59-72 (1988)

[49] D.L. Chen, B. Weiss and R. Stickler: The effective fatigue threshold: significance of the loading cycle below the crack opening load, International Journal of Fatigue, 16, 485-491 (1994)

[50] G.P. Cherepanov: Mechanics of Brittle Fracture. McGraw-Hill, New York, 1979 (Russian edition of 1974)

[51] G.P. Cherepanov: Fracture Mechanics of Composite Materials. Nauka, Moscow, 1983 (in Russian)

[52] T.W. Chooker and B.N. Leiss, Eds.: Corrosion Fatigue: Mechanics, Metallurgy, Electrochemistry, and Engineering. ASTM STP 801. ASTM Press, Philadelphia, 1983

[53] R.M. Christensen: A theory of crack growth in viscoelastic materials, Engineering Fracture Mechanics, 14, 215-225 (1981)

[54] A. Chudnovsky and S. Wu: Effect of crack-microcrack interactions on energy release rates, International Journal of Fracture, 44, 543-566 (1990)

[55] B. Cottrell and J.R. Rice: Slightly curved cracks, International Journal of Fracture, 16, 155-169 (1980)

[56] O. Ditlevsen and K. Sobczyk: Random fatigue crack growth with retardations, Engineering Fracture Mechanics, 24, 861-878 (1986)

[57] A.G. Evans and J.W. Hutchinson: On the mechanics of delamination and spalling in compressed films, International Journal of Solids and Structures, 20(5), 455-466 (1984)

[58] F.P. Ford: Status of research on environmentally assisted cracking in LWR pressure vessel stress, Journal of Pressure Vessel Technology, ASME, 110, 113-128 (1988)

[59] N.E. Frost, J.K. Marsh and L.P. Pook: Metal Fatigue. Clarendon Press, Oxford, 1974

[60] E. Gdoutos: Problems of Mixed Mode Crack Propagation. Martinus Nijhoff, The Hague, 1984

[61] R.V. Goldstein and V.M. Yentov: Qualitative Methods in Continuum Mechanics. Nauka, Moscow, 1989 (in Russian)

[62] D. Gross: Bruchmechanik. Springer-Verlag, Berlin, 1990 (in German)

[63] A. Harboletz, B. Weiss and R. Stickler: Fatigue threshold in metallic materials – a review, in Handbook of Fatigue Crack Propagation in Metallic Structures, A. Carpinteri, Ed., Elsevier, Amsterdam, 847-882, 1994

[64] P. Harlow, S. Phoenix: The chain of bundles probability model for the strength of fibrous materials, Journal of Composite Materials, 12, 195-213 (1978)

[65] D.R. Hayhurst, P.R. Brown and G.J. Morrisson: The role of continuum damage in creep crack growth, Philosophical Transactions, Royal Society of London, A311, 1516, 131-158 (1984)

[66] K. Hellan: Introduction to Fracture Mechanics. McGraw-Hill, New York, 1984

[67] C.T. Herakovich and Yu.M. Tarnopolsky, Eds.: Handbook of Composites, Structures and Design. North-Holland, Amsterdam, 1989

[68] R.W. Hertzberg: Deformation and Fracture Mechanics of Engineering Materials. Wiley, New York, 1989

[69] R.W. Hertzberg and J.A. Manson: Fatigue of Engineering Plastics. Academic Press, New York, 1980

[70] J.P. Hirth and H.H. Johnson: Hydrogen problems in energy related technology, International Journal of Corrosion, 32, 3-15 (1976)

[71] J.W. Hutchinson: Nonlinear Fracture Mechanics. Technical University Denmark, Lyngby, 1979

[72] J.W. Hutchinson: Micro-mechanics of Damage in Deformation and Fracture. Technical University Denmark, Lyngby, 1987

[73] L.M. Kachanov: Introduction in Continuum Damage Mechanics. Martinus Nijhoff, Dordrecht, 1986

[74] M. Kachanov: Elastic solids with many cracks: a simple method of analysis, International Journal of Solids and Structures, 23, 23-43 (1987)

[75] T. Kanninen and G. Popelar: Advanced Fracture Mechanics. Pergamon Press, Oxford, 1987

[76] B.L. Karihaloo: Fracture Mechanics and Structural Concrete, Addison Wesley Longman, New York, 1995

[77] S.A. Kislyakov: Stability and growth of delaminations in cyllindrical composite shell under compression, Mekhanika Kompozitnych Materialov, N 4, 653-657 (1985) (in Russian)

[78] H. Kitagava: Introduction to fracture mechanics of fatigue, in Handbook of Fatigue Crack Propagation in Metallic Structures, A. Carpinteri, Ed., Elsevier, Amsterdam, 47-106 (1994)

[79] M. Klisnil and H. Lucaš: Fatigue of Metallic Materials. Academia, Prague, 1980

[80] W.G. Knauss: Delayed failure – the Griffith problem for linearly visco-elastic materials, International Journal of Fracture, 6, 7-20 (1970)

[81] W.G. Knauss: The mechanics of polymer fracture, Applied Mechanics Reviews, 26, 1-17 (1973)

[82] J.F. Knott: Fundamentals in Fracture Mechanics. Butterworth, London, 1979

[83] S. Kocańda: Fatigue Fracture of Metals. Scientific Engineering Publishers, Warsaw, 1985 (in Polish)

[84] V.P. Kogayev: Strength Analysis under Stresses Varying in Time. Mashinostroyeniye, Moscow, 1993 (in Russian)

[85] D. Krajcinovic: Continuum damage mechanics: when and how, International Journal of Damage Mechanics, 4, 217-229 (1995)

[86] D. Krajcinovic: Damage mechanics. Elsevier Science, Amsterdam, 1996

[87] D. Krajcinovic and J. Lemaitre: Continuum Damage Mechanics – Theory and Applications. Springer, Berlin, 1987

[88]   A.S. Krausz and K. Krausz: Fracture Kinetics of Crack Growth. Kluwer, Dordrecht, 1985

[89]   J.D. Landes and J.A.Begley: A fracture mechanics approach to creep crack growth, in Mechanics of Crack Growth, ASTM STP 590, ASTM Press, Philadelphia, 128-148 (1976)

[90]   B.N. Leiss, M.F.Kanninen, A.T.Hopper, J.Ahmad, and D.Broek: Critical review of the fatigue growth of short cracks, Engineering Fracture Mechanics, 23, 883-898 (1986)

[91]   J. Lemaitre: A Course on Damage Mechanics. Springer-Verlag, Berlin, 1996

[92]   H. Liebowitz, Ed.: Fracture. An Advanced Treatise. In 6 volumes, Academic Press, New York, 1968

[93]   W.B. Lisagor, T.W. Chooker and B.N. Leiss, Eds.: Environmentally Assisted Cracking: Science and Engineering. ASTM STP 1049, ASTM Press, Philadelphia, 1990

[94]   H.W. Liu: A review of crack growth analyses. Theoretical and Applied Fracture Mechanics, 16, 91-108 (1991)

[95]   V. Lufarda and D. Krajchinovic: Damage tensors and the crack density distribution, International Journal of Solids and Structures, 30, 2859-2877 (1993)

[96]   A.J. McEvily, Ed.: Atlas of Stress-Corrosion and Corrosion Fatigue Cracks. ASM International Metals Park, 1990

[97]   H.O. Madsen, S. Krenk and N.C. Lind: Methods of Structural Safety. Prentice-Hall, Englewood Cliffs, 1986

[98]   S.K. Maiti and R.A. Smith: Comparision of the criteria for the mixed mode brittle fracture based on the pre-instability stress-strain field, International Journal of Fracture, 23, 281-296 (1983), 24, 5-22 (1984)

[99]   J.Y. Mann: Bibliography of the fatigue of materials, components and structures 1838-1950. Pergamon Press, Oxford, 1970

[100] R.H. Martin, Ed.: Composite Materials. Fatigue and Fracture, ASTM STP 1230. ASTM Press, Philadelphia, 5, 1995

[101] J. Mazars and G. Pijaudier-Cabot: Continuum damage theory – application to concrete, Journal of Engineering Mechanics, ASME, 115, 345-365 (1989)

[102] K.J. Miller and E.R. de los Rios, Eds.: Short Fatigue Cracks, ESIS 13. Institute of Mechanical Engineers, London, 1992

[103] B.V. Minakov: Crack growth prediction in thermically subjected disks in the presence of microdamage and creep deformations, Problemy Mashinostroyeniya i Nadezhnosti Mashin, N 4, 41-48 (1993) (in Russian)

[104] E.M. Morozov and G.P. Nikishkov: Finite Element Method in Fracture Mechanics. Nauka, Moscow, 1980 (in Russian)

[105] T. Mura: Micromechanics of Defects in Solids. Martinus Nijhoff, Dordrecht, 1987

[106] Yu. Murakami, Ed.: Stress Intensity Factors. Handbook (in 3 volumes), Pergamon Press, Oxford, 1987, 1992

[107] G.Kh. Murzakhanov and V.N. Shchugorev: Effect of secondary cracks on stability and growth of delaminations in composite structures, Mekhanika Kompozitnykh Materialov, N 6, 1120-1124 (1988) (in Russian)

[108] S.V. Nefedov: Analysis of growth of elliptical delaminations in composites under sustained loading, Mekhanika Kompozitnykh Materialov, 5, 827-833 (1988)

[109] D.N. Nesin: Low-cycle fatigue of composites with interlayer defects, Mekhanika Kompozitnych Materialov, N 1, 144-146 (1985) (in Russian)

[110] H. Neuber: Kerbspannungslehre. Springer-Verlag, Berlin, 1958 (in German)

[111] J.C. Newman and W. Elber, Eds.: Mechanics of crack closure, ASTM STP 982. ASTM Press, Philadelphia, 1988

[112] V.V. Panasyuk, Ed.: Fracture Mechanics and Strength of Materials. Handbook Manual. In 4 volumes, Naukova Dumka, Kiev, 1988-1990 (in Russian)

[113] P.E. Paris: Fatigue. An Interdisciplinary Approach. Syracuse University Press, Syracuse, 1964

[114] V.Z. Parton and E.M.Morozov: Mechanics of Elasto-Plastic Fracture. Nauka, Moscow, 1985 (in Russian)

[115] R.E. Peterson: Stress Concentration Factors. John Wiley and Sons, New York, 1974

[116] J. Petit, D.L. Davidson, S. Suresh and P. Rabbe, Eds.: Fatigue Crack Growth Under Variable Amplitude Loading. Elsevier, New York, 1988

[117] R. Pippan, H.P. Stuwe and K. Golos: A comparison of different methods to determine the threshold of fatigue crack propagation, International Journal of Fatigue, 16, 579-582 (1994)

[118] G. Pluvinage: Mecanique Elastoplastic de la Rupture. Cepad, Toulouse, 1989 (in French)

[119] A.R.S. Ponter and D.R. Hayhurst, Eds.: Creep in Structures. Springer-Verlag, Berlin, 1981

[120] Yu.N. Rabotnov: Mechanics of Deformable Solids. Nauka, Moscow, 1979 (in Russian)

[121] M. Ramulu and A.S. Kobayashi: Numerical and experimental study of mixed mode fatigue crack propagation, in Handbook of Fatigue Crack Propagation in Metallic Structures, A. Carpinteri, Ed., Elsevier, Amsterdam, 1073-1123 (1994)

[122] J.N. Reddy: Mechanics of Laminated Composite Plates: Theory and Analysis. CRC Press, Boca Raton, 1997

[123] K.L. Reifsnider: Fatigue behavior of composite materials, International Journal of Fracture, 16, 563-583 (1980)

[124] J.R. Rice: Conserved integrals and energetic forces, in Fundamentals of Deformation and Fracture, Cambridge University Press, Cambridge, 33-56 (1985)

[125] J.R. Rice: First-order variation of elastic fields due to variation in location of a planar crack front, Journal of Applied Mechanics, ASME, 52, 571-579 (1985)

[126] H. Riedel: Fracture at High Temperature. Springer-Verlag, Berlin, 1987

[127] M.M. Rocha, G.I. Schüeller and H. Okamura: The fitting of one- and two-dimensional fatigue crack growth laws, Engineering Fracture Mechanics, 44, 473-480 (1993)

[128] O.N. Romaniv and G.N. Nikiforchin: Mechanics of Corrosion Fracture of Structural Alloys. Metallurgiya, Moscow, 1986 (in Russian).

[129] J. Schijve: Four lectures on fatigue crack growth, Engineering Fracture Mechanics, 11, 167-271 (1979)

[130] R.A. Shapery: A theory of crack initiation and growth in visco-elastic media, International Journal of Fracture, 11, 141-159, 369-388, 549-562 (1975)

[131] V.V. Shchugorev: Delaminations under the combined action of tearing and inter-layer shear, Mekhanika Kompozitnych Materialov, N 3, 539-542 (1987)

[132] I. Sheinman, G.A. Kardomateas and A.A. Pelegri: Delamination growth during pre- and post-buckling phases of delaminated composite laminates, International Journal of Solids and Structures, 35 (1-2), 19-31 (1998)

[133] C.S. Shin, K.C. Man and C.M. Wang: A practical method to estimate the stress concentration of notches, International Journal of Fatigue, 16, 242-256 (1994)

[134] G.C. Sih and H. Liebowitz: Mathematical theories of brittle fracture, in Fracture. An Advanced Treatise. Academic Press, New York, 2, 67-190 (1969)

[135] G.C. Sih, Ed.: Handbook of Stress Intensity Factors. Lehigh University, Bethlehem, 1973

[136] G.C. Sih and B.M. Barthelemy: Mixed mode fatigue crack growth prediction. Engineering Fracture Mechanics, 13, 365-377 (1980)

[137] K. Sobczyk and B.F. Spencer: Random Fatigue: from Data to Theory. Academic Press, New York, 1992

[138] W.W. Stinchomb and N.E. Ashbaugh, Eds.: Composite Materials. Fatigue and Fracture, ASTM STP 1156. ASTM Press, Philadelphia, 4, 1993

[139] B. Storåkers: Nonlinear aspects of delaminations in structural members, in Theoretical and Applied Mechanics, P. Germain, M. Pian, and D. Caillerie, Eds., Elsevier, Oxford, 315-336 (1989)

[140] S. Suresh: Fatigue of Materials. Cambridge University Press, Cambridge, 1991

[141] H. Tada, P.C. Paris and G.R. Irwin: The stress analysis of cracks. Handbook. Del Research Corporation, St. Louis, 1985

[142] R. Talreja: Continuum modelling of damage in ceramic matrics materials, Mechanics of Materials, 12, 165-180 (1991)

[143] V.P. Tamuzs and V.D. Protasov, Eds.: Fracture of Structures of Composite Materials. Zinatne, Riga, 1986 (in Russian)

[144] D. Taylor: Fatigue Thresholds. Butterworth, London-Boston, 1989

[145] A. Tetelman and A. McEvily Jr.: Fracture of Structural Materials. John Wiley, New York, 1967

[146] S.P. Timoshenko: History of Strength of Materials. McGraw-Hill, New York, 1953

[147] V.T. Troshchenko and L.A. Sosnovsky: Fatigue Resistance of Metals and Alloys. Handbook in 2 volumes. Naukova Dumka, Kiev, 1987 (in Russian)

[148] W.-T. Tsai and L.R. Dharani: Non-self-similar fiber structure in unidirectional composites, Engineering Fracture Mechanics, 44(1), 43-49 (1993)

[149] S.W. Tsai and H.T. Hahn: Introduction to Composite Materials. Technomics, Lancaster, Basel, 1980

[150] A.A. Vakulenko and M.L. Kachanov: Continuum theory of cracked media, Izvestiya AN SSSR, Mekhanika Tverdogo Tela, N 4, 159-166 (1971) (in Russian)

[151] D.A. Virkler, B.M. Hillberry and P.K. Goel: The statistical nature of fatigue crack propagation, Journal of Engineering Materials and Technology, ASME, 101, 148-153 (1979)

[152] M.P. Wnuk: The nature of fracture in relation to the total potential energy, British Journal of Applied Physics, ser. 2, 1(2), 217-236 (1968)

[153] M.P. Wnuk, Ed.: Nonlinear Fracture Mechanics. CISM Course 314. Springer, Wien-New York, 1990

[154] M.P. Wnuk and W.G. Knauss: Delayed fracture in viscoelastic-plastic solids, International Journal of Solids and Structures, 6(7), 995-1009 (1970)

[155] J.G. Williams: Fracture Mechanics of Polymers. Ellis Horwood, Chichester, 1984

[156] C.H. Wu: Maximum-energy-release-rate criterion applied to a tension-compression specimen with a crack, Journal of Elasticity, 8, 235-257 (1978)

[157] C. Zweben and B. Rosen: A statistical theory of material strength with application to composite materials, Journal of Mechanics and Physics of Solids, 18 (3), 189-206 (1970)

SUBJECT INDEX

AUTHOR INDEX